GEOGRAPHICAL IMAGINATIONS

DEREK GREGORY

BLACKWELL
Cambridge MA & Oxford UK

First published 1994

Reprinted 1994, 1996, 1997, 1998

Blackwell Publishers Inc.
350 Main Street
Malden, Massachusetts 02148, USA

Blackwell Publishers Ltd
108 Cowley Road
Oxford OX4 1JF, UK

Library of Congress Cataloging in Publication Data
Gregory, Derek, 1951–
Geographical imaginations/Derek Gregory.
p. cm.
Includes bibliographical references and index.
ISBN 0–631–18329–9 (hbk) — ISBN 0–631–18331–0 (pbk)
1. Geography—Philosophy. Cultural studies. I. Title.
G70.G72 1993 93–9601
910'.01—dc20 CIP

British Library Cataloguing in Publication Data
A CIP catalogue record for this book is available from the British Library

Typeset in Garamond on 10.5/12pt
by Pure Tech Corporation, Pondicherry, India
Printed in the United States of America

This book is printed on acid-free paper

Contents

Figures

Acknowledgements

I am grateful to the following for permission to reproduce material: William Heinemann, Ltd., and Harcourt Brace Jovanovich, Inc., for the epigraph from Antoine de Saint-Exupéry, *Wind, sand, and stars* (translated by Lewis Gallantière); Blackwell Publishers for the epigraph from Neil Smith, *Uneven development: Nature, capital and the production of space*; HarperCollins for the epigraph from Michel Tournier, *Friday or The Other Island*; Pion, Ltd., for the epigraph from Rosalyn Deutsche, "Boys town"; Penguin Books for the epigraph from William Boyd, *An Ice-cream War*; Blackwell Publishers for the epigraph from Peter Haggett, *The geographer's art*; Harcourt Brace Jovanovich for the epigraph from Italo Calvino, *Invisible cities*; Pion, Ltd., for the epigraph from Allen Scott and Edward Soja, "Los Angeles: Capital of the late twentieth century"; the Association of American Geographers for the epigraph from Pierce Lewis, "Beyond description"; Hamish Hamilton, Ltd., for the epigraph from Raymond Chandler, *The high window*; Random House for the epigraph from Elizabeth Wilson, *Hallucinations: Life in the postmodern city*; Heinemann Publishing for the epigraph from Graham Swift, *Waterland*; the Association of American Geographers for the epigraph from Carl Sauer, "Foreword to historical geography"; Random House and Jonathan Cape for the epigraph from James Joyce, *Ulysses*; Blackwell Publishers for the epigraph from David Harvey, *The condition of postmodernity: An enquiry into the origins of cultural change*; Penguin Books for the epigraph from Émile Zola, *L'Assommoir* (translated by L. W. Tancock); the University of Minnesota Press for the epigraph from Kristin Ross, *The emergence of social space: Rimbaud and the Paris Commune*.

Chapter 3 is a revised and extended version of an essay that first appeared in *Strategies: A journal of theory, culture, and politics*; Chapter 4 is a revised and extended version of an essay that first appeared in *Geografiska Annaler*. I am grateful to the editors and publishers for allowing me to use that material here.

I am also indebted to my good friends Michael Dear, Allan Pred, Edward Soja, and Michael Watts for permission to reproduce illustrations from their own work.

Preface

When I left the University of Cambridge in 1989 to come to the University of British Columbia, I had in my baggage the manuscript of a book which I had called *The Geographical Imagination*; and since that book was very different from this one an explanation of how I came to abandon that draft (I threw it away at the end of my first term) and to write *Geographical Imaginations* might help to make sense of what follows.

I now realize – and writing this book has helped me realize – that moving to Canada required me to think about three issues which, if I thought about them at all in England, I had never foregrounded, and yet I now had to come to terms with them not as purely intellectual concerns but as matters of everyday practice. In the first place, I had to confront a colonial legacy that faced in two directions. On the one side was the continuing presence of European pasts: the frayed ties, now as much cultural as political or economic, which bind Anglophone Canada to Britain and Francophone Canada to France, and also the tangled braids that crisscross the border between Canada and the United States. On the other side was the inscription of European power and knowledge (and ignorance) on the lives and lands of native peoples. In the second place, I found myself living in an avowedly multicultural society – with its share of racial tensions, prejudices, and discriminations as well as its enrichments, vibrancies, and differences – and teaching at a university where many people could trace their roots to quite other continents, and most often to an Asia that could not be reduced to Europe's Other. And in the third place, I was working at a university where questions of gender and sexuality were taken more seriously than I had been accustomed, and in the wake of the hideous massacre of women students at the École Polytechnique in Montréal those questions took on a new and agonizing seriousness.

All of these things have helped to make me aware of my own "otherness," troublingly and imperfectly, and in many ways the essays in this book represent an attempt to think these issues through, to come to terms with my transplantation from Europe to North America and to understand the continuing importance of a European horizon of meaning in my own work. These are not purely personal preoccupations, for in this increasingly interconnected world the predicament of culture, as James Clifford calls it, touches all of us in myriad ways. If I think it has a special salience in geography, it is because I have started to understand my own situatedness and to think about its implications in a discipline that has had as one of its central concerns an understanding of other people and other places. "Discipline" is a double-edged term, of course, and it cuts in two directions: by this I mean to imply not a set of sovereign concepts, still less any rigorous policing operation, but instead a more diffuse (though nonetheless deep-seated) acknowledgment of the importance of place and space that shapes the ways many of us approach our work and our lives. This habit of mind is rooted in all sorts of experiences, inside and outside the academy, and it grows in different ways in different places. But such a metaphor is also duplicitous, for it usually grafts geography onto the classical tree of knowledge – systematic, hierarchical, grounded – so that its cultivators can scrutinize its fruit, fuss over its pruning, and worry about its felling. But it may be more appropriate to think instead of the nomadic tracks and multiplicities of the rhizome (and here I borrow from Gilles Deleuze and Félix Guattari): to open up our geographies to interruptions and displacements, to attend to other ways of traveling, and to follow new lines of flight. In this connection I have been fortunate in being able to join a remarkably talented and wonderfully diverse group of geographers in Vancouver. It is a pleasure to thank the colleagues, graduate students, and friends who have sustained me on this journey (and to note how often those colleagues and graduate students *are* good friends). I owe a particular and continuing debt to Trevor Barnes, Nick Blomley, Alison Blunt, Michael Brown, Noel Castree, Daniel Clayton, Robyn Dowling, Cole Harris, Marwan Hassan, Daniel Hiebert, Brian Klinkenberg, David Ley, Terry McGee, Cathy Nesmith, Tim Oke, Geraldine Pratt, Olav Slaymaker, Matthew Sparke, Lynn Stewart, Bruce Willems-Braun, Jonathan Wills, and Graeme Wynn.

In thinking about the predicament of positionality in this crossroads city, I have become aware of many writers who insist that it is both impossible and illegitimate to speak for or even about others; but as a teacher of geography I believe I have a responsibility to enlarge the horizons of the classroom and the seminar. I know that I cannot claim to do so from some Archimedian promontory; I know, too, that there are dangers in doing so – of being invasive, appropriative – and I do not pretend to have any answers to these anxieties. But the consequences of not doing so, of locking ourselves in our own worlds, seem to me far more troubling. I put the problem in pedagogic terms because I have always done research in order to teach. I know there will be readers who will wonder at the "relevance" of these concerns, particularly those who have a rather different view of research and its rewards, but my hope is that the ideas I discuss here – and the practices of critical inquiry I seek to foster – might inform not only the conduct of research but also the practice of teaching.

I have organized this book as a set of essays that consider diverse geographical imaginations, not "the" geographical imagination – a distinction for which I will be eternally grateful to Denis Cosgrove – but they do nonetheless spiral around a common set of themes dealing with power, knowledge, and spatiality. The phrasing is Michel Foucault's, but it is obvious to me how much my interest in these questions owes to David Harvey, whose work is centrally present in every one of these essays and to whom I owe more than I can adequately express. His *Social justice and the city* was published at the end of my first year as a postgraduate student, which was also the start of my first year as a lecturer at Cambridge. I can still remember the shock. This was the first book in geography that I knew I didn't understand. Trained in spatial science, I was accustomed to technical difficulty, but conceptual difficulty was an altogether different order of things, particularly when it required an engagement with social theory and a recognition of ethical and political responsibility. It was in that book, too, that Harvey first wrote about his "geographical imagination" and registered a series of claims about the importance of place and space in the conduct and constitution of social life. Those ideas have continued to guide – and on occasion to provoke – my own work; they still seem to me some of the most creative interventions in the discourse of geography. But I am acutely aware of another thematic that is present only in the margins of this book. I am conscious of the question of "nature" – living in this vast, beautiful, and sometimes terrifying land, how could I not be? – but for the most part I have concerned myself with the politics of spatiality. This is not because I think environmental questions are unimportant, and I welcome the contemporary interest in political ecology. But, rather like the outsider in Richard Rorty's *Philosophy and the Mirror of Nature* who keeps wanting to talk about something else even though the conversation has moved on, I still want to talk about spatiality: I do not think its questions have been resolved, and I am convinced that they are connected in profound ways to what is now often called the culture – and cultural politics – of nature.

I have implied that this has been a nomadic project, and I have been encouraged in my wanderings by any number of other friends outside Vancouver, including Alan Baker, Mark Billinge, Nicholas Clifford, Stuart Corbridge, Michael Dear, Rosalyn Deutsche, Felix Driver, Nick Entrikin, Tony Giddens, Peter Gould, Michael Heffernan, Peter Jackson, Ron Johnston, John Paul Jones, Gerry Kearns, Doreen Massey, John Langton, Linda McDowell, Gunnar Olsson, Chris Philo, John Pickles, Allan Pred, Gillian Rose, Edward Soja, Susan Smith, David Stoddart, Michael Storper, Nigel Thrift, Dick Walker, and Michael Watts. I am particularly grateful to David Livingstone and Neil Smith, who commented in vigorous and constructive detail on a first version of the book, to Paul Jance, who converted my rough sketches into finished diagrams, Stephanie Argeros-Magean, who edited the manuscript with great skill and sensitivity, and above all to John Davey, who has taken a characteristically warm and intelligent interest in the project.

Derek Gregory
Vancouver

PART I

Strange Lessons in Deep Space

Introduction

But what a strange lesson in geography I was given! Guillaumet did not teach Spain to me, he made the country my friend. . . . The details that we drew up from oblivion, from their inconceivable remoteness, no geographer had been concerned to explore. Because it washed the banks of great cities, the Ebro River was of interest to map-makers. But what had they to do with that brook running secretly through the water-weeds to the west of Motril, that brook nourishing a mere score or two of flowers?

"Careful of that brook: it breaks up the whole field. Mark it on your map." Ah, I was to remember that serpent in the grass near Motril! . . . And those thirty valorous sheep ready to charge me on the slope of a hill!

Little by little, under the lamp, the Spain of my map became a sort of fairyland. The crosses I marked to indicate safety zones and traps were so many buoys and beacons. I charted the farmer, the thirty sheep, the brook. And, exactly where she stood, I set a buoy to mark the shepherdess forgotten by the geographers.

Antoine de Saint-Exupéry, *Wind, sand, and stars*

The twentieth century has ushered in the discovery of deep space, *or at least its social construction, and yet it is only as the century draws to a close that this fundamental discovery is becoming apparent. . . . Deep space is quintessentially social space; it is physical extent infused with social intent.*

Neil Smith, *Uneven development*

When I began to draft the essays in Part I, my intention was to introduce the history of geography to readers outside the discipline and, at the same time, to call into question some of the ways in which that history had been written by those inside the discipline. As I worked on this material, however, the distinction became increasingly problematic. Traces of the original purpose are probably still present, but I now think of these essays as interventions in a discourse rather than a discipline. They sketch out two narratives.

The first concerns what one might call the "socialization" of human geography (and I hope the inevitable distortions of such a shorthand will be forgiven). In one sense, clearly, human geography has always been socialized. Even the rigid naturalism of environmental determinism or the unyielding physicalism of spatial science was a social construction. But it is only recently, I think, that many of its practitioners have come to reflect, critically and systematically, on the connections between social practice and human geography. In doing so, they (we) have learned "strange lessons": insights into human life on earth which, on occasion, soar into the sky like the aviator in Saint-Exupéry's novella (and I will have more to say about these aerial views in due course) but which are also as often grounded in the profoundly existential significance of place, space, and landscape to people like Guillamet and the anonymous shepherdess (forgotten in more ways than one).

The second narrative approaches from a different direction. Many of those working in the other humanities and social sciences have also become interested in questions of place, space, and landscape – in the ontological significance of what Neil Smith calls "deep space." Their studies have multiple origins, but they treat the production of social space, of human spatiality, in new and immensely productive ways. In philosophy, for example, I think of various rereadings of Martin Heidegger's texts and the geographies written into many of Michel Foucault's histories of the present; in historical materialism, I think of the work of Fredric Jameson and Henri Lefebvre; in feminism and poststructuralism, of bell hooks and Donna Haraway; and in cultural studies and postcolonialism, of Edward Said and Gayatri Chakravorty Spivak. The list is incomplete and heteroclite, like the "Chinese encyclopedia" with which Foucault opens *The order of things*, and I do not mean to imply that any of these proposals should be accepted uncritically. But the startling juxtapositions in a brief roll call like this conjure up a sense of political and intellectual ebullition that is, I think, unprecedented. It is out of the fusion between these two narratives that the discourse I have in mind is generated.[1]

I can perhaps describe this in another way. At the opening of the last decade, Clifford Geertz spoke of a "refiguration of social thought" in which the boundaries between formal intellectual inquiry and imaginative writing were becoming blurred. As the "closet physicists" of social science were returned to the closet, Geertz declared, so social life was increasingly conceived as a game, a drama, or a text. And, as usual, he had no doubt what this meant: "All this fiddling around with the properties of composition, inquiry,

[1] Cf. Edmunds Bunkse, "Saint-Exupéry's geography lesson: Art and science in the creation and cultivation of landscape values," *Annals of the Association of American Geographers* 80 (1990) pp. 96–108; Neil Smith, "Afterword: The beginning of geography," in his *Uneven development: Nature, capital and the production of space* (Oxford: Blackwell Publishers, 2nd ed., 1990) pp. 160–78.

and explanation represents, of course, a radical alteration in the sociological imagination."[2] So it does. But what Geertz failed to notice (or did not think significant) was that all three metaphors concern constructed *spaces* in which human action literally takes *place*. I hold no brief for any of these particular ways of thinking about social life, let me say, but I do think it highly unlikely that those spatial implications would have been unremarked by the end of that decade. For these "strange lessons in deep space" have more than metaphorical significance. Indeed, one of the most compelling aims of the project that Smith describes is to transcend the partitions between "metaphorical space" and "material space." And since I have used two metaphors – mapping and traveling – to think about much of what follows, I want to establish their materiality before I set out.

Maps of the intellectual landscape

In Chapters 1 and 2 I construct a "map" of the intellectual landscape by following the path shown in figure 1, on which I have plotted the relations between human geography and a number of other disciplines in the humanities and the social sciences. In the first essay I move down the lefthand side

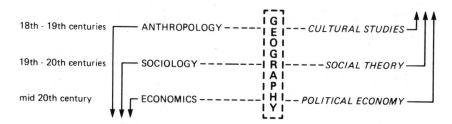

Figure 1 Maps of an intellectual landscape.

of the diagram, because I think that in many ways modern human geography has been defined through a series of strategic encounters with anthropology in the eighteenth century, sociology at the turn of the nineteenth and twentieth centuries, and economics in the middle of the twentieth century. I pay particular attention to three episodes which, taken together, bear directly on a specifically modern constellation of power, knowledge, and spatiality in which visualization occupies a central place – what I call "geography and the world-as-exhibition." In the second essay I move up the

---

[2] Clifford Geertz, "Blurred genres: The refiguration of social thought," in his *Local knowledge: further essays in interpretive anthropology* (New York: Basic Books, 1983) pp. 19–35; the essay was first published in 1980.

righthand side of the figure and suggest that the critique of modern spatial science in the closing decades of the twentieth century has involved a reactivation of the preceding dialogues in reverse order, from political economy through social theory to cultural studies. In doing so, the assumptions and privileges that inhere within the world-as-exhibition have been called into question, so it seems to me, and the complexity and contingency of human spatiality has produced what I describe as "geography and the cartographic anxiety."

This is by no means a disinterested representation. History is never innocent; it is always "history-for," and intellectual histories are no different. They are ways of locating claims within traditions that seek to establish them as authoritative and legitimate, and also ways of positioning claims in opposition to other traditions and so establishing their own authority and legitimacy by negation. I don't think it much matters whether these stories are the metanarratives that excite Lyotard's postmodern rage or the more modest *petits récits* that he endorses: All of them function as rhetorical devices. They are all strategies that seek to persuade readers of the cogency of their leading propositions. The same is true of my "map." Its objectivity is a serious fiction that represents a particular intellectual landscape from a particular point of view. As Felix Driver puts it:

Representing geography's past is inevitably an act of the present, however much we attempt to commune with the past. Indeed, the idea of mapping the historical landscape depends on the construction of perspective, a view from the present, around which the panoramas are made to revolve.[3]

I imagine it is hardly necessary to add that in confessing all this I have not once abandoned the rhetorical field; after all, the confessional has its own poetics of persuasion. And yet, as I have implied, the metaphoric of mapping – of panorama and perspective – is itself problematic. According to two critics,

This notion of map-able space involves a specific epistemic topography: a landscape, a form of knowing or seeing which denies its structuring by the gaze of white male bourgeois knowers on Other knowns. It limits the possibility of critique by refusing to acknowledge other kinds of space.[4]

But I want to ask: Always? Everywhere? I understand (and share in) this cartographic anxiety, but when these objections are put in this particular

---

[3] Felix Driver, "Geography's empire: histories of geographical knowledge," *Environment and Planning D: Society and Space* 10 (1992) pp. 23–40; the quotation is from p. 36.

[4] Stephen Pile and Gillian Rose, "All or nothing? Politics and critique in the modernism/postmodernism debate," *Environment and Planning D: Society and Space* 10 (1992) pp. 123–36; the quotation is from p. 131.

form they seem to me to accept cartography's own historiography even as they contest it. It is perfectly true that historians have usually presented cartography as the Survey of Reason, a narrative journey of progress from darkness to enlightenment, in the course of which maps become supposedly more "accurate" and more "objective." But it is also true that there is now a critical historiography, which has established the implication of maps in the constitution of systems of power-knowledge and, through the work of Brian Harley in particular, has suggested ways of deconstructing their technologies of power.[5] In doing so, it has become apparent that mapping is necessarily situated, embodied, partial: *like all other practices of representation*. This makes it misleading to counterpose metaphors of mapping that supposedly always and everywhere "refuse to acknowledge other kinds of space" with other metaphors that somehow ineluctably do. There is no reason to suppose that "location," for example, automatically challenges the supremacy of the knowing male gaze (and the history of location theory and spatial science provides a compelling argument for exactly the opposite). Given these critical historiographies, it is surely presumptuous to claim that "images of maps, landscapes, and spaces" are always advanced as "unproblematic" by those who use them, while images of location, position, and geometry – all of which are advanced in the essay from which I draw these admonitions – are not. What entitles the critics to assert their own privilege in this way? My point is that *all* metaphors are problematic and that "translation terms" like these – seemingly general terms used in strategic and contingent ways, as James Clifford calls them – get us a certain distance *and* fall apart.[6]

For my part, I believe it is possible to use images of maps, landscapes, and spaces *and also* images of location, position, and geometry in ways that challenge the Archimedian view of knowledge, in ways that insist that geographies of knowing make a difference. But these are not absolutes, and the differences that they are able to make depend on the specific ways in which they are used. Not only can conventional cartographic discourse be turned against itself, for example; not only can the mapping metaphoric be used ironically or parodically; but it is also possible to envisage other more open forms of cartographic discourse.[7]

[5] J. Brian Harley, "Maps, knowledge and power," in Denis Cosgrove and Stephen Daniels (eds.), *The iconography of landscape* (Cambridge: Cambridge University Press, 1988) pp. 277–312; *idem*, "Deconstructing the map," in Trevor Barnes and James Duncan (eds.), *Writing worlds: Discourse, text and metaphor in the representation of landscape* (London: Routledge, 1992) pp. 231–47.

[6] James Clifford, "Travelling cultures," in Lawrence Grossberg, Cary Nelson and Paula Treichler (eds.), *Cultural studies* (New York: Routledge, 1992) pp. 96–112.

[7] Graham Huggan, "Decolonizing the map: Post-colonialism, poststructuralism and the cartographic connection," *Ariel* 20 (1989) pp. 115–31. Huggan ties the possibility of more open discourses of cartography more closely than I would wish to the pluralizing, disseminating, deterritorializing "cartography" of Gilles Deleuze and Félix Guattari, *A thousand plateaus: Capitalism and schizophrenia* (Minneapolis: University of Minnesota Press, 1987); this was first published in France in 1980.

In constructing and reading my map, I need to identify both the "internal" and the "external" coordinates of human geography. The first of these tasks is scarcely unusual. There are many internalist histories of geography that focus on the individuals, schools, and traditions that are supposed to have played a part in the making of the discipline. But most of these assume that a discursive space was already reserved for geography; that one simply had to wait for the explorers, surveyors and settlers to appear and convert that immanent claim into a palpable reality. Such a procedure does little justice to the complexities of disciplinary emergence, and when I insist on the importance of the "internal" I do not mean to imply that the history of human geography can be reduced to a series of purely intellectual arguments.

On the contrary, geography (like cartography) has always been a thoroughly practical and deeply politicized discourse, and it continues to be marked by its origins. But I do think that its philosophico-theoretical contours need to be drawn with considerable care, because the topographies of human geography cannot be interpolated as so many responses to changes in the "real." To do so would be to make the same mistake as internalist histories, only in reverse: to think of geography as a discipline-in-waiting, whose formation is determined not so much by the internal logic of intellectual inquiry as the imperatives of an "external" reality. If the critique of realism has taught us anything, it is surely that the process of representation is constructive not mimetic, that it results in "something made," a "fiction" in the original sense of the word.[8] One does not have to endorse Rorty's pragmatism (to take a particularly audacious example) to see that his shattering of the pedestal on which analytical philosophy placed itself as quite literally "the mirror of nature" has a more general cogency: that discourse is not an unproblematic reflection of the world but is instead an intervention in the world.[9] The problem is thus to find some way of blurring the conventional distinctions between the "internal" and the "external," and at the very least of recognizing that what is *thought* of as internal or external is the product of a reciprocal process of constitution.

The predicament is compounded because those distinctions have been blurred in other ways too. The history of modern geography has often

---

[8] I owe the remark to Clifford Geertz, "Thick description: Toward an interpretive theory of culture," in his *The interpretation of cultures* (New York: Basic Books, 1973) pp. 3–30. His claim is more particular than mine. Ethnographies, so he says, are "fictions, in the sense that they are 'something made,' 'something fashioned' – the original meaning of *fictio* – not that they are false, unfactual, or merely 'as if' thought experiments" (p. 15).

[9] Richard Rorty, *Philosophy and the mirror of nature* (Princeton: Princeton University Press, 1980). What matters here is not so much the specifics of Rorty's position – about which I have considerable reservations – but the wider critique of analytical philosophy as the "cultural dominant" of the Western philosophical tradition. This is not confined to Rorty, of course, and a more complete survey is provided in Kenneth Baynes, James Bohman, and Thomas McCarthy (eds.) *After Philosophy: End or transformation?* (Cambridge: MIT Press, 1987).

been presented as an intellectual gazeteer. In the eighteenth century geography is supposedly a "European science"; in the nineteenth century distinctive schools of geography are identified with particular national traditions ("the French School," "the German School"); and in the twentieth century particular places are often used as markers in the intellectual landscape ("the Chicago School," "the Berkeley School," and more recently still even "the Los Angeles School"). But the global circulation of information and ideas – with all its inequalities, restrictions, and deformations – has made many of these parochialisms unusually problematic. No doubt we have retained some of them, and I suspect that we have invented a host of new ones, but there is nevertheless an important sense in which our "local knowledge" simply isn't local any more. Clifford argues that we now have to make sense of a world without stable vantage points; a world in which the observers and the observed are in ceaseless, fluid, and interactive motion; a world where "human ways of life increasingly influence, dominate, parody, translate, and subvert one another." All of this constitutes what he describes as the predicament of ethnographic modernity:

We ground things, now, on a moving earth. There is no longer any place of overview (mountaintop) from which to map human ways of life, no Archimedian point from which to represent the world. Mountains are in constant motion. So are islands: for one cannot occupy, unambiguously, a bounded cultural world from which to journey out and analyze other cultures.[10]

This is as pressing for intellectual activity as it is for any other sphere of social life, perhaps even more so, and with this in mind I also have to attend to the prospects and perils of "traveling theory."

Traveling theory

I have borrowed the phrase from Edward Said, who uses it to draw attention to the *situatedness* of theory: "Theory has to be grasped in the place and time out of which it emerges." Said also emphasizes that those situations are always overdetermined and constantly changing, and that no theory "exhausts the situation out of which it emerges or to which it is transported."[11] But what "theory"? and which "situation"? I suggest there are two overlapping motifs at work here. The first has been captured most

[10] James Clifford, "Introduction: partial truths," in James Clifford and George Marcus (eds.), *Writing culture: The poetics and politics of ethnography* (Berkeley: University of California Press, 1986) pp. 1–26; the quotation is from p. 22.

[11] Edward Said, "Traveling theory," in his *The World, the text and the critic* (London: Faber and Faber, 1984) pp. 226–47; the quotation is from pp. 241–42.

perceptively by Jonathan Culler in his description of contemporary theory as a genre. He explains it like this:

"Theory" is a genre because of the way its works function... [to] exceed the disciplinary framework within which they would normally be evaluated and which would help to identify their solid contributions to knowledge. To put it another way, what distinguishes the members of this genre is their ability to function not as demonstrations within the parameters of a discipline but as redescriptions that challenge disciplinary boundaries. The works we allude to as "theory" are those that have had the power to make the strange familiar and to make readers conceive of their own thinking, behavior and institutions in new ways. Though they may rely on familiar techniques of demonstration and argument, their force comes – and this is what places them in the genre I am identifying – not from the accepted procedures of a particular discipline but from the persuasive novelty of their redescriptions.[12]

Although I find the different arguments of Derrida and Foucault particularly compelling, I should say at once that, unlike Culler, my own interest is not circumscribed by poststructuralism. My focus in what follows will be on social theory more generally, which I conceive as a series of overlapping, contending and colliding discourses that seek, in various ways and for various purposes, to reflect explicitly on the constitution of social life and to make social practices intelligible. This is a minimalist definition, of course, and further discrimination is necessary. My particular concern is with the multiple discourses of *critical theory*: discourses that seek not only to make social life intelligible but also to make it *better*. This may seem a curiously anemic way of expressing myself, but when I think of the bloody consequences of many more traditional, supposedly disinterested modes of inquiry, I prefer this simple declaration of hope to those pious positivities. Critical theory is a large and fractured discursive space, by no means confined to the Frankfurt School and its legatees, but it is held in a state of common tension *by the interrogation of its own normativity*. And as part of that concern, many of the discourses of most moment challenge traditional ideas of what "theory" is (or might be), and think of their own function as one of interruption and intervention in the representation and negotiation of social life. It is in exactly this sense, too, that Culler thinks of them (and their "persuasive redescriptions") as transgressive.

Culler makes much of the capacity of theory to cross and call into question disciplinary boundaries. Now the intellectual division of labor has always been an untidy affair – which is, in part, why there have been so many new disciplines and interdisciplinary projects brought into being – but it is not completely arbitrary. It is always possible to provide reasons

---

[12] Jonathan Culler, *On deconstruction: Theory and criticism after structuralism* (London: Routledge, 1983) p. 9.

(historical reasons) for the boundaries being drawn this way rather than that. Once those boundaries are established, however, they usually become institutionalized. All the apparatus of the academy is mobilized to mark and, on occasion, to police them. But these divisions do not correspond to any natural breaks in the intellectual landscape; social life does not respect them and ideas flow across them.

It is this busy cross-border traffic that I have in mind when I talk about social theory as a *discourse*. This is not just another word for "conversation," or if it is then it is conversation in a greatly enlarged sense. For discourse refers to all the ways in which we communicate with one another, to that vast network of signs, symbols, and practices through which we make our world(s) meaningful to ourselves and to others. It is particularly helpful, I think, in clarifying the situatedness of theory: the contexts and casements that shape our local knowledges, however imperiously global their claims to know, and the practical consequences of understanding (and indeed being in) the world like this rather than like that. This state of affairs is not peculiar to the humanities and the social sciences. These ideas also bear directly on the natural sciences and, as Joseph Rouse has shown in a marvelous exposition of their protocols and procedures, even the laboratory sciences are grounded in specific sites and discursive practices whose various "local knowledges" are progressively and provisionally extended into other sites and other practices. As he also shows, to speak of discourse rather than discipline is not to escape the bonds between power and knowledge. On the contrary, to use this vocabulary is to reflect explicitly on those constellations and their distinctive regimes of truth.[13] In much the same way, and for many of the same reasons, I am more interested in the discourses of geography than in the discipline of geography. Geography in this expanded sense is not confined to any one discipline, or even to the specialized vocabularies of the academy; it travels instead through social practices at large and is implicated in myriad topographies of power and knowledge. We routinely make sense of places, spaces, and landscapes in our everyday lives – in different ways and for different purposes – and these "popular geographies" are as important to the conduct of social life as are our understandings of (say) biography and history.[14]

---

[13] Joseph Rouse, *Knowledge and power: Toward a political philosophy of science* (Ithaca: Cornell University Press, 1987).

[14] The example is not a casual one. Mills once defined what he called "the sociological imagination" as an ability to reveal the multiple intersections between biography and history in our own present. But he made it plain that he did not mean this in any strictly disciplinary sense and that the habit of mind could be found across the whole field of the humanities and the social sciences. See C. Wright Mills, *The sociological imagination* (New York: Oxford University Press, 1959). One of my own concerns is to bring this sociological imagination, in all its various forms, into dialogue with those "geographical imaginations" that also escape the usual disciplinary enclosures.

But when we are required to think critically and systematically about social life and social space, we usually need to distance ourselves from those commonplace, taken-for-granted assumptions. We can never suspend them altogether, and our reflections will make sense, to ourselves and to other people, only if they retain some connection with the ordinary meanings that are embedded in the day-to-day negotiations of lifeworlds. But we need to interrogate those "common sense" understandings: We need to make them answer to other questions, to have them speak to other audiences, to make them visible from other perspectives. And we also need to show how they engage with one another; how they connect or collide in complexes of action and reaction in place and over space to transform the tremulous geographies of modernity.

This has become one of the central tasks of social theory, which I insist is not the possession of any one discipline – not sociology, not even the social sciences – but is, rather, that medium within which anyone who seeks to account for social life must work. I say "must work" because empiricism is not an option. The facts do not and never will speak for themselves, and no one in the humanities or the social sciences can escape working with a medium that seeks to *make* social life intelligible and to *challenge* the matter-of-factness of "the facts." And I say "working with" because social theory does not come ready-made. As I have said, it provides a series of partial, often problematic and always situated knowledges that require constant reworking as they are made to engage with different positions and places. Conceived thus, social theory, like geography, is a "traveling discourse," marked by its various origins and moving from one site to another. For this reason, I think of working with social theory as spinning a *multiple* hermeneutic between its *different* sites rather than any "double hermeneutic" that confines social theory to a single site.[15]

But this is to move toward a second motif and an altogether different sense of traveling and transgression that has to do with the globalization of intellectual cultures. The trope of traveling, of tracking "roots" and "routes," has become an established genre of intellectual inquiry – and of contemporary writing more generally – and it is one that will occupy a central place in what follows. But it is important to understand that in its present form it does not so much redraw our maps of the intellectual landscape as call the very principles of mapping into question.[16] In many ways modern social theory still bears the marks of its Enlightenment origins, and its claims to know continue to respond to Kant's attempt to

---

[15] Cf. Anthony Giddens, *New rules of sociological method: A positive critique of interpretative sociologies* (London: Hutchinson, 1976) p. 162.

[16] Marilyn Strathern, "Or, rather, on not collecting Clifford," *Social analysis: Journal of cultural and social practice* 29 (1990) pp. 89–95; this is a critical commentary on James Clifford, *The predicament of culture: Twentieth-century ethnography, literature and art* (Cambridge: Harvard University Press, 1988).

install reason as the undisputed arbiter in all spheres of social life: in science, morality, and art. I suppose it is a characteristic of modern intellectuals – or "universal intellectuals," as Foucault once called them – to see themselves as legislators: as dealers in generalities rather than brokers in particulars, uniquely qualified to chart the course of society-in-general or society-as-totality.[17] The discourses of modern social theory are driven by an assertive generality, so to speak, in which, as Habermas puts it, "the transcendent moment of universality bursts every provinciality asunder." He accepts that these discourses are inevitably "carriers of context-bound everyday practice," embedded in a particular here and now, but he insists that they also typically claim to *erase* all particularities and to *transcend* time and space.[18] *It is precisely this claim that the metaphor of traveling calls into question.* But it does not replace it with its opposite. The objection is not so much that social theories are inescapably context-bound, but rather that the origins of "traveling theory" need to be scrupulously acknowledged because it will always be freighted with a host of assumptions, often derived from different and radically incommensurable sites, which may not – and usually should not – survive the journey intact. Traveling thus becomes a way of resisting the imperial ambitions of theory, of making those who work with it accountable for its movements, and of challenging what Donna Haraway calls "the politics of closure."[19]

And yet, as bell hooks points out in a moving reflection on these themes, holding on to the metaphor of travel can also be a way of clinging on to imperialism: or at any rate, it comes with its own baggage in which, as she says, it is not easy to find room for "the Middle Passage, the Trail of Tears, the landing of Chinese immigrants at Ellis Island, the forced relocation of Japanese-Americans, or the plight of the homeless." For hooks, as a black woman living in the United States, "to travel is to encounter the terrorizing force of white supremacy."[20]

Certainly, to read many of Clifford's essays is to be reminded of the privileges that accrue to the elites of Western intellectual culture. The freedom to move, to read, to write that he enjoys is a *situated* freedom, a "cosmopolitanism" that is, like my own, gendered, classed and (ironically) *located*. I doubt that Clifford is unaware of this. What disturbs him, I take

---

[17] Zygmunt Bauman, *Legislators and interpreters: On modernity, postmodernity, and intellectuals* (Cambridge: Polity Press, 1987).

[18] Jürgen Habermas, *The philosophical discourse of modernity* (Cambridge: Polity Press, 1987) pp. 322–23.

[19] Donna Haraway, "Situated knowledges: The science question in feminism and the privilege of partial perspective," in her *Simians, Cyborgs and Women: The reinvention of nature* (London: Routledge, 1991) pp. 183–201.

[20] bell hooks, "Representing whiteness in the black imagination," in Lawrence Grossberg, Cary Nelson, and Paula Treichler (eds.), *Cultural studies* (New York: Routledge, 1992) pp. 338–46; the quotations are from pp. 343–44.

it, is the assumption that particular classes of people are cosmopolitan ("travelers") while the rest are merely local ("natives"). He argues that this is merely the ideology of one, extremely powerful "traveling culture," and that there are indeed numerous other traveling cultures constituted through force as well as by privilege. And in his later writings in particular Clifford seems to me to provide a way of "traveling with maps" that involves more than redrawing and annotating them as he goes, his function more than the "scribe of our scribblings" patronized and pigeonholed by Paul Rabinow.[21] For Clifford's maps call into question their own enabling conventions *by* traveling: They worry away at their orientation, scale, and grid, their presences and absences, in ways that clarify the modalities by means of which, as he puts it himself, "cultural analysis constitutes its objects – societies, traditions, communities, identities – in spatial terms and through specific spatial practices of research."

These concerns are, perhaps, still privileges and luxuries; all spatial metaphors are no doubt compromised and tainted; and acknowledging these restrictions will not issue in an innocent inquiry conducted in a state of grace. But reminding oneself of the clinging mud of metaphor, of the mundanity and materiality of intellectual inquiry, is nonetheless a vital critical achievement.[22]

This is not the place to attempt a rigorous genealogy of human geography. What follows is a series of vignettes or, in the terms of my containing metaphors, fixes of position, which trace the emergence of a distinctive tradition of Western intellectual inquiry. I should perhaps make it clear that these do not constitute even the outlines of an alternative history. Such a project would have to attend much more scrupulously to archival matters than is possible here. All I seek to do is make a series of incisions into the conventional historiography of geography and show that its strategic episodes can be made to speak to many other histories. It should also be remembered that "human geography" did not emerge in any institutionalized sense until the closing decades of the nineteenth century; but its development appealed to much older traditions of inquiry through exactly the kind of legitimating devices that I mentioned earlier. For now, it will be enough to off-centre those appeals and call some of their assumptions into question.

---

[21] Clifford, "Traveling cultures," p. 108; Paul Rabinow, "Representations are social facts: Modernity and post-modernity in anthropology," in Clifford and Marcus, *Writing cultures*, pp. 234–61; the quotation is from p. 242.

[22] Clifford, "Traveling cultures," p. 97; Bruce Robbins, "Comparative cosmopolitanism," *Social Text* 31/32 (1992) pp. 169–86.

1

Geography and the World-as-Exhibition

I demand, I insist, that everything around me shall henceforth be measured, tested, certified, mathematical, and rational. One of my tasks must be to make a full survey of the island, its distances and its contours, and incorporate all these details in an accurate surveyor's map. I should like every plant to be labeled, every bird to be ringed, every animal to be branded. I shall not be content until this opaque and impenetrable place, filled with secret ferments and malignant stirrings, has been transformed into a calculated design, visible and intelligible to its very depths!

Michel Tournier, *Friday or the other island*

Distancing, mastering, objectifying – the voyeuristic look exercises control through a visualization which merges with a victimization of its object.

Rosalyn Deutsche, *Boys town*

Visualization

In this essay I offer a particular perspective on the constitution of modern geography. I have selected three episodes in its history which, at first sight, might seem to have little in common: the eighteenth-century European odyssey in the South Pacific; the celebrated regional geographies of late nineteenth- and early twentieth-century France; and the emergence of spatial science in Anglophone geography in the decades following World War II. The evidence from these episodes is heterogeneous, but when it is brought into the same frame it brings what I want to call *the problematic of visualization* into particularly clear focus. A number of writers have already drawn attention to vision as what Martin Jay calls "the master sense of the modern era": to the ubiquity of the visual tropes that (deliberately) stud my preceding sentences and to the gendering of the gaze that is there in Jay's own summary description.[1] But these claims assume a special significance in geography, because the discipline continued to privilege sight long after many others became more – well, circumspect. "Geography," writes one commentator, "is to such an extent a visual discipline that, uniquely among the social sciences, sight is almost certainly a prerequisite for its pursuit." This is hyperbole, of course, but the classical origins of geography are closely identified with the optical practices of cartography and geometry; its interests have often been assumed to lie in the landscape and the particular "way of seeing" that this implies; and its decidedly modern interest in theory invokes visualization not only covertly, through the Greek *thea* ("outward appearance") and *horao* ("to look closely") of "theory" itself, but also openly, through the display and analysis of spatial structures. None of these individual coordinates are confined to geography, but their joint intersection with its disciplinary trajectory does, I think, intensify and particularize its ocularcentrism: its characteristically visual appropriation of the world. In taking "the supremacy of the eye for granted," declares Tuan, its practitioners "move with the mainstream of modern culture."[2] Maybe so; but the visual thematic of modern culture is by no means an uncritical one, and I hope to show that it is possible to draw on ideas from art history and the history of science, from philosophy and critical theory, to interrogate geography's visual thematic.

Cook's Tour: anthropology and geography

I cut into the history of modern geography in the second half of the eighteenth century when, so David Stoddart claims, geography first emerged as a distinctively modern science. In doing so he is deliberately tweaking the noses of historians who have identified much earlier surfaces of emergence. He knows only too well that the standard histories of geography parade Eratosthenes, Strabo, and Ptolemy across their pages, followed by Hakluyt, Purchas, and Varenius, but he insists that these figures are remote from the concerns of modern readers. "Their contributions have meaning in the contexts only of their own time, not of ours."[3] Such a dismissal is contentious, of course, and it is intended to be so. An interest in these writers does not immediately imply any antiquarianism.

[1] Martin Jay, "Scopic regimes of modernity," in Scott Lash and Jonathan Friedman (eds.), *Modernity and identity* (Oxford: Blackwell Publishers, 1992) pp. 178–95.

[2] These remarks are drawn from D. C. D. Pocock, "Sight and knowledge," *Transactions of the Institute of British Geographers* 6 (1981) pp. 385–93 and Yi-Fu Tuan, "Sight and pictures," *Geographical Review* 69 (1979) pp. 413–22.

[3] David Stoddart, "Geography – a European science," in his *On geography and its history* (Oxford: Blackwell Publishers, 1986) pp. 28–40.

Livingstone has presented a compelling case for the importance of geography to the construction of modernity during the sixteenth and seventeenth centuries. On his reading, the Scientific Revolution of the seventeenth century may have depended, in complex but nonetheless crucial ways, on the voyages of discovery. Livingstone argues that these "first-hand encounters with the world" – which, like Stoddart, he regards as "the very stuff of geography" – "brought an immense cognitive and cultural challenge to tradition."[4] But I suspect Stoddart's response, and one which Livingstone anticipates in his own essay, would be to say that those early geographies just as often confirmed or reinforced the compulsions of tradition. Whatever force it may have had in other directions, geography – like other forms of knowledge in the sixteenth and seventeenth centuries – was also deeply implicated in magic and myth, cosmography shaded indiscriminately into astrology, and the shores of empirical science were still distant, blurred. On this reading, the difference between the sixteenth and the eighteenth centuries turns not so much on the distinction between myth and direct observation as on the different status accorded to information derived from the two. Those "other" knowledges are absent from the standard histories of geography too, but Stoddart's point is simply that they have no place in the *modern* discipline at all. His diagnostic question is equally simple: "When did *truth* become our central criterion?"

It is not, of course, as simple as all that, and I will want to suggest considerable caution about absolutizing "Truth" in this way. But what Stoddart has in mind is the emergence of geography as a quintessentially *empirical science*, and I want to begin by thinking through some of the implications of his argument.

Adventures in natural history

Stoddart's history is unorthodox. He dates the transformation of geography into an empirical science to 1769, the year in which Cook first entered the Pacific. The apparent precision of the date is deceptive, since in many ways the invocation of Cook is figurative: He is made to stand for a cluster of overlapping intellectual traditions. But in more conventional histories of geography he does not appear at all; or if he does, it is as little more than a bystander whose voyages are used to correct Ptolemy's errors and complete the outlines of the world map, so that they become narrowly *empirical* in significance rather than way stations en route to an empirical *science*. Only in Glacken's *Traces on the Rhodian shore* does Cook appear, as he does in Stoddart's account, as the superintendent of "a scientific undertaking, a

---

[4] David Livingstone, "Geography, tradition and the scientific revolution: An interpretative essay," *Transactions, Institute of British Geographers* 15 (1990) pp. 359–73; the quotation is from p. 364.

harbinger of the nineteenth-century scientific traveling of Humboldt, Darwin, and the *Challenger*," and as the bearer of a discourse that legitimated itself through "reliability in detail and authenticity."[5]

The ostensible purpose of Cook's first voyage (1768–1771) was to participate in an international project to observe the transit of Venus, an event that would not take place again until 1874. If the passage of the planet across the face of the sun could be observed at different stations around the world, it was hoped that astronomers would be able to calculate the distance between the earth and the sun, an achievement of considerable importance to both science and the art of navigation. Accordingly, Cook's instructions were to sail to Tahiti and set up a temporary observatory. But there was also a substantial interest both within the Royal Society and at court in having Cook explore and chart the South Pacific and, so it was hoped (still more fervently, I imagine), "discover" the hypothesized great southern continent, *terra australis incognita*. Such a mission had obvious implications for Britain as a maritime power, and this brief was duly incorporated into Cook's secret instructions.

Shortly before the *Endeavour* set sail, however, a young botanist, Joseph Banks, persuaded the Admiralty to allow him to accompany the expedition. Following the precedent set by Bougainville and other French voyagers, he brought with him two naturalists, both former students of the Swedish botanist Carl von Linné (Linnaeus), and two illustrators. Linnaeus was the author of *Systema naturae* (1735), one of the founding texts of natural history, and he soon learned of Banks's intentions. He was informed by one correspondent that:

No people ever went to sea better fitted out for the purpose of Natural History. They have got a fine library of Natural History; they have all sorts of machines for catching and preserving insects; all kinds of nets, trawls, drags and hooks for coral fishing; they have even a curious contrivance of a telescope, by which, put into the water, you can see the bottom at great depth.[6]

Similar provisions were made for subsequent voyages. Johann Forster and his son George accompanied Cook on his second voyage, while Banks himself became President of the Royal Society and acted as the veritable "custodian of the Cook model" for later expeditions.[7]

[5] Clarence Glacken, *Traces on the Rhodian shore: Nature and culture in western thought from ancient times to the end of the eighteenth century* (Berkeley: University of California Press, 1967) p. 620.

[6] J. C. Beaglehole, (ed.), *The Voyage of the* Endeavour, *1768–1771* (Cambridge: Cambridge University Press, 1955) p. cxxxvi. Beaglehole calls Linnaeus "the high priest of botanical studies" but I prefer another metaphor. Donald Worster seems much closer to the mark in describing him as one of the agents of "the empire of reason": see his *Nature's economy: A history of ecological ideas* (Cambridge: Cambridge University Press, 1985) p. 26 and *passim*.

[7] David Mackay, *In the wake of Cook: Exploration, science and empire, 1780–1801* (London: Croom Helm, 1985) p. 20.

Stoddart argues that the work of these teams of scientists, collectors, and illustrators displayed three features of decisive significance for the formation of geography as a distinctly modern, avowedly "objective" science: a concern for realism in description, for systematic classification in collection, and for the comparative method in explanation. But Stoddart presses this further. Neither Banks nor the Forsters confined themselves to plants and animals. They also took a keen interest in peoples, and in much the same way. Although Banks participated in many of the practices and rituals of the aboriginal peoples he encountered in Tahiti and elsewhere, he seems to have approached them "as he did any other species"; the Forsters were uninterested in even Banks's rudimentary attempts at participant observation, and preferred to hold their objects at a distance and attempt "above all else to compare and categorize the cultures they observed."[8] Hence what I take to be Stoddart's central claim: that it was "the extension of [these] scientific methods of observation, classification, and comparison to peoples and societies that made our own subject possible."[9]

This is a revisionist history, as I have already indicated, and Stoddart makes no bones about it:

My heroes are not the usual ones – the Ritters, Ratzels, Hettners, entombed by conventional wisdom. My geography springs from Forster, Darwin, Huxley: and it works.[10]

Yet one ought not to take this too literally. For all the disclaimers, the traditional pantheon continues to cast its shadow over Stoddart's narrative. "Humboldt was born in the year Cook first saw the Pacific," he remarks, "Ritter in the year he died there." Indeed, Banks and the Forsters are cast as the precursors of a tradition of field science (rather than "desk-bound scholasticism") that *culminated* in the work of Humboldt, who thus becomes "the inheritor of the great tradition of exploratory field science of the Enlightenment."[11] Stoddart plainly invokes Cook's voyage in order to

---

[8] Lynne Withey, *Voyages of discovery: Captain Cook and the exploration of the Pacific* (New York: William Morrow, 1987) pp. 111–12, 120, 290.

[9] Stoddart, "Geography" pp. 32–33, 35. It is one thing to claim that these developments made modern geography possible, but quite another to say that they also made it *distinctive* from other branches of knowledge, including anthropology and sociology (p. 29). I do not think this second, stronger thesis can be sustained, and in what follows I focus on the first.

[10] *Ibid.*, p. ix. Revisionist this may be in its way, but in other ways, as I will show in a moment, Stoddart follows the conventions of mainstream historiography: a celebration of heroes not heroines, and *European* heroes to boot. I focus primarily on that troublesome adjective, but for a preliminary discussion of the gendered construction of mainstream historiographies see Mona Domosh, "Toward a feminist historiography of geography," *Transactions, Institute of British Geographers* 16 (1991) pp. 95–104.

[11] Stoddart, *On Geography*, p. 36; *idem*, "Humboldt and the emergence of scientific geography"; I have quoted from the typescript because, as far as its author knows, the volume for which the essay was commissioned never appeared in print.

anchor geography within natural history and the natural sciences, but he also wants to establish that science, exploration and, indeed, "adventure" occupied (and, by implication, continue to occupy) a common space: to show that field science was (and remains) an indissoluble moment in the advance of science more generally. That sense of (manly?) adventure is not incidental to Stoddart's own investigations, and his history constantly invokes it. "On uninhabited Pacific atolls," he writes, "sailing along the barrier reefs of Australia and Belize, in the mangrove swamps of Bangladesh, on English coastal marshes, I have been concerned with making sense of nature." Adventure is an integral part of Stoddart's geography, but it exists in tension with his conception of science. For "adventure" surely requires the emotional investment and the affective response that he finds in Darwin's reaction to nature, which he sees as both a paradox and a complement to Darwin's view of science. "On the one hand, Darwin's reaction was emotional and integrative; he grasped like Humboldt, the harmony and wholeness of the Cosmos." Making sense of nature in *this* way was a response to the siren song of the romantic sublime, Stoddart suggests, and it was reflected in Darwin's vocabulary: He found "stillness, desolation, solitude," felt "wonder, awe, admiration." And yet "on the other hand, there was his analytical, instrumental, scientific approach, measuring and comparing, dissecting objects out from their background for closer investigation."[12]

I will return to this distinction in a moment, but I should first emphasize that there is no doubt about the importance of the *Endeavour* nor of the extraordinary significance of the South Pacific for the development of the natural sciences. Over the next century or so, as Smith notes, the region "provided a challenging new field of experience for Europeans, one which placed unprecedented pressure upon the Biblical creation theory and provided, simultaneously, a wealth of new evidence out of which was fashioned eventually the first scientifically credible theory of evolution."[13] Seen in this light, Stoddart's more particular purpose is to use Cook's voyage to establish the historical provenance of what he has long taken to be Darwin's seminal importance for modern geography.[14] Historians sailing in different waters, guided by different geographies, would evidently identify other routes and other ports-of-call. If the Napoleonic expedition to Egypt at

---

[12] Stoddart, "On geography," p. ix; *idem*, "Geography, exploration and discovery," *loc. cit.*, pp. 142–57; *idem*, "Grandeur in this view of life," *loc. cit.*, pp. 219–29.

[13] Bernard Smith, *European vision and the South Pacific* (New Haven: Yale University Press, 1985) p. viii.

[14] "Much of the geographical work of the past hundred years . . . has either explicitly or implicitly taken its inspiration from biology, and in particular from Darwin": David Stoddart, "Darwin's impact on geography," in his *On geography*, pp. 158–79. Other scholars prefer Lamarck; see J. A. Campbell and David Livingstone, "Neo-Lamarckism and the development of geography in the United States and Great Britain," *Transactions, Institute of British Geographers* 8 (1983) pp. 267–94.

the end of the eighteenth century was a "sort of first enabling experience for modern Orientalism," for example, it also provided a beacon for the reorientation of French geography.[15] Stoddart makes little of these ventures, and his "European science" is a largely English affair. Still, the importance of his argument lies in the connections he makes between Europe and the globalizing project of natural history. For it is those connections, it seems to me, that are directly implicated in the relations between geography and anthropology.

European science and its regime of truth

I should begin by clarifying "natural history." There are several standard historiographies, but I have found Foucault's account in *The order of things* particularly helpful. Foucault treats natural history as a "regional" space within what he calls the classical episteme. An episteme is, very roughly, a conceptual grid that provides conceptions of order, sign and language that allow a series of discursive practices to qualify as "knowledge." In Renaissance Europe, Foucault argues, the episteme was structured by resemblance, a way of thinking and being in the world in which there was no gap between "words" and "things," no difference in principle between signs on parchment and signs in nature. The world was known through a ramifying network of signatures, each one providing a glimpse into the design of the perfect whole. In the late seventeenth and eighteenth centuries, however, Foucault claims that a space opened between the two: as "words" were dissociated from "things" so resemblance yielded to *representation*. And it was within that gap that the discourse of natural history was constituted as part of a project to navigate the passage between the two or, as Foucault puts it, "to bring language as close as possible to the observing gaze, and the things observed as close as possible to words."

Since this concern occupies so much of this chapter, I ought to underline two of its most important implications. The first is the accent on vision and visibility. Natural history, says Foucault, "is nothing more than the nomination of the visible," revealing what is seen through what is said, and is concerned with the structure of the visible world, with "surfaces and lines" not with "functions and invisible tissue."[16] This matters because

---

[15] Edward Said, *Orientalism* (London: Penguin Books, 1985) p. 122; for a more detailed – and more nuanced – account, see Anne Godlewska, "Traditions, crisis and new paradigms in the rise of the modern French discipline of geography, 1760–1850," *Annals of the Association of American Geographers* 79 (1989) pp. 192–213.

[16] Michel Foucault, *The order of things: An archaeology of the human sciences* (London: Tavistock, 1970); this was first published in France in 1966.

by this means "the idea of a natural 'history' takes on a meaning close to the original Greek sense of a 'seeing'."[17] So, for Linnaeus, the rationality of nature could be disclosed through a taxonomy that classified plants by the visual characteristics of their reproductive parts. Only by this means, Goldsmith later wrote in his *History of the Earth*, by system and method, could one "hope to dissipate the glare, if I may so express it, which arises from a multiplicity of objects at once presenting themselves to the view." If this was the "Ariadne's thread" that would guide the botanist through the labyrinth, as Linnaeus argued, it was also a mapping of a logical order onto a supposedly "natural order": a sort of "optics" of plant morphology in which nature was seen as a table and *spatialized*.[18] The second implication follows directly: natural history is, in the modern sense of the word, thus resolutely *non*-historical.

The locus of this history is a *non-temporal* rectangle [the table] in which, stripped of all commentary, of all enveloping language, creatures present themselves one beside the other, their surfaces visible, grouped according to their common features and thus already virtually analyzed, and bearers of nothing but their own individual names.[19]

With these considerations in mind, Stoddart's account might be clarified in three main ways. In the first place, the "objective science" which is the lodestar of his history must not be confused with any naive realism. Stoddart accentuates the importance of measurement and calibration, of innovations in instrumentation and illustration, which enabled reports to be based on "direct observation." The power of these reports was further enhanced by the practice of collecting which, as Thomas has shown, attested in the most insistently material of ways to the fact of having visited remote places and observed novel phenomena. These specimens could then be inscribed within an emergent grid of intellectual authority in Europe: "Those who speculated without having engaged in direct observation were

---

[17] Garry Gutting, *Michel Foucault's archaeology of scientific reason* (Cambridge: Cambridge University Press, 1989) p. 163. And, indeed, Foucault's own work has as one of its central concerns a history of ways of seeing: John Rajchman, "Foucault's art of seeing," in his *Philosophical events: Essays of the 80s* (New York: Columbia University Press, 1991) pp. 68–102.

[18] Rajchman, "Foucault's art," pp. 75–76; James Larson, *Reason and experience: The representation of natural order in the work of Carl von Linné* (Berkeley: University of California Press, 1971). The language of mapping is not misplaced; Larson draws attention to Linnaeus's justification of the systematic arrangement of nature into five groups: "Geography passes from kingdom to canton through the intervening province, territory, and district; military science passes from legion to soldier by means of cohort, maniple, and squad" (p. 150). The analogy speaks directly to Foucault's triad of power, knowledge, and spatiality; see Michel Foucault, "Questions on geography," in his *Power/knowledge: Selected interviews and other writings, 1972–1977* (ed. Colin Gordon) (Brighton: Harvester Press, 1980) pp. 63–77.

[19] Foucault, *Order of things*, p. 131; emphasis added.

rendered unreliable by the special claims of a few to direct knowledge."[20] But these reports and observations were direct only in the (important) sense that they were produced by people who visited the lands they described; they continued to be mediated by European conceptual categories and European ways of seeing. I have used the plural of those terms because it seems necessary to resist essentializing "Europe" and implying a singular European gaze. As Smith shows with great care and subtlety, there was a complex dialogue between the contrasting traditions that the draftsmen and the artists on board the *Endeavour* brought to bear on their work.

The influence of the former on the latter seems to have been decisive, so much so that in the course of the early nineteenth century "the sciences of visible nature," as Smith calls geology, botany, zoology and (significantly) anthropology, "imposed their interests upon the graphic arts." But this new interest in accurate recording nonetheless continued to operate "within an intricate interplay of ideas in which, for the most part, European observers sought to come to grips with the [new] realities of the Pacific by interpreting them in familiar terms." This sort of argument does not turn on an extreme cultural relativism, but neither does it treat the process of representation as unitary or unproblematic. That process was rendered still more complicated on Cook's return to England, when the Admiralty hired a freelance writer, John Hawkesworth, to turn the journals and the log of the first voyage into a narrative that would appeal to a general audience. In order to do so, Hawkesworth drew freely on classical mythology and travelers' tales to embellish and dramatize the narrative, and the original illustrations were reworked into images that would be familiar to a European public.[21]

In the second place, the figure of Cook does not represent quite the scientific tradition for which Stoddart seeks paternity. According to Paul Carter, Cook's own journals stand in contrast to the records made by Banks and the other scientists who accompanied him. Cook's was (quite literally) a "geo-graphy," a writing of lands, which respected the intrinsic and singular spatiality of experience – the phrase is Carter's – rather than the spatiality of classification, whereas Banks produced a botany that was preoccupied with abstraction, with domesticating difference by caging it within a general taxonomic grid.[22] Although Cook's journal was the product of a great deal

[20] Nicholas Thomas, *Entangled objects: Exchange, material culture and colonialism in the Pacific* (Cambridge: Harvard University Press, 1991) p. 141. I use the conditional because collecting was not always conceived thus, and Thomas distinguishes between a mere "curiosity" which had no assertive intellectual framework within which the objects could be classified or hierarchized and "a more theoretical discourse," which did indeed define objects as scientific specimens within some ostensibly rational framework (pp. 140–41).

[21] Smith, *European vision*, pp. viii, 5, 333; Beaglehole, *Endeavour*, pp. ccxlii–ccliii; Withey, *Voyages*, pp. 175–76.

[22] Paul Carter, "An outline of names," in his *The road to Botany Bay: An essay in spatial history* (London: Faber, 1987) pp. 1–33.

Figure 2 Joseph Banks surrounded by trophies from the voyage of the Endeavour [original by Benjamin West, December 1771; mezzotint engraving by J. R. Smith, 1773].

of drafting and redrafting, and in the process he copied some of Banks's own notes, Carter insists that the distinction is one of substance:

Banks's interest in taxonomy quite excludes as part of his knowledge the circumstances of discovery. Knowledge, for Banks, is precisely what survives unimpaired the translation from soil to plate and Latin description. There is, in Banks's philosophy, no sense of limitation, no sense of what might have been missed, no sense of the particular as special. By a curious irony, even though he sets out to botanize on the supposition his botanical knowledge is incomplete, his knowledge is always complete: each object, found, translated into a scientific fact, and detached from its historical and geographical surroundings becomes a complete world in itself.... [Its] existence in a given, living space is lost in the moment of scientific discovery.[23]

The blank surround of the botanical plate was dead, Carter continues, but the blank spaces of Cook's maps and charts were active; and the lines, numbers, and names drawn on them preserved, in graphical form, the traces of particular encounters and the memories of particular experiences, and they also prefigured – made possible – the particulars of other journeyings. "The same calculations that enabled him to steer a course also enabled him to leave the coastlines he sighted where they were."[24]

In the face of Carter's enthusiasm for Cook, it needs to be remembered that it was *Cook* who sighted those coastlines and that the "spatiality of experience" that he respected was his *own*: that this was a way of drawing that world within his own, European horizon of meaning. But he did so in ways that were significantly different from Banks, and the distinction can be drawn in another, still more general way. In his lectures at Königsberg on geography, Kant emphasized that classification could proceed either logically or physically. It is the first of these that Stoddart has in mind when he writes so approvingly of the extension of the scientific method of classification to peoples and societies: It provides for what Kant called a "natural system" – "a system of nature" like that of Linnaean taxonomy – in which elements are placed in the same category on the basis of their *similarity*. But Kant insisted that history and geography proceeded differently; that they both relied on a physical classification that

---

[23] Carter, "An outline," pp. 21–22. If nature was to be read as a book, however, the grammar of Botany Bay was far from being immediately intelligible. The first president of the Linnaean Society confessed that in such situations the botanist "can scarcely meet with any fixed points from which to draw his analogies; and even those that appear most promising, are frequently in danger of misleading, instead of informing him. The whole tribes of plants, which at first sight seem familiar to his acquaintance, as occupying links in Nature's chain, on which he is accustomed to depend, prove, on nearer examination, total strangers." It was, of course, the function of Linnaean taxonomy to issue those "total strangers" with their identity papers. See Ross Gibson, *South of the West: Postcolonialism and the narrative construction of Australia* (Bloomington: Indiana University Press, 1992) pp. 24–25.

[24] Carter, *Botany Bay*, p. 23.

places elements together on the basis of their *proximity*. If there was a precursor of Stoddart's geography on board the *Endeavour*, then it must surely be Banks, not Cook, and the geography that he inaugurates is evidently very different from Kant's. The consequences of spatializing knowledge in this abstract, logical way (figure 2) are brought out very clearly, I think, in this passage from Mary Louise Pratt's aptly titled *Imperial Eyes*:

One by one the planet's life forms were to be drawn out of the tangled threads of their life surroundings and rewoven into European-based patterns of global unity and order. The (lettered, male, European) eye that held the system could familiarize ("naturalize") new sites/sights immediately upon contact, by incorporating them into the language of the system. The differences of distance factored themselves out of the picture . . .[25]

Within this optic, those differences were spatialized: They were placed in that "non-temporal rectangle," the table, which had no space for Kant's geography.

   In the third place, and following directly from these remarks, the generalizing science that is supposed to have provided the origin of modern geography (and much else) was no simple "extension." Placing other peoples and societies within its highly particular horizon of meaning had the most radical of consequences for the constitution of the human sciences and their conception of human subjects. The two "great devices" that Stoddart singles out for particular attention in bringing "the huge diversity of nature" within "the bounds of reason and comprehension" – namely, classification and comparison – were the central pinions of the classical episteme. Its claims to "truth" were not transhistorical, however, and neither was the tribunal to which it appealed. What displaced the classical episteme and inaugurated the modern episteme was, precisely, the incorporation of "man" within the conceptual grid of European knowledge. Here too Kant is a figure of decisive importance, and Foucault suggests that his *Critique of Pure Reason*, published 20 years after Cook's first voyage, marked "the threshold of our modernity" by drawing attention to the epistemic structuring of the world by the human subject. "As the archaeology of our thought easily shows," Foucault concluded in a striking and often repeated phrase, "man is an invention of recent date." What he meant was that the limits of representation were not breached until the years between 1775 and 1825: It was then that "man" was constituted as both an object of knowledge and a subject that knows, and in the process the space of European knowledge "toppled," "shattered," and dissolved *into a radically new configuration.*[26]

---

[25] Mary Louise Pratt, *Imperial eyes: Travel writing and transculturation* (London: Routledge) p. 31.

[26] Foucault, *Order of things*, pp. 221, 312, 386. The gendering of these passages is (in my case) deliberate and (in any case) entirely appropriate: see Genevieve Lloyd, *The man of reason: "Male" and "female" in Western philosophy* (London: Methuen, 1984).

Toward the end of the eighteenth century, the classical identity between thought and representation gave way to systematic reflection on the specificity of representation, to a characteristically modern emphasis on *reflexivity*. There are two thematics in Foucault's account that seem to bear directly on the protomodern discourses of anthropology and geography. The first was the investment of knowledge with historicity. European knowledge invented for itself "a depth in which what matters is no longer identities, distinctive characters, permanent tables with all their possible paths and routes, but great hidden forces on the basis of their primitive and inaccessible nucleus, origin, causality and history."[27] The consequences for anthropology and, I think, anthropogeography were dramatic. In the course of the nineteenth century both discourses pivoted around an axis "whereby differences residing in geographical space were [rotated] until they became differences residing in developmental historical time." As McGrane puts it, "*beyond* Europe was henceforth *before* Europe."[28]

The second thematic was the dissolution of knowledge as a homogeneous space. Foucault argues that once the unitary field of visibility and order was "opened up in depth," knowledge was no longer mapped as a single, general science of order but was instead distributed across a three-dimensional space. On the first side were the mathematical sciences; on the second side, the empirical sciences (biology, economics and linguistics); and on the third side, philosophical reflection. These axes triangulated the space of the new "human sciences" (including sociology and psychology), which existed in a state of constant, creative, and highly unstable tension with their enframing discourses (figure 3). In turn they attempted mathematical formalizations, they borrowed models from the empirical sciences, and they surrendered themselves to philosophical reflection.[29] The implications of this second set of claims for anthropology and human geography are more complicated. The position of anthropology was ambiguous: Foucault describes it as a "counter-science" that "ceaselessly 'unmakes' that very man who is creating and re-creating his positivity in the human sciences." Human geography had no place at all within the human sciences as Foucault conceived them, although it would not be difficult to trace the same epistemic displacements in its modern history: mathematization; biological, economic and linguistic analogues; philosophical reflection. My purpose is not to imply that Foucault's archaeology is unproblematic, however, still less that it can be used as a template from which to recover the archaeologies of anthropology or geography in any direct way: Foucault's own

[27] Foucault, *Order of things*, p. 251.

[28] Bernard McGrane, *Beyond anthropology: Society and the other* (New York: Columbia University Press, 1983) p. 94.

[29] Foucault, *Order of things*, pp. 347, 387.

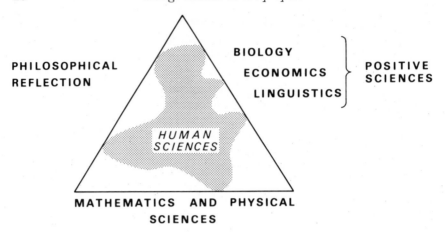

Figure 3 Foucault's archaeology of the human sciences.

view was quite the opposite.[30] But I do think that the disjuncture – the "threshold" – that he places between the classical and the modern calls into question any simple "extension" of the methods of the natural sciences "to people and societies."

Foucault's own account concerned the incorporation of European "man" within the grid of the human sciences. It was this process, so he claimed, that was centrally involved in the constitution of a modality of power that was characteristic of modern societies. He argued that the prevalent form of power in previous societies was *sovereign power*, a juridico-political model in which power "descended" from the heights to the depths, from the center to the margins. But in the course of the eighteenth century this was joined and jostled (but not replaced) by a profoundly nonsovereign power: an anonymous, polymorphous, and capillary *disciplinary power*, rooted in the depths and the margins, which established what Foucault called a society of normalization:

As power becomes more anonymous and more functional, those on whom it is exercised tend to be more strongly individualized; it is exercised by surveillance rather than ceremonies, by observation rather than commemorative accounts, by comparative measures that have the "norm" as reference rather than genealogies giving ancestors as points of reference.[31]

Disciplinary power is inherently spatializing, Foucault suggests, in its gaze, its grids, its architectures. It is also intrinsically productive: It produces

---

[30] Michel Foucault, "Questions on geography," in his *Power/knowledge: Selected interviews and other writings, 1972–1977* (Brighton: Harvester, 1980) pp. 63–77.

[31] Michel Foucault, *Discipline and punish: The birth of the prison* (London: Penguin, 1977) p. 193; this was first published in France in 1975; see also "Two lectures" in his *Power/knowledge*, pp. 78–108.

human bodies as human subjects, a process that Foucault calls *assujetissement* to convey both subjection and "subjectification." These are provocative suggestions and they have much to say about the constitution of a dispersed "spatial science" of sorts, already visible in the seventeenth century and emboldened in the course of the eighteenth and nineteenth centuries, concerned with what Foucault describes as an anatomo-politics of the human body and a bio-politics of the human population.[32]

These developments are, I suspect, of direct relevance to any genealogy of the discourse of geography. But Foucault's account of the human sciences was confined to Europe and, according to one otherwise sympathetic commentator, he seemed largely "unaware of the extent to which [these] ideas of discourse and discipline are assertively European" and of how they were used to administer, study, and reconstruct, and then to occupy, rule, and exploit "almost the whole of the non-European world." For this reason Said argued that much of what Foucault urged "makes greatest sense not as an ethnocentric model of how power is exercised in modern society, but as part of a much larger picture involving the relationship between Europe and the rest of the world."[33] Maybe so; other critics are more skeptical. But if one does move not only beyond Foucault's archaeology toward his genealogy but also beyond Europe, I am not sure that the distinction between sovereign and disciplinary power is then able to address the incorporation of *non-European* "man" into the table and the grid.

Many writers have argued that the European construction of the other as a figure of "nonreason" was an essential moment in the self-constitution of the European subject as the sovereign figure of reason and normality. And it is certainly possible to reverse the direction of the Linnaean gaze, as Pratt suggests, "looking at Europe from the imperial frontier" and connecting the scientific appropriation of the non-European world to processes of discipline and normalization within Europe. But suppose that gaze is held steady; suppose the imperial frontier is kept in view – what then? The self-image of late eighteenth-century science was undoubtedly one of innocence, and Pratt makes much of its planetary project as a discourse of "anticonquest" – a means whereby Europeans were promised intellectual possession of the world without physical dispossession of the people who inhabited it. What she calls the "conspicuous innocence" of the naturalist thus acquired meaning only in relation to the unnamed but

---

[32] Michel Foucault, *The history of sexuality: An introduction* (London: Penguin Books, 1978) pp. 139–45; this was first published in France in 1976.

[33] Edward Said, "Criticism between culture and system," in his *The world, the text, and the critic* (London: Faber, 1984) pp. 178–225; the quotation is from p. 222. It is for this reason that Said makes so much of the Napoleonic expedition. Its objective was "to render [Egypt] completely open, to make it totally accessible to *European* scrutiny" and thereby turn it "into a department of French learning": Said, *Orientalism*, p. 83.

nonetheless palpable "guilt of conquest." But the incorporation of non-European "man" into the table, the taxonomy and the grid effectively prised non-European people away from the land which they inhabited, and once they had been *textually* removed from the landscape, it was presumably easier to do so *physically* as well. It was not necessary to confine them to the formal categories of a European taxonomy since the rules of logical classification – of a "sameness" inscribed within an economy of difference – continued to underwrite less formal textual strategies.

As Pratt shows in a particularly powerful essay, for example, John Barrow's account of his early nineteenth-century travels through "the land of the Bushmen" in southern Africa was remarkable for the way in which the narrative of the journey and the description of the landscape were emptied of indigenous human presence. "The Bushmen" (the !Kung) are present in the next only as faint scratches on the surface of the land, traces of cultivation and signs of habitation, until they are eventually paraded around a separate ethnographic arena, a "bodyscape" altogether separate from the landscape.[34] But this is not sovereign power, I think, and neither is it – in this form – disciplinary power either.

In any event, from this perspective it is surely incontrovertible that the "extension" of the methods of natural history and, ultimately, of natural science to peoples and societies was powered by more than the sails of reason. These nominally scientific advances were at the same time, in complex and sometimes opaque ways, the spearheads of colonialism and conquest. For this reason I am puzzled by the contrast Stoddart seems to make between the European discovery of America in the fifteenth and sixteenth centuries and their discovery of Australia and the South Pacific in the eighteenth century.

In the first case, the questions raised were primarily theological and philosophical rather than scientific. . . . It proved impossible to subsume the cultures of the Andes and the Amazon within accepted European models: in consequence European accounts of the New World are overwhelmingly records of conquest and destruction.[35]

But matters were more complicated. The European discovery of the New World brought in its wake, by virtue of its "newness," what Pagden describes as a "metaphysical unease" which may have contributed directly to the eventual hegemony of the experimental method. Precisely because the New World could not be codified within the existing corpus of fixed texts, Kepler believed that Columbus had been instrumental in showing that "modern man had no choice but to be a rational experimentalist." This was a thoroughly modern reading of Columbus, however, and Pagden

---

[34] Pratt, *Imperial eyes*, pp. 34–36, 58–67.
[35] Stoddart, *On Geography*, p. 32.

suggests that Humboldt was nearer the historical mark when he saw Columbus as the embodiment of a new *moral* attitude toward nature and "human nature" and compared him to "the two great poets of the natural world, Buffon and Goethe."[36] But I am not sure that one has to choose between these interpretations, and it may be that this is the rock on which Stoddart's comparison founders too. Although Columbus was untutored in the sciences, Butzer argues that his ability "to observe, compare and describe" was of immense importance in establishing a tradition of scientific appropriation that extended from his son Fernando Colón's landscape taxonomy through Oviedo's biotic taxonomy to Acosta's attempt to establish "a scientific and ontological framework for the New World" in his *Historia natural* (1590). It seems equally likely that the moral dimensions of that intellectual encounter were far from straightforward. While he by no means diminishes the impress of domination and destruction, Greenblatt suggests that Columbus's voyage initiated "a century of intense wonder" – a project that involved not only "the colonizing of the marvelous" but, if one reads some texts in particular ways (and I hear those hesitations and qualifications too), also conferred upon the marvelous what he calls "a striking indeterminacy." His argument is a rich and subtle one, but it turns on the suggestion that "wonder" leads (and lead) in two opposite directions. One is the path to estrangement, objectification and possession; the other, less familiar, is the path to self-recognition and acceptance of alterity. As I read Greenblatt, even Columbus, on occasion, and as he followed the first, seems to have contemplated the second.[37]

These complications do not reverse Stoddart's judgment in the first case. But what of the second? The scientific project represented by Cook's voyages may not have been "interventionist in any immediate sense," as Thomas cautions, and it would be quite wrong to foreshorten the colonial past.[38] But it is scarcely possible to claim that the European encounter with the peoples of the South Pacific was disinterested or that the protocols of "objective science" left their world as it was. As the Comaroffs have recently reminded us:

The essence of colonization inheres less in political overrule than in seizing and transforming "others" by the very act of conceptualizing, inscribing and interacting with them on terms not of their choosing; in making them into pliant objects and

---

[36] Anthony Pagden, "The impact of the New World on the Old: The history of an idea," *Renaissance and modern studies* 30 (1986) pp. 1–11.

[37] Karl Butzer, "From Columbus to Acosta," *Annals of the Association of American Geographers* 82 (1992) pp. 543–65; Stephen Greenblatt, *Marvelous possessions: The wonder of the New World* (Chicago: University of Chicago Press, 1991) pp. 14, 24–25, 64–65, 135. Stoddart describes Oviedo and Acosta as "rare exceptions" but they were nonetheless immensely important and their natural histories were in many respects seminal works.

[38] Thomas, *Entangled objects*, p. 139.

silenced subjects of our scripts and scenarios; in assuming the capacity to "represent" them, the active verb itself conflating politics and poetics.[39]

In other ways, of course, "they" were neither pliant nor silenced, and in the early phases of the colonial encounter in particular indigenous peoples could be "no less powerful and no less able to appropriate than the whites who imagine[d] themselves as intruders."[40] This is an important qualification, but others would sharpen the point still further and insist on the intrinsic ambiguity of inscription and "possession." Thus:

Possessing Tahiti was a complicated affair. Indeed, who possessed whom? Native and Stranger each possessed the other in their interpretation of the other. They possessed one another in an ethnographic moment that got transcribed into text and symbol. They each archived that text and symbol in their respective cultural institutions. They each made cargo of the things they collected from one another, put their cargo in their respective museums, remade the things they collected into new cultural artifacts. They entertained themselves with their histories of encounter. Because each reading of the text, each display of the symbol, each entertainment in the histories, each viewing of the cargo enlarged the original encounter, made a process of it, each possession of the other became a self-possession as well. Possessing the other, like possessing the past, is always full of delusions.[41]

So it is. But surely one of the most insidious delusions is to presume that this was an equal and symmetrical process.

Stoddart's "objective science" was objective – and no doubt, remains objective – in all of these ways, then, and yet its modalities of power lie beyond the compass of his account. To be sure, one must make allowances for the heterogeneity of European colonial discourses. If these aimed to construct a "mechanism of mastery" – and literally so, as the women displayed as the frontispiece of Stoddart's history of geography suggest – they did not speak with one voice and neither were they altogether indifferent to context. But their main thrust (and the sexual connotation is not inappropriate) can be stated quite directly. As one of those most affected by the expeditions put it 200 years later, on the occasion of the Cook Bicentennial, "Fuck your Captain Cook! He stole our land."[42] In stealing that land those who came after also stole biography, history, and

---

[39] Jean Comaroff and John Comaroff, *Of revelation and revolution: Christianity, colonialism and consciousness in South Africa*, vol. 1 (Chicago: University of Chicago Press, 1991) p. 15.

[40] Thomas, *Entangled objects*, p. 184.

[41] Greg Dening, "Possessing Tahiti," *Archaeology in Oceania* 21 (1986) pp. 103–18; the quotation is from p. 117.

[42] The remark is quoted in H. C. Brookfield, "On one geography and a Third World," *Transactions, Institute of British Geographers* 58 (1973) pp. 1–20; the quotation is from p. 10. Brookfield's repetition of the remark caused a fluttering in the colonial dovecotes: the language was deemed objectionable. For a survey of other, more objectionable aspects of the encounter, see Alan Moorehead, *The fatal*

identity. George Forster had warned as much: "If the knowledge of a few individuals can only be acquired at such price as the happiness of nations, it would be better for the discoverers, and the discovered, that the South Sea had still remained unknown to Europe and its restless inhabitants."[43]

In all of these ways, I think one must conclude that Stoddart's geography was plainly not only a distinctively "European science," as he says, but also a distinctly Euro*centric* science. Surprisingly, Stoddart's own historiography contributes to that sense of closure. Non-European traditions of geography are disallowed and even dispossessed of their own "intellectual structure" in ways that strikingly confirm Said's view of one of the essential motifs of European imaginative geography: "A line is drawn between two continents. Europe is powerful and articulate; Asia is defeated and distant."[44] And yet, as the previous paragraphs have shown, the trajectory of this "modern" geography, this "European science," cannot be separated from those other societies and the "people without history."[45] But this requires a way of remapping those spaces of power-knowledge or, better, of exploring the interconnections between power, knowledge, and spatiality. As I have suggested, some of the most seminal cross-fertilizations during this period were those between anthropology and geography and what became known as anthropogeography is a tradition of basic importance to the formation of the modern discipline. This is necessarily to use "anthropology" and "geography" in highly general, diffuse, and overlapping ways. Kant lectured on both of them at Königsberg for over 30 years and believed that together they formed the totality of pragmatic, empirical knowledge of the world.[46] But neither of them was formalized, institutionalized, and professionalized in a modern, disciplinary sense within the Western academy until the closing decades of the nineteenth century. Anthropogeography continued to be of great moment; one reads its concerns in Ratzel and also in Sauer. Yet it was the encounter with another infant discipline, sociology, that added a new dimension to geography as an "objective science."

---

impact: An account of the invasion of the South Pacific, 1767–1840 (London: Penguin, 1966). The representation of "the land" as feminine, to be mastered by a powerful masculine reason, is a common trope of colonial discourse and I discuss it in more detail below, pp. 129–32.

[43] Withey, *Voyages*, p. 230.

[44] Said, *Orientalism*, p. 57. There were powerful, articulate Arabic and Chinese traditions of geography.

[45] The phrase is, of course, ironic; I have borrowed it from Eric Wolf, *Europe and the people without history* (Berkeley: University of California Press, 1982). See also Johannes Fabian, *Time and the other: How anthropology makes its object* (New York: Columbia University Press, 1983) and McGrane, *Beyond anthropology*. All of these texts focus on anthropology; geography is only beginning to develop a critical historiography of comparable sophistication: see Felix Driver, "Geography's empire: histories of geographical knowledge," *Environment and Planning D: Society and Space* 10 (1992) pp. 23–40.

[46] See J. A. May, *Kant's concept of geography and its relation to recent geographical thought* (Toronto: University of Toronto Press, 1970) pp. 107–31.

Borders: sociology and geography

I have said enough to be able to work more directly toward geography
and the world-as-exhibition: to consider the possibility that, by the closing
decades of the nineteenth century, it was a characteristic of European ways
of knowing to render things as objects to be viewed and, as Timothy Mitchell
puts it, to "set the world up as a picture ... [and arrange] it before an
audience as an object on display, to be viewed, investigated, and experienced." This machinery of representation was not confined to the zoo, the
museum, and the exhibition – all European icons of the nineteenth century
– but was, so Mitchell claims, constitutive of European modernity at large.[47]

Enframing the world

To grasp the implications of this way of knowing, Mitchell draws upon
one of Heidegger's later essays, which refers to modernity as "the age of
the world picture," by which he said he did not mean a picture of the
world "but the world conceived and grasped as a picture." To think of a
picture in this way implies both a setting of the world in place before
oneself, as an object over and against the viewing subject, and a making
of the world intelligible as a systematic order through a process of *enframing*.
But Heidegger was advancing a series of more or less ontological claims.
Although he traced enframing back to Descartes in the seventeenth century,
this was a purely nominal gesture: More fundamentally, it was supposed
to represent "the apocalyptic culmination of over two millennia of *Seinsvergessenheit*."[48] I do not propose to follow that abstract history here; but
Heidegger's argument can be made to intersect in complex and historically
more specific ways with the genealogy of linear perspective.

   This had its origins in Brunelleschi's marvelous experiments during the
Italian Renaissance, but it was more than an optics of artistic representation;
more, even, than the visible surface of a technology of power implicated
in the calculations of the merchant, the surveyor, and the general.[49] It also

---

[47] See two works by Timothy Mitchell: "The world as exhibition," *Comparative studies in society
and history* 31 (1989) pp. 217–36; and *Colonizing Egypt* (Cambridge: Cambridge University Press,
1988). What follows is deeply indebted to Mitchell's brilliant account, but I have some reservations
about it and discuss these in more detail below, pp. 175–76, 180–81.

[48] Richard Wolin, *The politics of Being: The political thought of Martin Heidegger* (New York: Columbia
University Press, 1990) pp. 161–62. *Seinsvergessenheit* derived from Heidegger's theory of *Seinsgeschichte*:
an avowedly antihumanist project that sought to disclose the authentic destiny of Being *(Dasein)*
which, so he claimed, had been suppressed (forgotten, *vergessen*) in the course of human history.

[49] Martin Kemp, *The science of art: Optical themes in western art from Brunelleschi to Seurat* (New Haven:
Yale University Press, 1990); Denis Cosgrove, "Prospect, perspective and the evolution of the
landscape idea," *Transactions of the Institute of British Geographers* 10 (1985) pp. 45–62.

Figure 4 Interior with a Geographer *[Jan Vermeer, 1669].*

formed part of a series of Renaissance codes that were involved, crucially, in the development of the camera obscura in the seventeenth century: "codes through which a visual world [was] constructed according to systematized constants and from which any inconsistencies and irregularities [were] banished to ensure the formation of a homogeneous, unified and fully legible space."[50] Crary shows that the camera obscura assumed a

[50] Jonathan Crary, "Modernizing vision," in Hal Foster (ed.), *Vision and visuality* (Seattle: Bay Press, 1988) pp. 29–44; the quotation is from p. 33. For a fuller discussion, see *idem*, "The camera obscura and its subject," in his *Techniques of the observer: On vision and modernity in the nineteenth century* (Cambridge: MIT Press, 1990) pp. 25–66.

preeminent importance in delimiting and defining the relations between observer and world, and it was this, so he claims, that provided both the foundation and the framework for an objectivist epistemology which, as Heidegger suggested, was set out by Descartes. It was also embodied in Vermeer's painting, *Interior with a Geographer* (c. 1669), which shows the scholar at work in a shadowy study, poring over a chart flooded with light from the casement (figure 4). The camera – the study – becomes the site within which an orderly projection of the world is made available for inspection by the mind, and Crary sees the portrayal of the geographer as a particularly lucid evocation of the autonomous subject "that has appropriated for itself the capacity for intellectually mastering the infinite existence of bodies in space."[51] By extension, we can surely say that it was much the same optic that framed Stoddart's "objective science" at the end of the eighteenth century.

Mitchell follows these lines of inquiry into the nineteenth century in order to show that the process of enframing was historically implicated in the operation of colonial, or better "colonizing" power. "Colonizing" is to be preferred, I think, because the constellation of power-knowledge that he has in mind was not confined to Europe's colonies (though, as he shows with considerable sophistication, it was undoubtedly important there too). But precisely because it put in place an inside and an outside, a center and a margin, this was a way of seeing the world as a differentiated, integrated, hierarchically ordered *whole*. What matters is not so much the visual metaphoric as the sense of systematicity, because this is the same unity – the same "field of projection," Crary calls it – provided by the model of the camera obscura. Seen like that, one can better understand Heidegger's claim that "the fundamental event of the modern age is the conquest of the world as picture." For it is through the process of enframing that "man contends for the position in which he can be that particular being who gives the measure and draws up the guidelines for everything that is."[52] Within this modern optic, the "certainty of truth" is made to turn on the need to establish a distance between observer and observed. From that

---

[51] Crary, *Techniques of the observer*, pp. 43–47. It is no accident that in a parallel painting Vermeer depicted the astronomer in the same situation, poring over a celestial globe. The two canvases are sometimes confused, but then, as Crary remarks, in both paintings Vermeer worked with the same figure: "Rather than opposed by the objects of their study, the earth and the heavens, the geographer and the astronomer engage in a common enterprise of observing aspects of a single indivisible exterior" (p. 46).

[52] Martin Heidegger, "The age of the world picture," in his *The question concerning technology, and other essays* (New York: Harper and Row, 1977) pp. 115–54; the essay was originally written in 1938. What is decisive about this, Heidegger argues, is that "Man himself expressly takes up this position as one constituted by himself ... and that he makes it secure as the solid footing for the possible development of humanity. Now for the first time is there any such thing as a 'position' of man" (p. 132). The affinities between these claims and Foucault's subsequent elaboration of the modern episteme are, I think, obvious (see above, pp. 26–28).

position (from that perspective) order may be *dis*-covered and *re*-presented. As Mitchell explains:

Without a separation of the self from a picture . . . it becomes impossible to grasp "the whole." The experience of the world as a picture set up before a subject is linked to the unusual conception of the world as an enframed totality, something that forms a structure or system.[53]

What makes such a conception so unusual is that the process of enframing on which it relies conjures up a framework that seems to exist apart from, and prior to, the objects it contains – a framework that appears "as order itself, conceived in no other terms than the order of what was orderless, the coordinator of what was discontinuous." This is a highly particular way of thinking about – and, indeed, being in – the world, so Mitchell argues, which is peculiar to European modernity. Indeed, non-Occidental visitors to the world exhibitions at the close of the nineteenth century saw them as emblematic of "the strange character of the West, a place where one was continually pressed into service as a spectator by a world ordered so as to represent." Mitchell seizes on this *aperçu* to rework Heidegger's original characterization and bring it into a sharper fin-de-siècle focus. Trading on Benjamin's suggestive sketch of the world exhibitions at the end of the nineteenth century as "sites of pilgrimage to the commodity fetish," in which "the character of a commodity [was extended] to the universe," he describes this particular constellation of power-knowledge not as "the world as picture" but "the world-as-exhibition."[54]

Exhibitions and tableaux

What have these, perhaps forbiddingly abstract considerations to do with geography? In the first place, they can be given the most concrete of forms.

---

[53] Mitchell is here invoking Flaubert's attempt to render the "bewildering chaos of colours" that was (to a European sensibility) nineteenth-century Egypt, but his argument is a general one. In much the same way Said (*Orientalism*, pp. 185–86) notes that Flaubert is interested "not only in the content of what he sees but . . . in *how* he sees, the way by which the Orient, sometimes horribly but always attractively, seems to present itself to him"; he also observes that the vision of the Orient "as spectacle, or *tableau vivant*" (p. 158) was by no means uncommon.

[54] Mitchell, *Colonizing Egypt*, pp. 13–14; Walter Benjamin, "Paris, capital of the nineteenth century," in his *Reflections: Essays, aphorisms, autobiographical writings* (ed. Peter Demetz) (New York: Schocken Books, 1986) pp. 146–62. Benjamin's essay was first published in 1935. What Mitchell's substitution does not make sufficiently clear, however, is the way in which the process of enframing constitutes not only its object *but also its observer*. Partly for that reason, I suspect, it also fails to draw attention to the gendering of the gaze. This rebounds not only Mitchell's specific analysis of colonial Egypt but also on his more general discussion of fin-de-siècle Europe; see also Griselda Pollock, "Modernity and the spaces of femininity," in her *Vision and difference: Femininity, feminism and the histories of art* (London: Routledge, 1988) pp. 50–90.

Figure 5   Reclus's proposed Great Globe at the Place d'Alma, Paris [Galeron, 1897].

Mitchell opens his account at the World Exhibition in Paris in 1889, which coincided with the International Congress of Geographical Sciences presided over by Ferdinand de Lesseps, the architect of the Suez Canal. One of the most imposing exhibits was the enormous Villard-Cotard globe, over 40 meters in circumference, showing mountains, oceans, and cities with what visitors took to be an astonishing verisimilitude. For the World Exhibition of 1900 the geographer Elisée Reclus proposed the construction of a still larger globe. Visitors would travel around its circumference on a spiral staircase or tramway, observing its finely sculptured relief at close quarters, and then move inside where a vast panorama would be projected across its vault (figure 5).[55] Although these globes are powerful expressions of the world-as-exhibition, what matters is not so much their particular iconography but the wider constellation in which they are set: a spectacular geography in which the world itself appeared as an exhibition.

One ought not to be quite so literal about things, however, and I want to suggest in the second (and more interesting) place that much the same

---

[55] G. Dunbar, "Elisée Reclus and the Great Globe," *Scottish geographical magazine* 90 (1974) pp. 57–66. These were by no means the first "great globes," but it was not until the nineteenth century that globes were planned so large that they had to be set outdoors. Two *géoramas* were displayed in Paris in 1823 and 1844, and in 1851 James Wyld, Geographer to the Queen, constructed a globe in Leicester Square as a private venture to capitalize on the tourists flocking to London for the Great Exhibition.

machinery of representation was deployed during the institutionalization of the French school of human geography at the turn of the century. Consider, for a moment, Vidal de la Blache's *Tableau de la géographie de la France*. In this, probably the classic monograph of the French school, written when Vidal was in his late fifties, the country is represented as a picture: one composed of "a multiplicity of shades," "a wealth of tones" – the landscapes of the different localities or *pays* – in which "all discordant tints melt into a series of graduated shades."[56] One critic has suggested that the very title of the book invoked, and was perhaps intended to invoke, "the geographer as landscape painter." For Vidal imposed "solely visual criteria" on his renderings, treating landscape (as he said himself) as "what the eye embraces with a look," as "that part of the country that nature offers up to the eye that looks at it."[57] Others have commented on geography's visual obsession before – its graphos, not logos – but in Vidal's case the gaze was doubly purposive. He was at pains to present France as a coherent and in some sense a *complete* composition.

France is one of those [countries] that took shape earliest. While in the more continental parts of Europe, the great countries of the future, Scythia and Germany, were only looming in semi-darkness, the outlines of France were already discernible.

The emboldening of those outlines – the framing of France, as it were – was by no means inevitable. Its telos was realized through collective human action, whose achievement was both revealed in and confirmed by the very *legibility* of the landscape – the visual expression that made the country the image of its people.

A country is a storehouse of dormant energies, laid up in the germ by Nature but depending for employment upon man. It is man who reveals a country's individuality by moulding it to his own use. He establishes a connection between unrelated features, substituting for the random effects of local circumstances a systematic co-operation of forces. Only then does a country acquire a specific character differentiating it from others, till at length it becomes, as it were, a medal struck in the likeness of a people.[58]

The parallel with Michelet's *Tableau* is striking. This had been published in 1833 as a section of his *Histoire de la France*, but in 1875 it was expanded

[56] Paul Vidal de la Blache, *Tableau de la géographie de la France* (Paris: Hachette, 1903). The early sections have been translated into English as *The personality of France* (London: Christophers, 1928); my quotations are from pp. 57–58, 69.

[57] Kristin Ross, *The emergence of social space, Rimbaud and the Paris Commune* (Minneapolis: University of Minnesota Press, 1988) pp. 86–87; Béatrice Giblin, "Le paysage, le terrain et les géographes," *Hérodote* 9 (1978) pp. 74–89. Ross also draws attention to Vidal's use of sexual metaphors in these passages and his implicit feminization of landscape and "nature." I return to this below.

[58] Vidal, *Personality*, p. 14.

and reprinted separately under the title *Tableau de la France: géographie physique, politique et morale*. Like Vidal, Michelet conducted his readers on a tour organized around a consciously visual itinerary; its formal apparatus was in fact derived from Hugo's "Bird's Eye View of Paris." Michelet saw this (in his case medieval) panorama as a vast and chaotic landscape whose inarticulate diversity was imposed by "Nature": "Division triumphs; each point of space asserts its independence."[59] Still more significantly:

> Michelet characterizes each of the provinces proleptically, by looking forward toward the personality it will reveal through time. The course of the journey, an outward-inward spiral from the peripheries to the heart (Paris) is a narrative of coming-to-self-consciousness, a sort of geographical *Bildungsroman* in which the nation-to-be finally "finds" itself (in Paris, naturally).[60]

Modernity thus witnessed what Michelet called "the condensation of France into oneness"; it stood revealed as "a person."[61]

Seen like this, human geography becomes complicit in the process of enframing. This was obviously not inaugurated by Vidal, and his affirmation of the principle of "terrestrial unity" – of "the conception of the earth as a whole, whose parts are co-ordinated"[62] – places him within a long and distinguished tradition of geographical thought. One can find the same vision of the world as an ordered totality in two of the founding texts of modern geography, Humboldt's *Cosmos* (1845–1862) and Ritter's *Erdkunde* (1817–1859), and in many other, extradisciplinary sites too. Vidal was undoubtedly familiar with Ritter's work, so much so that one historian of geography regards Ritter as the source of inspiration for some of Vidal's central insights.[63] But what is more important for my present purposes is the way in which both Humboldt and Ritter (and the tradition which they came to represent) invoked the world as totality through a sensibility that was at once aesthetic and scientific.[64]

---

[59] Stephen Kippur, *Jules Michelet: A study of mind and sensibility* (Albany: State University of New York Press, 1981) p. 62 and *passim*. Michelet's views on "Nature" were set out in a series of subsequent natural histories, which are discussed in detail in Linda Orr, *Jules Michelet: Nature, history and language* (Ithaca: Cornell University Press, 1976). On Michelet's feminization of "Nature" – which parallels Vidal's – see also Jeanne Calo, *La création de la femme chez Michelet* (Paris: Librairie Nizet, 1975) pp. 234–48.

[60] Hans Kellner, "Narrativity in history: Post-structuralism and since," *History and Theory* 26 (Beiheft) (1987) pp. 1–29; the quotation is from p. 10.

[61] Kippur, *Michelet*, pp. 72–73.

[62] Paul Vidal de la Blache, "Le principe de la géographie générale," *Annales de géographie* 5 (1896) pp. 129–42.

[63] Paul Claval, "The historical dimension of French geography," *Journal of historical geography* 10 (1984) pp. 229–45; *idem*, "Ritter and French geography," in Manfred Büttner (ed.), *Carl Ritter* (Paderborn, Germany: Schönigh, 1980) pp. 189–200.

[64] See, for example, Edmunds Bunksé, "Humboldt and an aesthetic tradition in geography," *Geographical Review* 71 (1981) pp. 127–46. Stoddart also draws attention to Humboldt's attempt to

I also want to emphasize, still more importantly, Vidal's more particular version of this global thesis – that is to say, the way in which he draws upon a palette of distinctive landscapes and projects them onto a coherent national canvas. For Vidal, one might say, as for Michelet, power over nature was synonymous with power over space. The progressive interactions between local societies and local ecologies that produced the mosaic of distinctive *pays* also put in place, less obviously but no less indelibly, a system of *circulation* between those different regions. Vidal thought that such a system had first emerged during the Roman occupation of Gaul, but its purpose was to connect the Mediterranean to the Atlantic. The system whose outlines were visible by the end of the eighteenth century was significantly different. It was marked by an imperfect, incomplete but nonetheless determined centralization. The principal roads now converged upon Paris so that the city sat there "like a spider in the center of its web." Unlike Michelet, however, Vidal insisted that this did not mean that individual regions had lost their distinctiveness, only that these differences were now systematically articulated through a political rather than a purely physical space: *"le système s'est nationalisé."*[65] This was an issue of considerable contemporary concern as well, of course, and one historian has claimed that it was in the decades between 1870 and 1914 that one could see "the disintegration of local cultures by modernity and their absorption into the dominant civilization of Paris."[66]

---

articulate the coherence of the cosmos but prefers to think of him as an instrumentalist, an experimentalist, and a rigorous proponent of the comparative method. It is easy to see how all this squares with Stoddart's own vision of geography as an objective science (see above, pp. 17–19). He does concede that Humboldt's "philosophical framework" was profoundly influenced by German *Naturphilosophie* and Romanticism, but he insists that these "ideas of beauty, harmony, and aesthetic experience lie uneasily alongside instrumental measurement and quantitative analysis." But I take the the burden of Mitchell's remarks to be that both the "aesthetic" and the "scientific" invoke the same model of the world-as-picture or (later) the world-as-exhibition. See Stoddart, "Humboldt." He makes much the same point about Darwin in his "Grandeur in this view of life," pp. 221–23. On the one side, as I noted earlier, Darwin was supposed to have rendered the landscape through the optic of romanticism; on the other side, he adopted "an analytical, instrumental, scientific approach" in which "the visually apprehended landscape became a conceptualized structure not limited to what could be directly observed." Stoddart admits that in both cases Darwin was still, in a sense, seeing. "Throughout all his works we are constantly invited to *see* with him, first with the eye the landscape he observes, then with the mind the insights he develops." This strikes me as a particularly lucid statement of the world-as-picture, whose essence was the enframed systematicity that both these modes of apprehension, the "aesthetic" and the "scientific," shared in full and equal measure.

[65] Vidal, *Tableau*, pp. 377–83.

[66] Eugen Weber, *Peasants into Frenchmen: The modernization of rural France 1870–1914* (London: Chatto and Windus, 1976) p. 486; for a critique see Roger Magraw, "Peasants into Frenchmen?" in his *France 1815–1914: The bourgeois century* (London: Fontana, 1983) pp. 318–53. The last project of that great *Annaliste* Fernand Braudel was to reopen Vidal's investigation of "the identity of France": to describe the diversity of *pays* – "a hundred, a thousand different Frances of long ago" –

Be that as it may, in France as elsewhere during this period, nationalism was a moving force in the admission of geography to the academy. Many commentators have since complained that Vidal's vision of geography was confined to a contemplation of the timeless landscapes of rural France – "a geography of permanences" – amounting to little more than a tableau of the ancien régime, which "naturalized," in the most obdurately physical of senses, the spatial foundations of a nationalist ideology. The traces of peasant culture in each *pays* were presented as a collective still life; the animation of the French Revolution, the Industrial Revolution, or the colonialism of the Third Republic were conspicuous only by their absence. Nothing was allowed to disturb the harmony of the hexagon.[67] The nationalism, even the chauvinism of Vidal's geography is as plain as a pikestaff, but this reading of it is unduly myopic. It is perfectly true that the *Tableau* was intended as an introduction to Lavisse's *Histoire de la France depuis les origines jusqu'à la Révolution* and so, of course, its focus was on the ancien régime. It is also the case that its landscapes were barely humanized so that, as Ross remarks, where people appear at all, they are made to do so "in such a way as to reinforce the natural harmony of the region: the native, the peasant is *part of* the landscape, in a synecdochic relationship of décor."[68] Yet Vidal's oeuvre is not closed around the *Tableau* and I suggest that his accounts of post-Revolutionary France require a more careful inspection.

Territorial claims

Let me turn, therefore, to Vidal's last book, *France de l'est*, which was written in the middle of World War I, soon after his son had been killed in combat, and published just one year before Vidal himself died. During the war, Vidal joined a number of French geographers who served on a series of special commissions established under the auspices of the Société de Géographie de Paris in order to advise the government and the general staff on the aims and conduct of the war. One of the most important was concerned with the determination of France's eastern frontier and the prospects for geopolitical realignment in a postwar Europe. Its chairman

---

and to discover the "connecting tissues" that bound them together "into a unitary whole, an overall design." See Fernand Braudel, *The identity of France*. Vol. 1: *History and environment* (London: Fontana, 1989) pp. 23, 31–32, 263–64.

[67] Ross, *Social space*, p. 87; Yves Lacoste, *La géographie, ça sert, d'abord, à faire la guerre* (Paris: La Découverte, 1978) pp. 44–48. Vidal's conclusion to the *Tableau*, p. 386, spoke directly to this last point: "When a gust of wind disturbs the surface of a clear pond, everything trembles and collides; but, presently, the image of the bottom reappears. A close study of what is fixed and permanent in the geographical conditions of France must more than ever be our guide."

[68] *Ibid.*, p. 87.

was Lavisse and its secretary was Vidal. I think it is possible to show that *France de l'est* was a direct outcome of Vidal's work on behalf of the political and military authorities in Paris and that it was also all of a piece with his own conception of human geography. Indeed, Heffernan insists that:

Vidal was a much more political and ideological writer than previous commentators have recognized.... Even if he meant what he often said about the political innocence of his scholarship – as, for example, at the start of *France de l'est* – to ignore the political message which pervades his work is to fail to understand the real power of Vidal and the French school of geography and to misinterpret its relations to French intellectual life under the Third Republic.[69]

In *France de l'est* Vidal sought to show · that Alsace-Lorraine was a distinctive *pays* – that its cultural landscape set it apart from other districts – but that it had been bound to France in other registers for generations: politico-juridically since the fifteenth century, economically since the eighteenth century. The revolution then fanned the flames of a regional consciousness so intense, so democratic, Vidal argued, that it finally (and the implication is irrevocably) soldered the region to the nation. Industrialization did nothing to weaken those bonds; in the course of the nineteenth century it radically changed the basis on which *genres de vie* were produced – and for that matter modes of regional production more generally – but it did so in ways that were fully continuous with the preexisting order. In the past, and quintessentially during the ancien régime, regions had been formed around the organic core of a peasant culture and economy, through the suturing of *paysan, paysage*, and *pays*, but modern regions crystallized around the materialities of industrial production. In Vidal's view, this was entirely symmetrical: the vitality of modern industry depended on the formation of coherent regional metropoles. Yet in the wake of the Franco-Prussian war of 1870–71, part of Alsace-Lorraine had been occupied and annexed by a victorious Germany. As a result, part of what had been a thriving industrial complex was now a mere satellite to the Ruhr. Not only was the coherence of the region fractured and its resource base ruptured; since regional difference was now articulated through a national space, the integrity of the nation (in every sense of that phrase) was also compromised. Vidal's conclusion was unequivocal and it drove his regional geography onto the wider terrain of geopolitics. Vidal believed that there was literally all the difference in the world between German expansionism within Europe (which he insisted was illegitimate) and French colonialism beyond Europe (which he regarded as wholly admirable). It was perfectly acceptable for France and Britain to find their *champs d'expansion* in Africa and Asia, he argued, because this would contribute to the growth and consolidation

---

[69] Michael Heffernan, personal communication.

of the liberal-democratic state; but it was quite wrong for Germany to threaten its neighbors and direct its ambitions toward *une Europe asservie*, as it had done in the case of Alsace-Lorraine, because this signaled a dramatic departure from the political model of the European Enlightenment.[70]

What can be concluded from all this? Vidal's portrayal of *France de l'est* is plainly no less nationalistic than his *Tableau géographique de la France*. If anything, his nationalism is even more pronounced. But the materials with which he works have altered: All the episodes that critics claim to be absent from his problematic – the Revolution, industrialization, colonialism – and that were deeply implicated in the constitution of modernity are here vividly present. And yet there is little sense of the contradictions, struggles, and tensions that were so crucially involved in those transformations. Vidal contains them by the imposition of a national frame that projects historicism onto a geographical canvas and thereby provides his narrative with its meaning and orientation. This enables him to describe the production of regions as the progressive realization of potentialities that were inscribed within what he takes to be the *essence* of France.[71] It is this historico-geographical trajectory – this destiny – which is then so abruptly blocked by events lying outside the frame: on the other side of the Rhine. It is only in this sense that I can make anything of Lacoste's claim that *France de l'est* reveals a new Vidal concerned "not with a space given once and for all" but with "contested spaces."[72] For that space is surely *constituted inside* the national frame but *contested outside* it: Vidal provides no conceptual ground for disclosing the spaces of contestation *within Alsace-Lorraine*. I want to suggest that this vision was blinkered by more than the animus of national politics, however, and that Vidal's observations were also shaped by the shifting positions of geography and sociology within the contemporary constellation of power-knowledge.

These were by no means independent considerations. As a first approximation one could say that it was Durkheim's sociology rather than Vidal's geography that was most concerned to expose the internal tensions that threatened the precarious order of the Third Republic. But this was no simple division of intellectual labor; Durkheim had read Ratzel and in consequence had something to say about geopolitics too. He was himself

---

[70] Paul Vidal de la Blache, *France de l'est* (Paris: Armand Colin, 1917) pp. 51, 60, 79, 163–64, 197–99.

[71] Vidal would not have used a phrase like "the production of regions" but the vocabulary of realization would not have been remotely foreign to him. He once suggested that geography "is interested in the events of history in so far as these bring to life and to light, in the countries where they take place, qualities and potentialities that without them would remain latent." Paul Vidal de la Blache, "Des caractères distinctifs de la géographie," *Annales de géographie* 22 (1913) pp. 289–99; the quotation is from p. 299.

[72] Yves Lacoste, "A bas Vidal...Viva Vidal!" *Hérodote* 16 (1979) pp. 68–81.

a native of Alsace-Lorraine and during the Franco-Prussian war, when he was only twelve years old, his hometown was occupied by Prussian troops. Forty years later he did not anticipate the outbreak of a new war against Germany with any enthusiasm, and neither did many of his contemporaries, but when it came his sympathies were fully with France. Unlike Vidal, Durkheim was no nationalist, but he believed that the imperial regimes of Prussia and Austria were tottering – "unnatural aggregates," he called them, "established and maintained by force" – and after the battle of the Marne in September 1914 he was firmly convinced that "the geography of Europe [would] be remade on a rational and moral basis."[73] But these remarks, and Durkheim's wartime writings more generally, need to be put in the context of his project as a whole. Although it would be a mistake to reduce his work to a series of political interventions, the crisis of liberalism within the Third Republic posed sociology's central problem with a particular urgency: namely, "what are the social forms capable of realizing the ideals of [liberty] and equality generated by the transition from the traditional order?"[74] It is doubly helpful to put the question in these terms, I think, because they show that Durkheim was concerned to establish the *contrast* between the traditional and the modern – much more so than Vidal, whose conceptual structure remained more or less intact through the transition – and because their appeal to the principles of the Revolution and the republic suggests that Durkheim's interest in establishing sociology as a stabilizing science, one that could indeed anticipate the emergence of a social order founded on the "rational and moral basis" of consensus, did not absolve him from the analysis of *conflict*. This too distinguishes him from Vidal, whose conception of the region – even in *France de l'est* – implied "a unitary society at one with its natural milieu and united in its collective will to exist."[75]

For all these differences, however, Durkheim's interest in the "geography" of postwar Europe was more than a turn of phrase: He had his own territorial ambitions within disciplinary space. Some commentators have described this as a "sociological imperialism," and so it was, but I want to connect these considerations more directly to Mitchell's thesis in order to show that, in quite other senses, both Durkheim's sociology and Vidal's geography were colonial, "colonizing" discourses. It will then be possible to be more precise about the divisions that were put in place between them within the *same* constellation of power-knowledge.

---

[73] Steven Lukes, *Émile Durkheim: His life and work* (London: Allen Lane, 1973) pp. 39–41; 547–52; the quotation is from p. 548.

[74] Anthony Giddens, "Durkheim's political sociology," in his *Studies in social and political theory* (London: Hutchinson, 1977) pp. 235–72; the quotation is from p. 251. This paragraph also relies on Steven Seidman, *Liberalism and the origins of European social theory* (Berkeley: University of California Press, 1983) pp. 153–78.

[75] Ross, *Social space*, p. 87.

As I have said, the military humiliation of France in 1871 was much on people's minds 40 years later. In the meantime, however, the French colonial empire had expanded dramatically and many commentators – Vidal among them – thought that this would be the saving of France. Certainly geography was no bystander in that movement. Its institutionalization as an academic discipline, in France as elsewhere, was prompted to some degree by the imperatives of colonialism and imperialism.[76] Vidal himself spent long periods in the Balkans and Turkey, in Syria and Palestine (he was in fact present at the opening of the Suez Canal) and, like many of his contemporaries, he regarded colonialism as "the crowning glory of our era."[77] The world-as-exhibition was a primary instrument of that accession, as Mitchell suggests, and geography was caught up in the spectacle of both, but it seems to me that discourses of this kind not only contributed to the colonization of the *pays d'outre-mer*, they also "colonized" the West itself. I am thinking in particular of Young's suggestion that we ought to "reposition European systems of knowledge so as to demonstrate the long history of their operation as the effect of the colonial other."[78] This is usually taken to refer to the ways in which the West conceived of "the Other" as a means of confirming its own distinctive identity (and superiority). I have no doubt of the importance of such a strategy, and it would not be difficult to read its privileges in the geographies or the sociologies of the day, but what I have in mind here is the way in which a particular constellation of power and knowledge was put in place in the colonies and also, so to speak, "repatriated" to the metropolis where it became accepted as the single office through which legitimate and supposedly disinterested claims to truth could be registered.

The spaces of science

It is nonetheless true that the "scopic regime" – the term is Metz's[79] – implied by the world-as-exhibition was not monolithic. The city of Paris in which (from 1877) Vidal mapped out the principles of his human geography

---

[76] For the general argument, see Brian Hudson, "The new geography and the new imperialism, 1870–1918," *Antipode* 9 (1977) pp. 12–19; Driver, "Geography's empire." On France, see Michael Heffernan, "A science of empire? The French geographical movement and forms of French imperialism," in Anne Godlewska and Neil Smith (eds.), *Geography and Empire: Critical studies in the history of geography* (Oxford: Blackwell Publishers, in press).

[77] See, for example, Ross, *Social space*, pp. 93–95; Paul Rabinow, *French modern: Norms and forms of the social environment* (Cambridge: MIT Press, 1989) pp. 139–42 and *passim*. These enthusiasms were not confined to the center and right. Reclus was no defender of imperialism but he nonetheless welcomed the ways in which Europeans were acting as "torch-bearers of learning and harbingers of ideas" and "Europeanization" was fusing local histories into one universal history. See Marie Fleming, *The geography of freedom: The odyssey of Elisée Reclus* (Montreal: Black Rose Books, 1988) pp. 181–84.

[78] Robert Young, *White mythologies: Writing history and the West* (London: Routledge, 1990) p. 119.

[79] See Jay, "Scopic regimes."

and where (from 1902) Durkheim established the precepts of his sociology was also and at the same time the site of a whirlwind of experimentation in the arts. This ebullient modernism, in all its various forms, drew attention to the process of enframing as something constructed, conventional, and, as a consequence, emphatically not "natural." The modernist project was far from homogeneous, of course, but what its various movements had in common was a displacement and a decentering of those conventional, taken-for-granted, above all *respectable* representations. Its critical salvos thus contributed to what Lunn calls the "intellectual bombardment of liberal certainties."[80] In short, modernism called into question, in radically unsettling ways, the "essential unity" that was focal to the world-as-exhibition.[81]

Mitchell ignores the avant-garde, and in some ways necessarily so: He is concerned with a particular, dominant modality of power-knowledge, and that is precisely why his thesis fits Vidal's work so well. For the most part the discipline of geography stood apart from those experimentations. If the figure of Vidal conjured up the geographer as painter, then his work was closer, in spirit if not always in substance, to the Salon paintings that hung on the walls of the Sorbonne than to the disconcerting canvases of the Impressionists; if he was known as *poète Vidal* when he was a student at the École Normale, then his mature style was safely Parnassian rather than shockingly *symboliste*.[82] One could perhaps press this disciplinary stringency still further. It has been suggested that Reclus, whose weekly *Nouvelle géographie universelle* enjoyed considerable public success, was marginalized by the academic establishment not so much because of his popularism or his political radicalism (his anarchism and his involvement with the Commune) as for the self-consciously artistic form of his work.[83] That may be so, but aesthetics and politics were never far apart and modernism was plainly

---

[80] Eugene Lunn, *Marxism and modernism* (London: Verso, 1985) p. 39.

[81] See, for example, the extremely interesting discussion in T. J. Clark, *The painting of modern life: Paris in the art of Manet and his followers* (Princeton: Princeton University Press, 1984). Clark pays considerable attention to the emergence of what he calls a "spectacular society" during the 1870s. He uses the term *spectacle* to point to "the ways in which the city (and social life in general) was presented as a unity in the later nineteenth century, as a separate something made to be looked at – an image, a pantomime, a panorama" (p. 63). The Impressionists drew attention to the artifice of these presentations, often ironically, and disrupted the "essential unity" of the conventional representations. For further reflections on the complex historical geographies of scopic regimes, see Crary, *Techniques of the observer*, pp. 67–136. His main argument is that both fin-de-siècle experimentation (at the margins) and the continuation of a nominally "realist" tradition (at the center) were made possible by a single series of changes that were put in place much earlier in the nineteenth century and that installed what he calls "a visual modernity" (p. 91).

[82] I owe the poetic analogy to Ross, *Social space*, pp. 86–87, who describes Parnassian culture as both "high bourgeois" (p. 43) and ocularcentric: excessively dependent "on the eye" (p. 83).

[83] Richard Lafaille, "En lisant Reclus," *Annales de géographie* 548 (1989) pp. 445–59. In fact his publisher insisted he reach out beyond the academy: "It is a literary book [I want from you]," Templier urged, "a sort of poem in which the Earth is hero." See Gary Dunbar, *Élisée Reclus: Historian of nature* (Hamden, Conn: Archon Books, 1978) p. 74 and *passim*.

involved in the crisis of liberalism. Whatever the merits of these comparisons, however, there is no doubt that the discipline of human geography – academic, institutionalized, establishment geography – was advertised as a *science* and not an art.

This was true of many other disciplines in the *Nouvelle Sorbonne* at the fin de siècle. That institution was regarded by its critics as the creature of republicans, its professors charged with providing a moral and "scientific" alternative to traditional clerical education. Few rose to the challenge as enthusiastically or energetically as Durkheim, who took great pains to distinguish his science of sociology from the imprecise and supposedly internalized accounts furnished by a canonical "literature."[84] He set out the program for this new science in 1895 in his *Rules of sociological method*. Its fundamental requirement was the injunction to see social phenomena *comme les choses* – "as things" – by which he meant, he said, that social phenomena shared the properties of things in the sense that they were resistant to human desire or will, so that their scientific investigation had to be an external, "objectivist" one.[85] But this did not limit social inquiry to a dull empiricism. For Durkheim, "the facticities of all the sciences are constituted by a given visible field of phenomena which are expressions of the realities that underlie it." What distinguished sociology from other sciences was that its "facticities" ("social facts") consisted of collective *representations* of the social – of "society" – whose *objective structure* surfaced through them.[86] This focus on the "representational" nature of the social thus coincided with what Mitchell takes to be "the more and more 'exhibitional' nature of modernity."[87] But Durkheim also believed that the systematicity of the social depended on its spatial structure: on what he called its "social morphology." For this reason, any account of the constitution of social life would have to incorporate many of the propositions of human geography, which Durkheim regarded as one of the "fragmentary sciences" that had to be drawn out of their isolation to contribute to an inclusively social science. Indeed, he claimed that it was only through such a plenary gesture that those disciplines would be able to grasp the objectivity and systematicity of society as a totality; only thus that they could ever be "social in anything but name."[88]

---

[84] Wolf Lepenies, "Agathon and others: Literature and sociology in France at the turn of the century," in his *Between literature and science: The rise of sociology* (Cambridge: Cambridge University Press, 1988) pp. 47–90; Lukes, *Émile Durkheim*, p. 75.

[85] See Mike Gane, *On Durkheim's* Rules of sociological method (London: Routledge, 1988) pp. 30–38. Durkheim's sociology was not closed around the *Rules*, of course, but I have confined my discussion to elements that remained central to his project as a whole.

[86] Paul Hirst, *Durkheim, Bernard, and epistemology* (London: Routledge and Kegan Paul, 1975) pp. 90–103; the quotation is from p. 93.

[87] For Mitchell's discussion of Durkheim, see his *Colonizing Egypt*, pp. 120–26; the quotation is from p. 126.

[88] See, for example, Durkheim's reviews of Ratzel's *Politische Geographie* in *L'année sociologique* 2 (1897–8) pp. 522–32 and *Anthropogeographie* in *L'année sociologique* 3 (1898–9) pp. 550–58. He

All of this was an advertisement for what Hirst calls "a vacant space awaiting its science."[89] But Durkheim made it very clear that this space was *irreducible* and that its integrity could not (and would not) be compromised by the sort of geography that insisted on embedding local cultures in local ecologies. For Durkheim, social morphology was a social construction; for Vidal, the production of regions was the result of routines and practices that reached into both "society" and "nature." The differences are stark, but I am not sure that there was ever much of a debate between Durkheim and Vidal.[90] Although some historians have done their best to stage one, it was, I think, largely an indirect affair, mediated through their different readings of Ratzel. Vidal was much more sympathetic to Ratzel and his persistent organicism than Durkheim, and his concept of the *genre de vie* can be read as a local translation of Ratzel's master-concept of *Lebensraum* ("living space"). Both terms conveyed not only a territorial imperative – "just as a plant needs room to spread and for its seeds to bear fruit," is how Vidal put it – but also (and in the same biophysical vocabulary) an essential unity between society and nature.[91] Certainly the Vidalians upheld the independent integrity of a human geography predicated not only on the horizontal relations of society over space but also on the vertical relations between society and nature. Whatever Durkheim may have thought, however, this did not involve them in any covert reductionism, still less in an environmental determinism. Local ecologies were supposed to provide a portfolio of options, as it were, within which local cultures could exercise considerable collective discretion. Vidal's human geography thus disclosed a "field of compromise," as Brunhes once put it, and represented the human being as "part of living creation" – the words are Vidal's – who could act upon nature "only through her and by her."[92] But

generalized these claims (with Paul Fauçonnet) in his "Sociologie et sciences sociales," *Revue philosophique* 55 (1903) pp. 465–97; the quotation is from p. 495.

[89] Hirst, *Durkheim*, p. 81.

[90] Cf. Vincent Berdoulay, "The Vidal-Durkheim debate," in David Ley and Marwyn Samuels (eds.) *Humanistic geography: Prospects and problems* (Chicago: Maaroufa Press, 1978) pp. 77–90; Catherine Rhein, "La géographie: Discipline scolaire et/ou science sociale? 1860–1920," *Revue française de sociologie* 23 (1982) pp. 223–51.

[91] Paul Vidal de la Blache, "Les genres de vie dans la géographie humaine," *Annales de géographie* 20 (1911) pp. 193–212; 289–304; the quotation is from p. 290; Anne Buttimer, "Anthropogeography and social morphology," in her *Society and milieu in the French geographic tradition* (Washington: Association of American Geographers, 1971) pp. 27–40. In Ratzel's case, ironically, this was used to underwrite the German expansionism that had previously cost France Alsace-Lorraine: see Mark Bassin, "Imperialism and the nation state in Friedrich Ratzel's political geography," *Progress in human geography* 11 (1987) pp. 473–95 and Woodruff Smith, "*Lebensraum* – theory and politics in human geography," in his *Politics and the sciences of culture in Germany, 1840–1920* (Oxford: Oxford University Press, 1991), pp. 219–33.

[92] The gendering of this passage – and of *la géographie humaine* more generally – requires a more careful consideration than I can provide here. It is not an artifact of the gendered construction

the distance from Durkheim's project was still considerable, and these claims evidently diminished the prospect of a unitary social science that would find its field wholly within the objectivity of the social.

The materialism of Vidal's geography is often overlooked, particularly by those who have tried to present the French school as a precursor to an idealist or interpretative geography, but so too is the sense in which this doctrine of "possibilism" (as Febvre called it) cast human geography within the framework of naturalism. Vidal made it plain that geography was "the scientific study of places and not of people."[93] This did not mean that he refused to people particular places – quite the contrary – only that he set his face against the individual and the isolated event: against what Braudel, one of his greatest admirers, later dismissed as *l'histoire événementielle*. Instead, Vidal's geography was directed toward the collective and the conjectural as these were inscribed in "typical" landscapes. Neither did his commitment to a recognizably scientific geography imply any commitment to the discovery of "laws" – he thought Ratzel's attempt to do so was at best premature, and the accounts offered by his own human geography were often singular, indeterminate, and even ill-defined – but it did rely (and centrally so) on a concept of *contingence* that was derived from the physical sciences.[94]

Although both Durkheim and Vidal presented their projects as "objective sciences," they clearly occupied different booths in the world-as-exhibition. Durkheim drew upon the natural sciences, but his was a methodological interest, a necessary first step along the path toward establishing sociology

of the French language, however, and I am reminded of the French sculptor Louis-Ernest Barrias, who produced several statues at the end of the nineteenth century portraying "Nature" as a woman removing her veil, exposing her breasts and "revealing her secrets to science"; Carolyn Merchant, *The death of nature: Women, ecology and the scientific revolution* (San Francisco: Harper and Row, 1989) pp. 189–91. When Merchant remarks how these representations transform "Nature" from "an active teacher and parent" to "a mindless, submissive body" (p. 190), I am left wondering about the patriarchal subtext of possibilism.

[93] Paul Vidal de la Blache, "Les caractères distinctifs de la géographie," *Annales de géographie* 22 (1913) pp. 289–99.

[94] This claim is developed in detail in Fred Lukermann, "The *Calcul des Probabilités* and the *École française de géographie*," *Canadian Geographer* 9 (1965) pp. 128–37. It is also accepted, at least in outline, by Rhein, *Emergence*, who draws attention to the Vidalians' disinterest in the social sciences and to Vidal's own "naturalistic" conception of the discipline. A different reading is provided by Howard Andrews, "The Durkheimians and human geography: Some contextual problems in the sociology of knowledge," *Transactions, Institute of British Geographers* 9 (1984) pp. 315–36. But this does not much alter my central conclusion. Andrews insists on the close connections between Vidal's conception of geography and the practice of history (rather than natural science) – a connection that I agree is extremely important though if Lepenies, *Between literature and science*, p. 64, is right, historians were by no means immune from the "scientization" of intellectual inquiry – but in any case Andrews uses the proximity of geography and history to establish "the differences separating the Durkheimians from the philosophical underpinnings of human geography as espoused by Vidal and his disciples." In other words, whichever way Vidal's *oeuvre* is approached, his geography remains at a distance from sociology.

as a science; the second step turned on respecting the substantive differences between the natural and the social so that sociology could be established as an autonomous *social* science. I suspect that Vidal may have been even closer to the methodology of the natural sciences than Durkheim, though this was usually expressed in a much less formal manner, but the two parted company most decisively when Vidal insisted on crossing what Durkheim saw as the essential substantive divide between the natural and the social. Yet to do otherwise would have been to "suspend society in the air," as Ratzel put it in his own response to Durkheim, and would have left geography literally groundless. Vidal insisted that geography was not (and could not be) a purely social science, therefore, and his attempt to distance human geography from sociology was more than a matter of disciplinary politics: It was founded on serious intellectual differences.

That sense of separateness was to become the enduring legacy of the fin-de-siècle encounter between geography and sociology. This determined distinctiveness – what two groups of critics were later to call, with somewhat different inflections, a doctrine of "exceptionalism" – became the prevailing, largely unexamined orthodoxy of both Anglophone and Francophone geography until World War II. I say "became" advisedly, because the differences separating Durkheim and Vidal – both figures for larger discursive formations – were indeed both serious and intellectual. Yet once they were frozen into disciplinary boundaries, they were patroled and policed; and once identity papers had to be produced for inspection, travelers were threatened with exile or internment: there was precious little space to question those differences, still less to transcend them. In this light one begins to suspect that the real contribution of Hartshorne's account of *The nature of geography* is not so much its emphasis on areal differentiation as its insistence on disciplinary differentiation.[95]

I should make it plain that this was not because Hartshorne was indebted to Vidal or the French school of geography more generally. On the contrary, as Sauer pointed out at the time and as Butzer has reminded us since, Hartshorne's views were developed through a highly selective exegesis of a German intellectual tradition. His approval of Hettner (in particular) was unrestrained, but the regional geography that he constructed was purged of both the physico-ecological and the cultural-historical implications that were indelibly present in Hettner and, for that matter, Vidal.[96] As a result, geography's window on the world-as-exhibition was narrowed:

[95] Richard Hartshorne, *The nature of geography* (Lancaster, Pa: Association of American Geographers, 1939).

[96] Carl Sauer described this period as "the Great Retreat" in his "Foreword to historical geography," *Annals of the Association of American Geographers* 31 (1941) pp. 1–24; see also Karl Butzer, "Hartshorne, Hettner and *The nature of geography*," in J. Nicholas Entrikin and Stanley Brunn (eds.), *Reflections on Richard Hartshorne's* The Nature of Geography (Washington: Association of American Geographers, 1989) pp. 35–52.

in the camera obscura of *The nature of geography*, only the "areal expression" of phenomena could be registered. This metaphor is appropriate partly because it mirrors Hartshorne's own sensibilities. His geography was "a particular point of view" on the world; it was an objective science that could be distinguished from the undisciplined subjectivities of art by its accuracy, certainty, systematicity; its distinctive perspective was that of areal differentiation. But the metaphor is appropriate for another reason too. The image captured by the camera obscura was, in the most obvious of senses, two-dimensional – and so too, I suggest, was Hartshorne's geography. Indeed, in his subsequent *Perspective on the nature of geography* he described geography as one of the three "spatial sciences." The other two, significantly, were astronomy and geophysics.[97]

Frontiers: economics and geography

I now turn to perhaps the apotheosis of the world-as-exhibition: the constitution of geography as a formal spatial science. For the most part, historians and contemporaries alike have chosen to represent this as a sharp break from Hartshorne's views, but I regard this as substantially mistaken. The underlying continuities, though less visible, are the more powerful. The contrast has usually been presented in terms of the neo-Kantian distinction between idiographic and nomothetic modes of concept-formation, but in ways that both misrepresent Hartshorne's argument and misunderstand the original formulation. *The nature of geography* undoubtedly gave a particular force to the idiographic – to a conception of geography directed toward an elucidation of the specificity of place, region, and spatial configuration – but it did not rule the nomothetic out of court. Still more important, the philosophical architects of the distinction, Rickert and Windelband, never made it an ontological one: idiographic concept-formation was a part of both the *Naturwissenschaften* and the *Geisteswissenschaften*.[98] There is no reason to suppose that the interest in the idiographic foreclosed on the possibility of constituting geography as a science, therefore, and I now propose to show how that mission was pursued with a renewed vigor in the postwar decades.

---

[97] Richard Hartshorne, *Perspective on the Nature of Geography* (Washington: Association of American Geographers, 1959) p. 177; see also in the same book, pp. 140–42.

[98] See J. Nicholas Entrikin, *The betweenness of place: Towards a geography of modernity* (London: Macmillan, 1991) pp. 93–98. The *Naturwissenschaften* are the natural sciences but the *Geisteswissenschaften* correspond to neither the English "social sciences" nor the French "human sciences"; they include both the social sciences and the humanities, and a possible translation might be "cultural sciences."

I want to confine my discussion to three areas. First, I will suggest that the formalization of geography as spatial science deepened its involvement in what I have been calling the world-as-exhibition. Second, I will argue that insofar as this depended on an appropriation of neoclassical economics it only increased human geography's debt to the natural sciences. And third, I will claim that these developments implicated spatial science in what Habermas calls the colonization of the lifeworld.

Spatial science and abstract space

One of the central pinions of the world-as-exhibition was a conception of order that was produced by – and resided in – a structure that was supposed to be somehow separate from what it structured: A framework that seemed to precede and exist apart from the objects that it enframed.

[This] ordering creates the impression that the gaps between things are an abstraction, something that would exist whether or not the particular things were put there. This structural effect of something preexistent, nonparticular and nonmaterial is what is experienced as "order."[99]

This conception of order was also at the very center of spatial science. It was articulated most clearly by Peter Haggett in his seminal review of *Locational analysis in human geography*, first published in 1965 when this so-called "new geography" was near its zenith. "Order depends not on the geometry of the object we see," he affirmed, "but on the organizational framework in which we place it."[100] For Haggett, it was the framework itself that was geometric, and over the past 25 years or so he has sought to recover a geometrical tradition – a concern for "spatial structure" – that was focal to the classical conception of geography, at least as it was set out by Eratosthenes and Ptolemy, and to install it in a central place within the modern discipline. Accordingly, *Locational analysis* was organized around the decomposition of a regional system into a series of abstract geometries: movements, networks, nodes, hierarchies and surfaces.[101] In an elegant essay reflecting on this project, Haggett described his vision of geography as a "distant mirror," a way of seeing the world from a distance, which I find extremely revealing. For it recalls the strategies of all those European

---

[99] Mitchell, *Colonizing Egypt*, p. 79. I ought to note that Mitchell repeatedly describes this order as "nonspatial" since this appears to conflict with the case I am about to make: but his own analysis makes so much of geometricity that he is clearly using "nonspatial" as a synonym for "nonmaterial" or "conceptual."

[100] Peter Haggett, *Locational analysis in human geography* (London: Edward Arnold, 1965) p. 2.

[101] The stasis of this schema was recognized in the second edition by the addition of a chapter on diffusion, but its treatment remained essentially morphological: see Peter Haggett, Andrew Cliff, and Allan Frey, *Locational analysis in human geography* (London: Edward Arnold, 1977) pp. 231–58.

travelers in Mitchell's Egypt who could only "form a picture" of what they saw – who could only render the world intelligible – by placing themselves on "viewing platforms" (Telegraph Hill outside Cairo or the Great Pyramid at Giza), which provided them with what Mitchell calls "optical detachment": a position from which they could see without being seen.[102] The parallel with the surveillant eye of spatial science is, I think, exact.

I should emphasize that these visual preoccupations are not phantoms of my own devising. Haggett himself has repeatedly claimed that "of all sciences [geography] has traditionally placed greatest emphasis on seeing."[103] Such concerns are not confined to the sciences either, and Krauss has proposed a reading of what she takes to be modern art's obsessive interest in the grid that speaks equally directly to the silent world of spatial science.

Surfacing in pre-war Cubist painting and subsequently becoming ever more stringent and manifest, the grid announces, among other things, modern art's will to silence, its hostility to literature, to narrative, to discourse. As such, the grid has done its job with striking efficiency. The barrier it has lowered between the arts of vision and those of language has been almost totally successful in walling the visual arts into a realm of exclusive visuality and defending them against the intrusion of speech.[104]

It would not be difficult to project Krauss's discussion of the "flattened, geometricized [and] ordered" grids of modern art onto the abstract lattices of spatial science; and it may well be that part of the appeal of these models – "pictures of the world," as Haggett once called them – is indeed an aesthetic one. Haggett makes no secret of his delight at the sheer "beauty of geographical structures." This is, at one level, purely phenomenological: "the structural symmetry of the planet as viewed from outer space, the sequence of atolls in a Pacific island chain or the terraced flights of irrigated fields on a Philippine hillside continue to be awesomely beautiful." When those symmetries are not immediately apparent, however, Haggett's project turns on the possibility of making them appear – calling them into presence – on another level altogether by re-presenting the world in a different spatial framework: an abstract geometry. "As so often," he remarks, "symmetry and elegance go hand in hand with truth."[105]

---

[102] Peter Haggett, *The geographer's art* (Oxford: Blackwell Publishers, 1991) p. 5; Mitchell, *Colonizing Egypt*, pp. 21–28.

[103] Haggett, *Locational analysis*, p. 2; idem, *Geographer's art*, p. 19.

[104] Rosalind Krauss, "Grids," in her *The originality of the avant-garde and other modernist myths* (Cambridge: MIT Press, 1985) pp. 8–22; the quotation is from p. 9. Krauss is not essentializing vision and visuality, and she argues that the twentieth-century modernist grid is strikingly different in its function from the lattice used since the fifteenth century in the construction of linear perspective.

[105] Haggett, *Geographer's art*, p. 184. If this appears to tremble on the edges of aestheticism, the sensibility is by no means confined to spatial science. Sauer's interest in the morphology of landscape was inspired, at least in part, by Goethe's "science of forms" and he constantly

But those who have been most concerned to advance this conception of geography, including Haggett, have usually invoked the language of the hard sciences to do so: a strategy symptomatic of the postwar constellation of science and technology that contained at its core an ideological nucleus that virtually eliminated the distinction between the technical and the practical by enclosing the one within the categories of the other.[106] Spatial science promised to enlist both physical and human geography in joint explorations of spatial structure – to dissolve the divisions (if not the distinctions) between the two – and to equip the expedition with the baggage of the physical sciences. This included not only their techniques and technologies but also their discursive practices: the data-set, the passive voice, the Harvard reference system. The leaders of the expedition used Kuhn's apparatus to survey the intellectual landscape, and extrapolated from his studies of scientific revolutions in the physical sciences to claim a similarly paradigmatic (which is to say "scientific") status for their own "new" geography.[107] In one particularly extravagant flight of fancy, the spat between Hartshorne and Schaefer was compared to "an argument between Michelson and Newton."[108] Haggett himself repeatedly drew parallels between modern physics and modern geography in order to fix the position of his project. In his view, "two of the greatest intellectual achievements of twentieth-century science," relativity theory and quantum mechanics, were "concerned with spatial structures," and he used this congruence of interest to locate his geography midway between "the elegant worlds of the galaxy and the atom."

Equidistant from the stars and the atoms lie the familiar middle-ground distances of the geographer: here are spatial structures which come half-way between the worlds of Einstein and Bohr.[109]

---

accentuated the importance of an aesthetic response to landscape. See Carl Sauer, "The morphology of landscape," and his "The education of a geographer," in John Leighly (ed.), *Land and life: Selections from the writings of Carl Ortwin Sauer* (Berkeley: University of California Press, 1963) pp. 315–50, 389–404; these essays were first published in 1925 and 1956 respectively.

[106] Cf. Jürgen Habermas, *Technik und Wissenschaft als Ideologie* (Frankfurt am Main: Suhrkamp Verlag, 1969); this text is included in his *Toward a rational society* (London: Heinemann, 1971).

[107] Peter Haggett and Richard Chorley, "Models, paradigms and the new geography," in Richard Chorley and Peter Haggett (eds.), *Models in geography* (London: Methuen, 1967) pp. 9–41. In fact, Kuhn's thesis suggested that the physical sciences were considerably less "objectivist" than either their protagonists or their critics supposed (which is precisely why Popper, Lakatos, and others attacked Kuhn so ferociously). For a general discussion of Kuhn's hermeneutic method, see Richard Bernstein, *Beyond objectivism and relativism: Science, hermeneutics and praxis* (Oxford: Blackwell Publishers, 1983); for a more specific discussion, see Andrew Mair, "Thomas Kuhn and understanding geography," *Progress in human geography* 10 (1986) pp. 345–69.

[108] William Bunge, "Fred K. Schaefer and the science of geography," *Harvard Papers in Theoretical Geography*, Special Paper A (1968) p. 12.

[109] See Haggett's *Locational analysis* and his *Geographer's art*, p. 23. This is not a retrospective gloss – distance lending enchantment, so to speak – and neither was it idiosyncratic. In much the

Those worlds are not without their wonders, of course, and the intellectual curiosity of modern physicists is by no means bounded by the observatory or the laboratory. Here, for example, is Bohr talking with another physicist, Werner Heisenberg, about their visit to Kronberg Castle in Denmark:

Isn't it strange how this castle changes as soon as one imagines that Hamlet lived here? As scientists we believe that a castle consists only of stones and admire the way the architect put them together. The stones, the green roof with its patina, the wood carvings in the church, constitute the whole castle. None of this should be changed by the fact that Hamlet lived here, and yet it is changed completely. Suddenly the walls and the ramparts speak a quite different language.[110]

Haggett cites this passage too; but it was that "different language" that concerned many of the critics of spatial science. Indeed, the most succinct expression of the distance between the two was probably his own decision to strip the place-names from the maps in *Locational Analysis* and replace them with Greek symbols, a maneuver that literally "replaced" them in an abstract – geometric, generalizable – space.[111]

The physics of the space-economy

It was not difficult to give these characterizations a more particular form, and Haggett suggested that Lösch's dazzling investigations of *The economics of location*, which run like a leitmotif through the first edition of *Locational analysis*, derived from a way of seeing the space-economy "in the manner of a solid-state physicist, each settlement like an atom bonded together in a matrix of molecular interconnections" (figure 6).[112] Although Lösch would probably not have dissented from such a gloss on his work – he employed a specifically functional form of mathematics originally devised to describe the physical world, and physical analogies (waves, refractions) recur throughout the text[113] – his project was nonetheless intended as a contribution to economics.

same way Harvey suggested that the geographer was concerned with neither "the spatial patterning of crystals in a snow flake" nor "the spatial patterning of stars in the universe" but with a "regional resolution level": David Harvey, *Explanation in geography* (London: Edward Arnold, 1969) p. 484.

[110] Werner Heisenberg, *Physics and beyond: Encounters and conversations* (New York: Harper and Row, 1972) p. 51.

[111] Haggett, *Geographer's art*, pp. 58–59; *idem, Locational analysis*, p. xii. Did the Greek symbols also conjure up the classical tradition of a "mathematical" geography?

[112] Haggett, *Geographer's art*, p. 70; August Lösch, *The economics of location* (New Haven: Yale University Press, 1954). The title of the original German edition, which was published in 1940, gives a more accurate indication of the scope of the text: *Die raümliche Ordnung der Wirtschaft* or *The spatial organization of the economy*.

[113] See Peter Gould, "August Lösch as a child of his time," in Rolf Funck and Antoni Kuklinski (eds.), *Space-structure-economy: A tribute to August Lösch* (Karlsruhe, Germany: von Loeper, 1985) pp. 7–19.

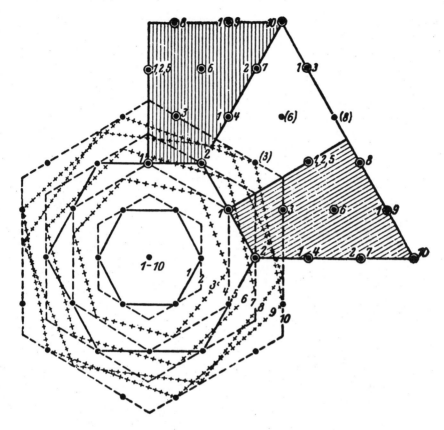

Figure 6 Lösch's "solid-state" landscape.

It is the importance of that economic base that I want to establish here. Lösch set out his stall in the preface, which he composed in the autumn of 1939. (This was soon after Hartshorne had completed his investigations of geography in Germany, but the only German geographer that Lösch made much of was Ratzel.) In his previous writings, most obviously in *Population cycles and business cycles*, Lösch had been concerned to wire demographic change to economic change, to trace the trajectory of the economy over time. Now he was proposing to examine "the relations between *space* and the economy." What followed was propadeutic, focusing on the geometries of a space-economy in equilibrium, but he believed that those theoretical foundations would eventually allow "all economic theory [to be] be reformulated thus from a spatial aspect."[114] His immediate purpose was to establish the conditions for a general equilibrium in space.

[114] Lösch, *Economics*, p. xiii. He identified three phases of theoretical economics: first, the formulation of a price theory that neglected time and space; second, the construction of theories

This was of the first importance: Lösch wanted to disclose, to make visible, the systematic order of a rational economic landscape – what he called "the rational and *therefore* natural order" (my emphasis) – because he was convinced that such a demonstration held out the prospect of domesticating the "illogical, irregular, lawless" forces that ravaged the "chaotic" reality of his own turbulent present.

Although Lösch was in many ways the high-priest of order – economics "must not become a science that describes chaos instead of preaching order," he declared – he insisted again and again that this had to be a *rational* order. If his investigations coincided with the "reactionary modernism" of the Weimar Republic and the Third Reich, they had very little in common with those "deformations of reason," as Habermas would call them.[115] In Lösch's view – and, here as elsewhere, the visual metaphoric has a particular force – the real had to submit to the power of the rational, and his empirical studies were designed to show "how strong the forces of order really are" and to reveal the contingency of "the discouraging impression of chaos under which we have suffered too long."[116]

The genealogy of location theory on which Lösch drew is far from simple. I cannot discuss those filiations in any detail, but I want to draw attention to the heterogeneity of three of its canonical texts. J. H. von Thünen's prospectus for *The isolated state* was derived from a critique of classical political economy which enabled him to formulate a series of models of agricultural land use, but his discussion of the politics of the "frontier wage" was informed by Hegel's *Philosophy of right*. Alfred Weber's general model of industrial location and agglomeration rested on little more than a physical analogy, but it also stood in a close (if sometimes tense) relation to his brother Max's sociology. Walter Christaller's formulation of central place theory was indebted to classical political economy, but he also (and more significantly) drew upon the marginalism of Menger

---

of interest and business cycles that incorporated time; and "now the third period [had] dawned, when space is seriously considered" (p. 93n).

[115] Jeffrey Herf, *Reactionary modernism: Technology, culture and politics in Weimar and the Third Reich* (Cambridge: Cambridge University Press, 1984). This requires a fuller discussion than I can provide here. Werner Sombart was certainly implicated in reactionary modernism, for example, and Lösch did make some use of Sombart's *Der moderne Kapitalismus* (1902): but he eschewed Sombart's critique of rationalization and his use of racial archetypes in *Die Juden und das Wirtschaftsleben* (1911). Yet in the second edition he also praised Walter Christaller, "Raumtheorie und Raumordnung," *Archiv für Wirtschaftsplanung* 1 (1941) pp. 116–35, as "excellent" (p. 350n) at a time when the political implications of Christaller's work for the Third Reich were hardly inconspicuous. Lösch himself drew attention to the importance of location theory for "the planning of the newly acquired eastern territories" (p. 351n).

[116] *Ibid.*, pp. 92–93, 216–20. Hence his criticism of Keynes, whose General Theory was "really based on phenomena of decadence in the economy" (p. 363); Marshall, in contrast, he thought "brilliant" (p. 364n). It was of course Marshall who brought supply as well as demand into the marginalist framework and in doing so effectively authored the neoclassical paradigm.

and Wieser and on the Younger Historical School of Sombart and (Max) Weber.[117]

It is necessary to emphasize the variety of sources on which these authors drew because virtually all the more developed versions of spatial science strained these same writings through a narrowly *economic* filter to produce partial or general models of an economic landscape or space-economy. I am not entirely sure why this should have been the case, though it was presumably connected to (it was certainly coincident with) a hegemonic liberalism. In any event, all those "commercial geographies" of the late nineteenth century had been invested in the construction of a specifically economic geography, which by the middle of the twentieth century had become the empirical core of human geography. And I suspect that many writers agreed with Harvey that economics was "the most successful of the social sciences in developing formal theory," so that in the absence of what he called "indigenous theory," a properly "theoretical geography" could most profitably draw upon economics.[118] It is perfectly true that Haggett himself showed little interest in theoretical questions – his main concerns have always been descriptive-predictive, I think, and it is by no means clear how much of even Lösch's theoretical armature he retained in *Locational analysis* – but the spatial order that other postwar writers supposed to be immanent in the human landscape was usually derived from neoclassical economics: from transcodings of its abstract calculus of supply and demand into an equally abstract "friction of distance."

Ironically, however, this maneuver merely strengthened the connections with the physical sciences. Although there were important differences between its proponents, the so-called "marginalist revolution" of the 1870s depended on concepts drawn directly and deliberately from the physical sciences and on methods borrowed from statistical mechanics (including, crucially, the technique of constrained maximization). There was no secret about any of this, and Walras even described his "pure" economics as a "physico-mathematical" science. So open and so systematic was the piracy that Mirowski mischievously suggests the entire episode is more properly described as the marginalist "annexation": except, of course, that it was more than a passing episode. The heir to marginalism was neoclassical

---

[117] J. H. von Thünen's *Der isolierte Staat in Beziehung auf Landwirtschaft und Nationalökonomie* was originally published in two volumes in 1826 and 1850; the first volume is available in English translation as Peter Hall (ed.), *Von Thünen's Isolated State* (London: Pergamon Press, 1966) and the second volume as W. B. Dempsey (ed.), *The frontier wage* (Chicago: University of Chicago Press, 1960). Alfred Weber's *Über den Standort der Industrien* was originally published in 1909 and is available in English translation as *Theory of the location of industries* (Chicago: University of Chicago Press, 1929). Walter Christaller's *Die zentralen Orte in Süddeutschland* was originally published in 1933 and is available in English translation as *Central places in southern Germany* (Englewood Cliffs, N.J.: Prentice-Hall, 1966).

[118] David Harvey, *Explanation*, p. 118.

economics, which continues to be modeled on a mid-nineteenth century, post-Newtonian physics and which, as Mirowski remarks, has in consequence been left stranded by the retreat of atomistic determinism (and much else besides) from twentieth-century physics.[119] One hundred years later one can read the same ironies in the texts of spatial science. In some cases the models are derived directly from physics – usually models of spatial interaction and spatial diffusion – and in other cases – usually models of location – the debts to the physical sciences are mediated by neoclassical economics. But in both cases the world they exhibit is enframed by the logic of highly constrained physical metaphors.[120]

Rationalization, spatiality, and the disciplinary society

If this produced a human geography that was conceptually estranged from the social, it was also one that in quite other ways impinged directly upon the social. At the heart of neoclassical location theory was a preoccupation with *rational* action and with *rational* landscapes, but "rationality" was invariably understood in a highly specific and purely formal sense. The technique of constrained maximization was so important to the progenitors of marginalism and neoclassicism because it appeared to allow them to theorize, in ostensibly scientific terms, about what Max Weber identified more generally as a purposive-rational (cognitive-instrumental) orientation to action, and to operationalize it in mathematical terms. Although Weber regarded such an orientation to action as characteristic of – and in some measure even constitutive of – the capitalist economy, rationality was not uniquely associated with capitalism in the original formulations of location theory. Both Alfred Weber and Christaller speculated on the replacement of the price mechanism as the regulator of the economy by what Christaller called the "organized reason" of the state under socialism (though neither of them had a very conventional conception of socialism). But with the passage to Lösch the formalization and, by implication, the naturalization of the neoclassical model ensured that the figure of *homo economicus*, the

---

[119] On the relations between neoclassical economics and physics, see two works by Philip Mirowski: "Physics and the marginalist revolution," *Cambridge Journal of Economics* 8 (1984) pp. 361–79 and "The ironies of physics envy," in his *More heat than light: Economics as social physics, physics as nature's economics* (Cambridge: Cambridge University Press, 1989) pp. 354–95.

[120] On the physicalism of Hägerstrand's diffusion models, see my "Suspended animation: The stasis of diffusion theory," in Derek Gregory and John Urry (eds.), *Social relations and spatial structures* (London: Macmillan, 1985) pp. 296–336; on the physicalism of neoclassical economic geography see Trevor Barnes, "*Homo economicus*, physical metaphors and universal models in economic geography," *Canadian Geographer*, 31 (1987) pp. 299–308 and his "Reading the texts of theoretical economic geography," in Trevor Barnes and James Duncan (eds.), *Writing worlds: Discourse, text and metaphor in the representation of landscape* (London: Routledge, 1992) pp. 118–35.

very embodiment of capitalist rationality, would assume an ever greater prominence. Although (Max) Weber drew attention to the historical specificity of a purposive-rational orientation to action, he apparently did not intend this as a criticism of marginalism.

Weber fully accepted that marginalism provided an adequate account of economic action in a capitalist society and even, at the economic level, of the specifically economic institutions of capitalist society. Thus Weber accepted Menger's account of the rational origins of money and of market exchange, and the marginalist conception of the economic institutions of capitalism as embodiments of economic rationality, as technical means adapted to the achievement of economic ends, and so as "facts," at least in relation to the ethical ideal of economic rationality.[121]

Be that as it may, what Weber did object to was the institutionalization of this "ideal" as a singular and transcendent orientation to action, because it culminated in the "iron cage" of capitalist modernity, and marginalism was surely implicated in its construction. The "fate of our times," Weber wrote in one of his last essays, is characterized by "the 'disenchantment of the world'." Insofar as the desire for mystery and mystification survived, he argued, it was transferred to the repetitive routine of the everyday ("which itself becomes a functional metaphysic") or to the means and effects of calculation itself: to the functional instrumentalities of "sacred" science.[122] Marginalism and the neoclassical calculus are both versions of that functionalist metaphysic.

This story has been told many times and in many ways, most recently in Habermas's critique of functionalist reason (which itself draws critically upon Weber's work). To appreciate the radically unsettling nature of accounts such as these, as I have implied, it is necessary to grasp their intrinsic reflexivity. They are describing events in which the human and social sciences were themselves deeply implicated. In the present context, this means recognizing the sense in which neoclassical economics and its extensions into human geography have been *constitutive* of the modern world. Whatever one thinks of the intellectual merits of its program, it would be impossible to deny that many of the assumptions and concepts of the neoclassical calculus have entered popular discourse and shaped public policy, and that through its teaching, research, and publishing

[121] Simon Clarke, *Marx, marginalism and modern sociology: From Adam Smith to Max Weber* (London: Macmillan, 1982) p. 204; see also Alan Sica, *Weber, irrationality, and social order* (Berkeley: University of California Press, 1988) p. 181. Clarke's analysis is perceptive but he claims too much (or too little) when he suggests that one of Weber's primary concerns was to establish boundaries between marginalist economics and "sociology." I doubt that Weber's thought is so readily pinned to the pages of any disciplinary text.

[122] Lawrence Scaff, *Fleeing the iron cage: Culture, politics and modernity in the thought of Max Weber* (Berkeley: University of California Press, 1989) pp. 224, 227.

programs, neoclassicism has naturalized and universalized a particular conception of economic process.

This ideological function – a market triumphalism – has been immensely consequential, and not only in North America and Western Europe. It has effectively endorsed the commodification of successive spheres of social life and contributed to the hyper-rationalization of the lifeworld – in a specifically and narrowly instrumental sense – to such a degree, indeed, that Habermas speaks of the *colonization* of the lifeworld.[123] In much the same way, I think, though in a minor key, spatial science installed a discursive regime – a nexus of power and knowledge – which both suppressed what Habermas would have called its "constitutive interest," the ways in which its claims were embedded in particular social structures and dispositions, and also sought to represent its contingent logic as a singular, universal, and enduring rationality that could not be meaningfully questioned precisely because it too was conducted under the sign of a supposedly *dis*-interested "science."[124] Like neoclassicism, this vision of human geography tacitly endorsed the existing structures of social life, which were left in almost complete darkness, and its illuminations of the world-as-exhibition functioned as strategic moments in the contemporary colonization of the lifeworld.

But the critical momentum in narratives of this sort is almost invariably carried by the unmasking of a particular model of *science*, and I do not think this properly registers the subterranean complicity of *spatial* science in the process of enframing. The spatiality of social life was a matter of complete indifference to both marginalism and neoclassicism. Marshall had insisted that the role of time was "more fundamental than that of space" in accounting for the complexities of market behavior and later economists continued to believe in what Isard called this "wonderland of no spatial dimensions."[125] Yet the submergence of spatiality was in some ways perhaps the most ideological function of all. As I have noted, Foucault seems to suggest that a dispersed and anonymous system of spatial sciences emerged in the eighteenth century – quite apart from the "bibliographic dinosaurs" that haunt the pages of the usual histories of geography – as part of a

---

[123] Jürgen Habermas, *The theory of communicative action. Vol. II: The critique of functionalist reason* (Cambridge: Polity Press, 1987) pp. 303–403; the original was published in Germany in 1981. Two qualifications are necessary. First, commodification is only one (extremely important) dimension of the colonization of the lifeworld; secondly, the metaphor of "colonization" should not obscure the ethnocentrism of Habermas's own theorization.

[124] Habermas first presented his model of "constitutive interests" in his *Knowledge and human interests* (London: Heinemann, 1972); the original was published in Germany in 1968. Since then he has considerably modified his account, but the general importance of contextualizing claims to knowledge – of specifying their "politics of location" – remains unchallenged by his development of the theory of communicative action.

[125] Walter Isard, *Location and space-economy* (Cambridge: MIT Press, 1956) pp. 25–26.

generalized medico-administrative system of knowledge that was deeply implicated in the formation of a disciplinary society and its constitutive technologies of surveillance, regulation, and control. Its characteristic figure is the Panopticon, from which Foucault observes:

Our society is not one of spectacle, but of surveillance; under the surface of images, one invests bodies in depth; behind the great abstraction of exchange, there continues the meticulous, concrete training of useful forces; the circuits of communication are the supports of an accumulation and a centralization of knowledge; the play of signs defines the anchorages of power; it is not that the beautiful totality of the individual is amputated, repressed, altered by our social order, it is rather that the individual is carefully fabricated in it, according to a whole technique of forces and bodies.[126]

By the middle decades of the twentieth century, so it seems to me, human geography had centralized and formalized its contribution to that project: the disciplinary knowledge of spatial science was systematically connected to disciplinary power. The construction of this discursive triangle between power, knowledge, and spatiality effected a colonization of the lifeworld in which "space" was given both metaphorical and material resonance: as Foucault declared, "Space is fundamental in any exercise of power."[127] I am aware of the difficulties in moving from Habermas to Foucault in this way, and I do not mean to minimize them, but at this level the connections are, I think, compelling.

Spatial science has not stood still, of course, and more recently – with the emergence of Geographical Information Systems (GIS) at the core of a new technical geography – it seems that we may no longer be gazing at the world-as-exhibition but rather traveling *through* the world-as-exhibition. In a provocative discussion of these late twentieth-century political technologies, Poster suggests that contemporary changes in systems of information capture and transfer (the metaphors are startling in their associations: the Middle Passage conducted through cyberspace) are in the process of constituting a distinctively new "mode of information" that departs significantly from Foucault's general model of Panopticism. While the prospects for systematic surveillance have been considerably enhanced by developments in geodemographics and related fields, and may be enhanced still further if GIS can be integrated more effectively with spatial analysis, Poster's own view is that electronically wrapped language reconfigures the

---

[126] Foucault, *Discipline and punish*, p. 217; see also "The eye of power," in his *Power/knowledge*, pp. 146–65.

[127] Michel Foucault, "Space, knowledge, and power," in Paul Rabinow (ed.), *The Foucault Reader* (New York: Pantheon, 1984) pp. 239–56; the quotation is from p. 252. See also Felix Driver, "Power, space, and the body: A critical assessment of Foucault's *Discipline and Punish*," *Environment and Planning D: Society and Space* 3 (1985) pp. 425–46.

human subject's relation to the world in such a way that the power of that language now derives "not so much from representing something else *but from its internal, linguistic structure.*"[128] Olsson made much the same point about what he called the dialectics of spatial analysis some 20 years ago – that its discourse revealed more about the language it was conducted *in* than the world it was supposedly *about*[129] – but Poster's point is a still sharper one: that this new configuration *dissolves the distinction* between the real and the representation.

The object tends to become not the material world as represented in language but the flow of signifiers itself. In the mode of [electronically mediated] information it becomes increasingly difficult, or even pointless, for the subject to distinguish a "real" existing "behind" the flow of signifiers.[130]

This is not so much a dissolution of the world-as-exhibition, I suggest, as its apotheosis. For it conforms to three of the cardinal characteristics of the world-as-exhibition: what Mitchell identifies as its certainty; its para-doxicality ("its certainty exists as the seemingly determined relation between the representations and 'reality,' yet the real world turns out to consist only of further representations of this reality"); and its imbrication in systems of colonial, "colonizing" power.[131]

Yet Poster sees these developments somewhat differently, as a challenge to Western logocentrism and as a radicalization of deconstruction:

Computer writing is the quintessential postmodern linguistic activity. With its dispersal of the subject in nonlinear spatio-temporality, its immateriality, its dis-ruption of stable identity, computer writing institutes a factory of postmodern subjectivity, a machine for constituting non-identical subjects, an inscription of an other of Western culture into its most cherished manifestation. One might call it a monstrosity.[132]

But this promise of monsters – to paraphrase Donna Haraway – is born of a critique of logocentrism which, unlike Haraway's work, does not address the systematic privilege accorded to *vision* within Western modern-ity. This privilege is central to GIS and other, still more radical develop-

---

[128] Mark Poster, *The mode of information: Poststructuralism and social context* (Chicago: University of Chicago Press, 1990) p. 11 (my emphasis). On geodemographics see the chillingly enthusiastic discussion of the surveillance of individuals and "psychographics" in Peter Brown, "Exploring geodemographics," in Ian Masser and Michael Blakemore (eds.), *Handling geographical information: Methodology and potential applications* (London: Longman, 1991) pp. 221–58; on spatial analysis and GIS see Stan Openshaw, "A spatial analysis research agenda," pp. 18–37.

[129] Gunnar Olsson, "The dialectics of spatial analysis," *Antipode* 6(3) (1974) pp. 50–62. See also his *Lines of power/limits of language* (Minneapolis: University of Minnesota Press, 1991) p. 59.

[130] Poster, *Mode of information*, pp. 14–15.

[131] Mitchell, *Colonizing Egypt*, p. 13.

[132] Poster, *Mode of information*, p. 128.

ments in virtual reality and cyberspace. The power to display complex data-sets in three dimensions, to rotate, manipulate, and track across their terrains, and to collapse continental and even global landscapes onto video screens was revealed with agonizing clarity during the Gulf War of 1990–91, a conflict which Smith describes as "the first full-scale GIS war."[133]

The possibilities of these global data bases and displays dwarf Reclus's Great Globe and they extend far beyond the early, noninteractive images of the earth from space. The first, and still most famous image of the whole, unshadowed globe was captured in December 1972, and Cosgrove suggests that this was a "key icon" that destabilized previous global mappings by "decentering" Europe and the Atlantic and "privileging" the South. But this interpretation overlooks the platform from which the imaging and transmission took place. For this is NASA photograph 22727 and what it shows, surely, is the North inspecting the South.[134] My point is simply that even these high-tech global images that construct the world-as-exhibition in such a dazzling display have to be produced from *somewhere*. The subsequent development of GIS has hidden its viewing platforms even more effectively, however, and much of the discussion continues to treat GIS as a detached "science" or as a response to the abstract "logic" of the market and its supposedly natural commodification of information. In doing so a rhetoric of concealment is deployed that passes over these configurations of power-knowledge in virtual silence.[135]

It is precisely this ideology of abstraction and detachment, the hidden platform, that Haraway calls into question through her discussion of *situated knowledges*. She challenges the modern decorporealization of vision, which she describes in resonantly Foucauldian terms as the gaze that "mythically inscribes all the marked bodies, that makes the unmarked claim the power to see and not be seen, to represent while escaping representation," and argues that this myth is carried forward and embedded within the visualizing practices of late twentieth-century technoscience.

Vision in this technological feast becomes unregulated gluttony; all perspective gives way to infinitely mobile vision, which no longer seems just mythically about

[133] Neil Smith, "History and philosophy of geography: Real wars, theory wars," *Progress in human geography* 16 (1992) pp. 257–71; see also Robert Stam, "Mobilizing fictions: The Gulf War, the media and the recruitment of the spectator," *Public Culture* 4 (1992) pp. 101–26.

[134] On global data bases, see Helen Mounsey and Roger Tomlinson (ed.), *Building data bases for global science* (London: Taylor and Francis, 1988); on NASA photograph 22727, see Denis Cosgrove, "New world orders," in Chris Philo (ed.), *New words, new worlds: Reconceptualising social and cultural geography* (Lampeter, Wales: IBG Social and Cultural Geography Study Group, 1992) pp. 125–30.

[135] See, for example, Michael Goodchild, "Geographical information science," *International journal of geographical information systems* 9 (1992) pp. 31–45; S. Openshaw and J. B. Goddard, "Some implications of the commodification of information and the emerging information economy for applied geographical analysis in the United Kingdom," *Environment and Planning A* 19 (1987) pp. 1423–39.

the god-trick of seeing everything from nowhere, but to have put the myth into ordinary practice.[136]

Haraway is not talking about geography or GIS either, but the parallels are much closer. The NASA photograph exemplifies exactly that "god-trick of seeing everything from nowhere," and subsequent developments in remote sensing and GIS have extended the reach of human vision still further. When this "mobile vision" is set loose in cyberspace (which Haraway describes as "the virtual reality of paranoia") the prospects for dissolving the distinction between the real and the representation are vastly enlarged.

The term "cyberspace" has its origins in the science fiction of William Gibson, but it is now increasingly used to describe a series of creative technocultural interventions that blur the usual boundaries between fact and fiction by conjuring up a world of information sustained by computers and global telecommunication circuits. According to one of its architects and protagonists, cyberspace is a "technosimulated space" which

involves a reversal of the current mode of interaction with computerized information. At present such information is external to us. The idea of cyberspace subverts that relation; we are now within information.[137]

That sense of being "within" is greatly enhanced by the visualizing practices of more advanced geographical information systems and interactive tele-communications systems, which at once set the world at a distance – accessed from a platform, seen through a window, displayed on a screen – and yet also promise to place the spectator in motion inside the spectacle. This is a radicalized vision of modernity, one that gives a wholly new dimension to Marx's image of a world in which "all that is solid melts into air," and to some thinkers it heralds the very end of modernity. Indeed, if the weight of my earlier sentence is shifted, from being *within* to *being* within, then one arrives at something very much like Vattimo's sense of a postmodern enframing in which, as he puts it,

---

[136] Donna Haraway, "Situated knowledges: The science question in feminism and the privilege of partial perspective," in her *Simians, cyborgs and women: The reinvention of nature* (London: Routledge, 1992) pp. 183–201; the quotations are from pp. 188–89. See also her "The promises of monsters: A regenerative politics for inappropriate/d others," in Lawrence Grossberg, Cary Nelson, and Paula Treichler (eds.), *Cultural Studies* (New York: Routledge, 1992) pp. 295–337. There are important parallels with Henri Lefebvre's critique of the visualizing logic of capitalist modernity; see my Chapter 6, this volume.

[137] Marcos Novak, "Liquid architectures in cyberspace," in Michael Benedikt (ed.), *Cyberspace: First steps* (Cambridge: MIT Press, 1991) pp. 225–54; the quotation is from p. 225. In the same book, see also Allucquere Rosanne Stone, "Will the real body please stand up? Boundary stories about virtual cultures," pp. 81–118.

The world of images of the world, the true world, as Nietzsche says, becomes a fable; or, to use the Heideggerian term, *Sage*. Hermeneutics is the philosophy of this world in which being is given in the form of weakening or dissolution.[138]

Similarly, though not identically, Baudrillard suggests that the appropriate response to cultural implosions of this kind – to "being within" – is to renounce the metaphysics of the real altogether. Unlike Vattimo, however, he grasps the intrinsic, strategic significance of concepts of spatiality. In a revealing metaphor playing on the Borgesian fable of a map so detailed that it covered the territory it represented, Baudrillard claims that "it is no longer a question of either maps or territory." "Something has disappeared," he declares, namely "the sovereign difference between them." In his view, modern society has transcended the rigidities of Foucault's society of surveillance to the point where it is *constituted through* electronic images and becomes a society of simulation, a "hyperreality" that is "the product of an irradiating synthesis of combinatory models in a hyperspace without atmosphere."[139]

In Gibson's own work cyberspace is turned into what Ross describes as "the heady cartographic fantasy of the powerful," aestheticized until it dissolves into a technological sublime, and his imagery of "console cowboys" rustling data across the empty vastness of cyberspace is shot through with white masculinist ideology (so too, I think, is much of Baudrillard's most recent work).[140] It is also plain that access to this world is highly uneven in space and that its networks may be the global highways of a new system of colonial, colonizing power with a stubbornly material geography. Yet I do not want to imply that these more general techno-cultural developments necessarily issue in a dystopian future in which human creativity is stifled and all resistance crushed. Neither do I want to loose the demons of technophobia: There are other ways of navigating the hyperreal, other configurations of power-knowledge that are possible within this new technoculture. Some optimistic commentators point to the installation of "the machinery of countersurveillance," of public data networks, bulletin-board systems, and alternative information and media links; others envisage global data bases nourishing a democratic "planetary consciousness"

---

[138] Gianni Vattimo, *The transparent society* (Baltimore: Johns Hopkins University Press, 1992) p. 117. Vattimo develops the idea of "weak thought" in his *The end of modernity: Nihilism and hermeneutics in post-modern culture* (Cambridge, England: Polity Press, 1988); this was first published in Italian in 1985.

[139] Jean Baudrillard, "Simulacra and simulations," in his *Selected writings* (ed. Mark Poster) (Cambridge, England: Polity Press, 1988) pp. 166–84; in the same book, see also "Fatal strategies," pp. 185–206.

[140] Andrew Ross, "Cyberpunk in boystown," in his *Strange weather: Culture, science and technology in the age of limits* (London: Verso, 1991) pp. 137–67; cf. Michael Heim, "The erotic ontology of cyberspace," in Benedikt (ed.), *Cyberspace*, pp. 59–80.

of vital importance to ecological politics (though at present the ideology of globalism still seems to be caught in the snares of traditional nationalisms).[141] But what I have in mind is described most tellingly by Haraway when she argues against a totalizing critique of vision and visualizing technologies.

There is no unmediated photograph or passive camera obscura in scientific accounts of bodies and machines; there are only highly specific visual possibilities, each with a wonderfully detailed, active, partial way of organizing worlds. All these pictures of the world should not be allegories of infinite mobility and interchangeability, but of elaborate specificity and difference and the loving care people might take to learn how to see faithfully from another's point of view.[142]

Where is human geography in this world of detail and difference? In one dimension, clearly, it is a long way from Haraway. Across the frontiers between human geography and computer science, mathematics and statistics lie three politico-intellectual constellations which construct the "reality"/ "representation" couple in different ways. But none of them has approached the richness, complexity and sensitivity of Haraway's project. In the center is spatial science which, in its most classical form, retains the distinction between reality and representation but which, through its consistent focus on modeling, does at least draw attention to the process of representation as a set of intrinsically creative, constructive practices. Hence, in part, the accent on the aesthetics of modeling, on the "elegance" of spatial models (though the metaphoric of "power" is never far away). Tracking forward and to one side of spatial science are explorations of cyberspace and hyperspace, which often abandon the distinction between reality and representation as the unwanted metaphysical baggage of modernity and chart instead a postmodern world of representations and simulations. But in doing so they too draw attention to the constitutive function of representation, to what Vattimo calls the "fabling" of the world.[143] On the other side of spatial science, however, are advances in GIS which seem to move in precisely the opposite direction: to assume that it is technically possible to hold up a mirror to the world and have direct and unproblematic access to "reality" through a new spatial optics. The question of representation, of regimes of truth and configurations of power, knowledge and spatiality, is simply never allowed to *become* a question.

---

[141] Peter Fitting, "The lessons of cyberpunk," in Constance Penley and Andrew Ross (eds.), *Technoculture* (Minneapolis: University of Minnesota Press, 1991) pp. 295–315; Ross, *Strange weather*, pp. 98–100.

[142] Haraway, "Situated knowledges," p. 190.

[143] Vattimo, *Transparent society*, pp. 24–25.

But there are other frontiers, other ways of navigating hyperspace, and in other dimensions I think human geography has started to move in the directions indicated by Haraway. And in doing so, as I will try to show in the next chapter, it has embarked on a trans-disciplinary voyage into "deep space."

2

Geography and the Cartographic Anxiety

Gabriel thought maps should be banned. They gave the world an order and a reasonableness which it didn't possess.

William Boyd, *An ice-cream war*

Much of geography is the art of the mappable.

Peter Haggett, *The geographer's art*

Descartes and deconstruction

The modern staging of the world-as-exhibition implied a commitment to foundationalism, and spatial science was no exception. It was, in effect, the cartography of objectivism, which claimed to disclose a fundamental and enduring geometry underlying the apparent diversity and heterogeneity of the world. In this essay I want to consider some of the paths that have been opened up by the *critique* of spatial science in the closing decades of the twentieth century. But I begin more than three hundred years before that, and more than a century before Cook's first voyage in the South Pacific, with a figure who is usually portrayed as the architect of foundationalism: the philosopher René Descartes (1596–1650). He was on a voyage of sorts too, searching for what he himself called a "firm and permanent structure in the sciences." In his second *Meditation* he observed the following:

Archimedes, in order that he might draw the terrestrial globe out of its place and transport it elsewhere, demanded only that one point should be fixed and immoveable; in the same way I shall have the right to conceive high hopes if I am happy enough to discover one thing only which is certain and indubitable.[1]

---

[1] René Descartes, *Meditations on first philosophy* (trans. Elizabeth Haldane and G. R. T. Ross) (Cambridge: Cambridge University Press, 1969) p. 144; this was originally published in Latin in 1641–2 and translated into French in 1647.

His own voyage discovered its continent of certainty through the so-called "Cartesian exclusion": it banished as radically other everything that could not be brought within the ring-fence (and hence sovereign rule) of Reason. This strategy was by no means confined to Descartes, and it was part of a more general philosophy of exclusion, of "a gesture which legitimizes its own claims to knowledge by devaluing whatever lies beyond its sovereign grasp."[2] What interests me here, however, is the use of these metaphors of exploration and discovery, of travel and territory, because they play an important part not only in the Cartesian project but in many of its successors equally preoccupied with charting the shores (and the shoals) of Reason.[3] In his first *Critique*, for example, Kant declared:

We have now not merely explored the territory of pure understanding, and carefully surveyed every part of it, but have also measured its extent, and assigned to everything in it its rightful place. This domain is an island, enclosed by nature itself within unalterable limits. It is the land of truth – enchanting name! – surrounded by a wide and stormy ocean, the native home of illusion, where many a fog bank and many a swiftly melting iceberg give the deceptive appearance of farther shores, deluding the adventurous seafarer ever anew with empty hopes, and engaging him in enterprises which he can never abandon and yet is unable to carry to completion.[4]

This speaks directly to the undertakings I discussed in the previous chapter and to Michel Tournier's marvelous epigraph, which I placed at its head: Robinson's demand that everything on "his" island (and the act of claiming possession, of asserting sovereignty in the name of Reason, is extremely significant) "henceforth be measured, testified, certified, mathematical and rational."

But these metaphors interest me most of all because, like Tournier's imaginative restaging of the Crusoe story, they also carry within them an edgy recognition of what Richard Bernstein (to whom I am indebted for this train of thought) calls the "Cartesian anxiety." Tournier's Robinson notes "with every day that passes [on the island] the collapse of whole sectors of that citadel of words within which our thought dwells and moves": "Those fixed points which thought uses for its progression, like

---

[2] Roy Boyne, "The Cartesian exclusion," in his *Foucault and Derrida: The other side of reason* (London: Unwin Hyman, 1990) pp. 35–52; Christopher Norris, "Transcendent fictions: imaginary discourse in Descartes and Husserl," in his *The contest of faculties: Philosophy and theory after deconstruction* (London: Methuen, 1985) pp. 97–122; the quotation is from p. 100.

[3] Georges van den Abbeele, "Cartesian co-ordinates," in his *Travel as metaphor: From Montaigne to Rousseau* (Minneapolis: University of Minnesota Press, 1991) pp. 39–61.

[4] Immanuel Kant, cited in Richard Bernstein, *Beyond objectivism and relativism: Science, hermeneutics and praxis* (Oxford: Blackwell Publishers, 1983) p. 18. The *Critique of pure reason* was originally published in 1781.

crossing a river on stepping-stones, are crumbling and vanishing beneath the surface."[5] In fact, Descartes himself wrote that it was "as if I had all of a sudden fallen into very deep water [and] I am so disconcerted that I can neither make certain of setting my feet on the bottom, nor can I swim and so support myself on the surface."[6] The other side of the Cartesian exclusion is thus a profound *un*-certainty.

The specter that hovers in the background of this journey is not just radical epistemological scepticism but the dread of madness and chaos where nothing is fixed, where we can neither touch bottom nor support ourselves on the surface. With a chilling clarity Descartes leads us with an apparent and ineluctable necessity to a grand and seductive Either/Or. *Either* there is some support for our being, a fixed foundation for our knowledge, *or* we cannot escape the forces of darkness that envelop us with madness, with intellectual and moral chaos.[7]

Bernstein reconstructs the dilemma only to refuse the choice, however, because he thinks it is both misleading and distorting. In his later writings he suggests that Derrida's practice of deconstruction provides one way of coming to terms with the predicament. Derrida's achievement is to show that the strange, the alien, the other are not massing outside the gates of Reason: They are already and, so to speak, *constitutively* inside. For deconstruction displaces the binary oppositions of both categorical and dialectical thought; it frustrates the attempt to draw a perimeter, to hold the line, by showing that all these boundary commissions and policing operations are inherently "undecidable": that their systems can be completed only by admitting terms that call that very completeness into question. The point is not that drawing boundaries is somehow impermissible – the play of difference that is language depends upon it – but that the permeability of those boundaries has to be constantly asserted; more than this, that the space in which they are drawn is not a simple plane. Each side folds over and implicates the other in its constitution. In Bernstein's view, therefore, Derrida's work is so powerful and disconcerting because he has "an uncanny (*unheimlich*) ability to show us that at the heart of what we take to be familiar, native, at home – where we think we can find our center – lurk (is concealed and repressed) what is unfamiliar, strange and uncanny."[8]

The same practices of deconstruction can be traced through the recent history of human geography, where such metaphors have a special salience.

---

[5] Michel Tournier, *Friday, or the other island* (London: Penguin, 1974) p. 58; this was originally published in French in 1967.

[6] Descartes, *Meditations*, p. 149.

[7] Bernstein, *Beyond objectivism*, p. 18; see also van den Abbeele, *Travel*, p. 61.

[8] Bernstein, "Serious play: the ethical-political horizon of Derrida," in his *The new constellation: The ethical-political horizons of modernity/postmodernity* (Cambridge: MIT Press, 1992) pp. 172–98; the quotation is from p. 174.

Here, too, they have been radically unsettling – which is why I have invoked the "cartographic anxiety" as the title of the present chapter. But, like Bernstein, I do not think this ushers in a radical relativism. Two examples will illustrate what I have in mind. The first is taken from a series of essays in which Olsson sketches what he calls "a cartography of thought," perhaps most effectively in his semiautobiographical reflections on three distinctive "lines of power": his own journey in human geography from the identity sign of positive science, through the "both/and" slash of dialectics to the horizontal "Saussurean bar" separating and joining signifier and signified. Anticipating a question to which he must be accustomed, and which (I presume) is usually intended as an objection, he asks: "Is this geography?" His answer is quite emphatic: "Of course it is! For what is geography if it is not the drawing and interpretation of lines [?]" As I read this essay, it is – like a number of Olsson's other performances – an attempt to retain the geometric obsessions of spatial science in order to turn them against themselves: to deconstruct spatial science by invoking "a world where lines are taken to their limits."[9] This involves him in often dazzling wor(l)dplay, and there is something about Olsson's facility in English and Swedish that sometimes makes me feel as though he is able to glide behind the back of language. But this does not mean that his analytics of space pirouettes into a giddy relativism. For he is interested, above all, in exploring and transgressing those established "limits," in showing what it is they do and what it is they serve, and his interventions seem to me essentially therapeutic critiques which raise a series of profoundly moral and ethical questions.[10] The performance is truly astonishing, and it is not influenced by Derrida alone. However, in one of his essays Olsson pays homage to Derrida's work in a passage that summarizes much of my argument:

It is easy to appreciate why Jacques Derrida's conception of grammatology rests so securely in the two disciplines of geometry and psychoanalysis. While the former teaches the techniques of making distinctions by drawing limiting lines, the latter warns that no line is to be trusted; the taken-for-granted is not in the content of the distinguished but rather in the marks of distinguishing.[11]

---

[9] Gunnar Olsson, "Lines of power," in his *Lines of power/limits of language* (Minneapolis: University of Minnesota Press, 1991) pp. 167–81; the quotation is from p. 181. See also his "Dematerialized" (pp. 5–8) and "Squaring" (pp. 183–96).

[10] See in particular *idem*, "The social space of silence" and "The eye and the index finger," *loc. cit.*, pp. 108–12 and pp. 129–45. There are limits to the therapy too; while Olsson increasingly foregrounds (his) masculinity, this often seems to involve a marked insensitivity to feminism.

[11] *Ibid.*, p. 139. If "grammatology" is the science of writing, then deconstruction is "Derridean grammatology": Christopher Norris, *Derrida* (Cambridge: Harvard University Press, 1987) p. 84. The *locus classicus*, which speaks directly to Olsson's argument, is Jacques Derrida, *Of grammatology* (trans. Gayatri Chakravorty Spivak) (Baltimore: Johns Hopkins University Press, 1976); this was first published in French in 1967.

My second example is taken from Harley's meditations on conventional cartography, which have been influenced still more directly by Derrida (and Foucault). Harley seeks to subvert what he sees as the mythical history of cartography, which presents itself as the progress of an objective science toward unmediated representation, toward a mapping that is somehow able to conjure up a world of pure presence. The rhetoric of scientificity, of scientific rules and procedures, has been central to this strategy.

> The primary effect of the scientific rules was to create a "standard" – a successful version of "normal science" – that enabled cartographers to build a wall around their citadel of the "true" map. Its central bastions were measurement and standardization, and beyond there was a "not cartography" land where lurked an army of inaccurate, heretical, subjective, valuative and ideologically distorted images.[12]

This is the same rhetorical apparatus that I described earlier, the continent of certainty and the island of truth, and the task of deconstruction is to show that the infidels occupy this citadel too. Thus Harley insists that cartography's "mask of a seemingly neutral science" hides and denies the modalities of power that are embedded in and enframed by the map text, and that ostensibly "scientific," "objective" maps cannot escape their (sometimes unwitting) complicity in ideology. "Deconstruction urges us to read between the lines of the map – 'in the margins of the text' – and through its tropes to discover the silences and contradictions that challenge the apparent honesty of the image." Far from this involving a relativism, a suspension of discrimination and judgment, this too is a call to subject the most insistently scientific and objective cartographic representations to critical scrutiny.[13]

Most of the discussion that follows will be less concerned with formal examples of deconstruction – and perhaps I should say that I do not mean to exempt either of these two from criticism – but my continued use of the term is not entirely inappropriate. Although I will be concerned with a more diffuse constellation of interventions, they have, I think, made the

---

[12] J. B. Harley, "Deconstructing the map," in Trevor Barnes and James Duncan (eds.), *Writing worlds: Discourse text and metaphor in the representation of landscape* (London: Routledge, 1992) pp. 231–47; the quotation is from p. 233. On what this strategy meant for the colonial reading of non-Western maps, see *idem*, "Rereading the maps of the Columbian encounter," *Annals of the Association of American Geographers* 82 (1992) pp. 522–42.

[13] Harley, "Deconstructing the map," p. 238. I should note that Harley misinterprets Derrida at one key point where he worries that the phrase "there is nothing outside the text" licenses a pure textualism and "defeats the idea of a social history of cartography" (p. 233). In fact, what Derrida meant by "*il n'y a pas de hors-texte*" – the phrase will be found in his *Grammatology*, p. 158 – is exactly what Harley wants to argue: that however "closed" a text might seem to be, it can never exclude what it claims to have placed "outside."

closures and certainties of the objectivist tradition within human geography increasingly suspect. Through a collective process of interrogation, an active displacement of the tacit paradigms of spatial science and its successors, a kind of strategic reversal has been put into effect, which now continually unsettles attempts to claim a synoptic completeness for the human geographical project. It did not begin quite like that. The critique of spatial science was at first an attempt to displace one set of foundations with another (historical materialism or humanism). But many of the most recent contributions have challenged the very idea of what Foucault once called a "canonical grid." This has not only made human geography much more interesting, but it has also brought to the foreground questions of ethics, morality, and politics in new and much less comforting ways.

I have tried to preserve something of that sense of vulnerability – of meanings anticipated, deferred, displaced, contested, echoed – by presenting the journey that follows as an itinerary that (I know) allows many more passages than those I am able to set out *en clair* – even though I do not always know what those other passages or are where they lead. But I am aware that any such survey, however partial, almost inevitably betrays a desire for completeness, "a search for common ground," at odds with what I have been trying to establish, and I urge the reader to keep prising open my statements and my silences.

I am also aware of another difficulty. Much of what I have written in the preceding chapter has been a matter of report – not of "record," I should say, since this is a history that I have told for particular purposes, a reconstruction and representation of a past in which I have placed myself as critic: in short, a history of the present – but, all the same, an account of events in which I was not directly involved. From here on, however, I have to make myself doubly present. As I proceed, it will become obvious that my previous sketches were made possible by the steps I am now taking. My readings of Cook and Banks, Vidal and Durkheim, Lösch and Haggett were all informed by arguments that were developed as critical responses to the objectivist tradition. But this is now a history in which I am personally involved, in various ways and to various degrees, and in what follows – even more than in what has gone before – I cannot pretend to occupy any detached, still less any privileged vantage point.

Let me therefore set out some preliminary and necessarily personal markers. In *Ideology, science and human geography* I sketched a critique of spatial science which, in outline, I still think was broadly correct.[14] I advanced three main proposals there that bear directly on the discussion that now follows. First, I continue to believe that a critical human geography ought to reject those strategies of representation that treat discourse as an unproblematic reflection of the world; that it should be predicated on the

---

[14] Derek Gregory, *Ideology, science and human geography* (London: Hutchinson, 1978).

constitutive, creative function of theoretical work; and that it must explicate those sociospatial structures that are both the conditions and consequences of human action. Whether I would now describe this as constituting a form of "structural explanation" is more doubtful. I originally intended the term to signal my hostility to empiricism – though emphatically not to empirical inquiry – and my reluctance to endorse the structural*ism* that was being canvassed as its main contender. At the time, I was particularly interested in structural Marxism and its extensions. I still think that many of those ideas are of the first importance, but I would now be much more skeptical about convening them within any conventional conception of totality.

Second, I still believe that reflexivity is an inescapable moment in any critical human geography; that it is vital to overcome that estrangement from people, places and landscapes that spatial science imposed upon the discipline; and that any critical human geography must attend to the ways in which meanings are spun around the *topoi* of different lifeworlds, threaded into social practices and woven into relations of power. But I would now be much less optimistic about achieving these redemptions (for that is surely what they are) through a dialogue between constitutive phenomenology and time-geography. I would now want to attend to the voices of feminism, poststructuralism and postcolonialism, and in consequence I would be considerably more critical of attempts to translate humanism and its highly particular conception of subjectivity into human geography.

Third, I still think of human geography as an irredeemably situated, positioned system of knowledge; I continue to believe that its moral and political commitments need to be carefully scrutinized; and I hope that a critical human geography can help to make social life not only intelligible but also *better*. Although I would no longer use Habermas's discussion of knowledge-constitutive interests to sharpen these points, his later writings continue his original project in a different register and are of direct relevance to many of my own concerns. I would no longer present socialism as an unproblematic, largely unexamined ideal, to be sure, but my commitments have not been changed by the fall of authoritarian regimes in Eastern Europe or by the dissolution of the Soviet Union. These events mark the collapse of communism but they do not herald the triumph of capitalism. Indeed, I continue to believe that the critique of capitalism has to be a major focus of critical inquiry: If anything, that commitment has been strengthened by the ways in which the "Second" and "Third" Worlds are being so rapaciously convened within its hideously unequal "One," always "First" World. But I am disturbed by the ethnocentrism pervading much of Habermas's work and I would want to place critiques of patriarchy, racism and other institutionalized systems of political, social, and cultural oppression alongside the critique of capitalism. Although these systems are often connected, and it is necessary to expose the mediations between them, they cannot be reduced to a single explanatory locus.

These aspirations are not idiosyncratic, but even in this reworked form they leave at least two issues unresolved. At the time, my primary concern was to clarify the implications of postpositivist philosophy and social theory for human geography, and for that reason I said very little about the other side of the dialogue. It seemed strategically more important to consider the humanization of geography than to open philosophical and theoretical reflection to questions of place, space, and landscape. I no longer think it desirable or even possible to limit the development of critical theory in this way, and the subsequent turn of events has shown the importance of opening (and keeping open) reciprocal lines of communication. By the same token, when I set out that original agenda I provided virtually no discussion of the relations between its three claims. Since then, however, it has become increasingly plain that attempts to articulate "structural explanation" and "reflexive explanation" within the discursive space of critical theory – to provide what I had termed a "committed explanation" – are likely to require an engagement with the intrinsic spatiality of social life. These concerns inform much of what follows.

Marks: political economy and human geography

The critique of spatial science was many-stranded, but as events have turned out the most productive objections of the late 1960s and early 1970s were, I think, derived from political economy, usually but by no means invariably from those versions drawing upon historical materialism. In saying this, I do not mean to imply that the development of a humanistic geography was unimportant. Its insistence on providing a space (and a voice) for human agency proved to be an essential moment in the critique not only of spatial science but also of structural Marxism. But structural Marxism was never as important in human geography as most of its critics there made out, and many of the central concerns of humanistic geography found a receptive audience in other traditions of historical materialism. And in saying *this* I do not mean to imply that structural Marxism was without merit or that humanistic geography was without fault.

I will begin with a series of questions that circle around meaning, intention, and agency – and which I convene through a tense debate between humanism and historical materialism – and return to them at the end, in the form of an equally tense encounter between Western Marxism and post-Marxism. These interrogatories will frame an account of the spatial analytics of the capitalist economy that was central to the critique of spatial science and was both theoretically productive and politically consequential.

Humanism, historical materialism, and the politics of theory

In my view the theoretical and the political are closely connected, and my reluctance to give priority to the contributions of humanistic geography is in large measure the result of many of its practioners' reluctance to engage more directly with theoretical concerns. Running through some of its more programmatic statements in the 1970s and even into the 1980s was a sense that "theory" was constraining rather than enabling (my own view is that it is both), and that a properly humanistic geography ought to be able to find common ground between its writers and its subjects through a more or less direct appeal to "experience." This was not the brute empiricism that underwrote some of the more single-minded exercises in spatial science – the claim that all scientific knowledge was grounded in the experience of an object-world – because humanistic geography invoked "experience" expressly to keep objectivism at bay. It marked the human subject with knowledges and skills and with emotions and feelings that were intrinsic to human agency and to human beings. These were not dangerous subjectivities that had to be kept outside the confines of responsible intellectual inquiry, therefore, but instead identified the very core of a properly human geography. I do not dissent from the importance of these concerns, and the conjunction of the practical-experimental and moral-existential meanings embedded in that single, difficult word "experience" requires the most careful analysis and reflection. But I do not think that this should be underwritten in ways that foreclose on theoretical questions: "Theory" is surely also a profoundly human capability.

There was considerable variation in the force and discrimination with which this suspicion toward theory – not toward particular theories, I stress, but toward theory-in-general – was advanced. It had counterparts within historical materialism, too, as I will subsequently show, but here I want to confine myself to three examples within humanistic geography. In his original essay setting out the parameters of humanistic geography, Yi-Fu Tuan associated this approach with sensibilities that were developed most deliberately in the humanities – the arts, history, literature, and philosophy – which were supposedly in tension with those developed in the sciences. If Tuan's subsequent essays are exemplars of what he had in mind, then his humanistic geography appears as a series of "ironic observations on familiar and exotic forms of geographical knowledge and experience."[15] It is quintessentially philosophical, in what Tuan calls "the old meaning of wisdom or an outlook on life and world," and indeed he describes his own approach as modeled on the ideal conversation:

[15] Yi-Fu Tuan, "Humanistic geography," *Annals of the Association of American Geographers* 66 (1976) pp. 266–76. For a thoughtful commentary, see Stephen Daniels, "Arguments for a humanistic geography," in R. J. Johnston (ed.), *The future of geography* (London: Methuen, 1985) pp. 143–58; the quotation is from p. 143.

In such a conversation, one person offers a theme – a point of view – which he clarifies with an example or two.... The listener then responds with a case of his own, to show that he has understood, or to show that the theme is capable of further development, or to show that it is problematical – that its application, for example, is less general than its proponent believes.[16]

This is not Habermas's ideal speech situation; it is advanced innocently, almost naively, without any of the theoretical apparatus or political claims that attend Habermas's account of the sphere of communicative action. Tuan describes his geographical imagination as "an attentive mode of inquiry, a vigorous engagement with the real," by means of which one pays heed to the world as it is in order to think carefully about what ought to be. This is necessary, Tuan believes, because the hegemony of analytical thought in the modern world means that "what we cannot say in an acceptable scientific language we tend to deny or forget." The purpose of his own project is thus to reflect on those everyday experiences and "surface phenomena" that are at once removed from the gaze of conventional (social) scientific inquiry and yet remain stubbornly present within the lifeworlds of its practitioners and within its own practices: simply, "to increase the burden of awareness."[17]

Scientists strive to stand far above their material, for a view from nowhere, with the hope that they will thereby be able to plunge well below the surfaces of reality. By contrast, cultural geographers-cum-storytellers stand only a little above their material and move only a little below the surfaces of reality in the hope of not losing sight of such surfaces, where nearly all human joys and sorrows unfold.[18]

But this phenomenology, if that is what it is, is not only a philosophy; it is also a philosophical anthropology. One of the most persistent motifs in Tuan's writings is an appeal to *common* sense and *common* experience: to a "we" whose inclusive address presumes a shared human condition to which "we" have access without theoretical mediation.[19] His humanistic geography is, in essence, a moral-aesthetic discourse; it is contemplative, at once reflective

---

[16] Yi-Fu Tuan, *Morality and imagination: Paradoxes of progress* (Madison: University of Wisconsin Press, 1989) p. ix. This is not invariably a dialogue between men; Tuan suggests that the ideal reader of his book would say " 'yes, but...' and write her own book" (p. x).

[17] Yi-Fu Tuan, *Space and place: The perspective of experience* (London: Edward Arnold, 1977) pp. 200–203.

[18] Yi-Fu Tuan, "Surface phenomena and aesthetic experience," *Annals of the Association of American Geographers* 79 (1989) pp. 233–41; the quotation is from p. 240.

[19] Tuan addressed phenomenology directly in his "Geography, phenomenology and the study of human nature," *Canadian Geographer* 15 (1971) pp. 181–92, but John Pickles criticized the polarization of phenomenology and empirical science to be found in that essay (and elsewhere) in his own *Phenomenology, science and geography: Spatiality and the human sciences* (Cambridge: Cambridge University Press, 1985) especially pp. 56–57.

and speculative, and yet – despite the model of the ideal conversation – at best studiously indifferent to the wider conversations that might be made possible *through the theoretical.*[20]

Tuan's writings are hardly typical, I realize, and those who have sought to develop a more systematic, analytical version of humanistic geography – though one which remains attentive to moral-aesthetic concerns – have been much less reluctant to fix their positions and enlarge their horizons through some sort of theoretical orientation. This is generally a circumspect affair, critical of the endless elaboration of theoretical abstractions for their own sake – of the involution of what Mills once called "Grand Theory" – and determined to develop what David Ley calls "conceptual heuristics where separation from everyday geographies is more modest." Although this essay was written thirty years after Mills's original critique of sociology, Ley's reflections on the limits to theory provide perhaps the closest parallel in human geography (though I should say that neither of them restricted their arguments to narrowly disciplinary enclosures). In his essay Ley identifies two concerns that I think characterize his work as a whole. In the first place:

Theory building has become increasingly insulated from the empirical world... [and theory] admired less for its ability to illuminate reality than for itself, as an intellectual product with an elegance and coherent logic, with a *beauty* of its own.

In the second place, this autonomy is connected to a fragmentation: the obsessive experimentation with different theoretical forms has produced "cacophony" and "confusion." Neither concern detracts from what Ley sees as "the compelling case in favour of a theoretical orientation in social science," but he says they do imply that "the triumph of theory has been too complete" since "reality is [now] made over in a purely formal language." His own preference is thus for a humanistic geography that reopens contact with the hermeneutics of everyday life and repositions theory "much closer to the ground." In this way it might be possible to pick one's way through the fragments and bring about an "enriched synthesis" in human geography.[21]

Ley has been probably the most consistent, certainly the most rigorous advocate of such a program, but where Mills was writing as a Marxist (of sorts) and his principal target was Parsons's structural functionalism, Ley has advanced his own program as (in part) a critical response to Harvey's

---

[20] *Idem, Morality and imagination op. cit.*, pp. 143–44.

[21] C. Wright Mills, *The sociological imagination* (New York: Oxford University Press, 1959) pp. 33–59; David Ley, "Fragmentation, coherence and limits to theory in human geography," in Audrey Kobayashi and Suzanne Mackenzie (eds.), *Remaking human geography* (Boston: Unwin Hyman, 1989) pp. 227–44; the extended quotation is from p. 235.

historico-geographical materialism. "The limits to theory" sounds very much like a counterpoint to Harvey's analysis of "the limits to capital," which Ley regards as one of the main sources of a "fixation upon theory" that is to be laid at the double-doors of (high) modernism and Marxism. For my part, I am not sure it is necessary to read Harvey in this way. Although he is often castigated for his supposed theoreticism, I am more impressed by his interest in both the theoretical *and* the empirical. That it is difficult to work the two together and that Harvey often accentuates the one rather than the other does not, in my view, diminish his commitment to a reciprocity between them. But Ley is adamant in this essay and elsewhere that Harvey's work exemplifies the "holding the world at a distance" implicit in the classical Greek project of *theoria* and explicit in his involvements in both spatial science and historico-geographical materialism. I will return to the continuities between the two in a moment; but Ley's own preference is for the local knowledges embedded in contemporary developments in the arts and architecture: for modest generalizations that he believes are sensitive to specific contexts, human proportions, and historical traditions and that return us somewhere close to Tuan's moral-aesthetic concerns but now in a different, supposedly *post* modern register.[22]

Such a cautious, even pragmatic view of theory is too much for other writers, however, and one of the most passionate indictments of these theoretical orientations and dispositions had already been furnished by Donald Meinig. In an elegant essay entitled "Geography as an art," Meinig was intensely sympathetic to what he saw as the "self-conscious drive to connect with that special body of knowledge, reflection and substance about human experience and human expression, about what it means to be a human being on this earth," namely the humanities. But he also contended that "almost all of this avowedly humanistic geography is analytical in intent" and that "most of it seems little more than an extension of science."[23] Meinig's humanistic geography was about the *evocation* of place – about giving voice to "inner feeling" – and it was distinguished by a compelling need to *share* those feelings in ways that "make effective connection with the lives of others." For this reason, he had no hesitation in proclaiming William Bunge's *Fitzgerald: Geography of a Revolution* as a "work of art" and, indeed, in dismissing the claim of its author that it was science: "Any reader will quickly realize it is not." For all its flaws – and Meinig made no secret of his disappointment at what he took to be Bunge's political prejudice and textual indiscipline – what nevertheless confirmed

---

[22] Ley, "Limits to theory," pp. 237, 243–44; David Harvey, *The limits to capital* (Oxford: Blackwell Publishers, 1982).

[23] Donald Meinig, "Geography as an art," *Transactions of the Institute of British Geographers* 8 (1983) pp. 314–28; the quotation is from p. 315.

Fitzgerald as art in his eyes was the sheer intensity of the experience and the passion surging through its pages.[24]

What was meant by "science" in critiques of this kind was, variously, the physical sciences, the experimental sciences, and the social sciences, and where these were framed in narrowly objectivist terms one can surely understand the force of the complaint. But "science" is not the anemic practice many of its critics (and protagonists) seem to think. These critiques often folded "science" more or less directly into "theory" so that the recoil from one entailed the dismissal of the other. The assumption, again, was that theory was unnecessary for – simply got in the way of – the attempt to articulate different experiences one with another that was central to the humanistic project. To Meinig, therefore, and to critics like him, geography had to be aligned with the creative, imaginative passion of art, not with the dismal disciplines of science and its theoretical devices, because only in that way would it be able to communicate and share those inner feelings: and only in that way could it approach the (hallowed) ground where it would merge "into 'literature.' "[25]

Although I hardly mean to disparage the importance of creative writing, or Meinig's own very considerable accomplishments, I do find his remarks disconcerting. They become delightfully ironic alongside Ley's critique of theoreticism as "a form of art for art's sake," and for that matter Harvey's critique of the aestheticization of politics, and these ironies speak directly to my own disquiet.[26] I will try to explain myself and connect my concerns directly to my continuing interest in historical materialism.

In the first place, I have no quarrel over the importance of evocation. I realize that most of us are not trained to be painters or poets but, like Pierce Lewis, I don't think we should boast about it. As a matter of fact, whenever I think of Lewis's own evocation of his childhood love affair with a special place, the sand dunes on the eastern shore of Lake Michigan, I never fail to be moved by it:

My love affair with those Michigan dunes . . . had everything to do with violent immediate sensations: the smell of October wind sweeping in from Lake Michigan, sun-hot sand that turned deliciously cool when your foot sank in, the sharp sting of sand blown hard against bare legs, the pale blur of sand pluming off the dune

[24] *Ibid.*, pp. 321–22; William Bunge, *Fitzgerald: Geography of a revolution* (New York: Schenkman, 1971). Although Bunge described this as "a humanist geography," he also insisted on its "disciplined objectivity" and, indeed, its "generalizability" (foreword).

[25] Meinig, "Geography as an art," p. 316. I suspect that Meinig has in mind a particular conception of "art," however, and I wonder what he would make of Olsson's invocation of "the geographer as a loving artist," of his execration of "the dogma of objectivity and scientific methodology" and (most of all) of the artistic practice that then follows: Olsson, *Lines of power*, p. 153.

[26] Ley, "Limits to theory," pp. 232–33; David Harvey, *The condition of postmodernity: An inquiry into the origins of cultural change* (Oxford: Blackwell Publishers, 1989).

crest against a porcelain-blue sky, Lake Michigan a muffled roar beyond the distant beach, a hazy froth of jade and white. As I try to shape words to evoke my feelings, I know why the Impressionists painted landscapes as they did – not literally, but as fragments of color, splashes of pigment, bits of shattered prismatic light. One is meant to feel those landscapes, not to analyze them. I loved those great dunes in my bones and flesh. It was only much later that I learned to love them in my mind as well.[27]

I feel the same, visceral affection for other places – including the dunes at Camber in England, Wreck Beach in Vancouver – and rage at my inability to evoke them in such vivid, salt-spangled prose. But Lewis's address turned on the need to go "beyond description." He made it very clear that description was no simple task, but he also insisted that "description, if it is any good, has to be good for something: it has to lead somewhere." And for this reason he talked not only about aesthetic description but what he called *intellectual* description. I am not sure that this is quite the right word, but it is not difficult to understand what Lewis was after: The ability to evoke people, places, and landscapes is an essential prerequisite to analysis but it is not a substitute for it.

I would put this somewhat differently, I think. A work like E. P. Thompson's *The making of the English working class* or *Whigs and Hunters* suggests that these moral-aesthetic sensibilities need to be inscribed *within* critical analysis; they are not thresholds or supplements to intellectual inquiry but essential moments within it. Thompson's own views on theory are well known, of course, and cannot be ignored (even if one disagrees with them, as I do): but his angry denunciations of "the poverty of theory" were surely directed against structural Marxism in particular and not theory in general. He constantly invokes the reciprocity between theory and evidence, and hardly abandons the field of historical materialism to a brute empiricism. Thompson draws deeply from the wells of the humanities, but his history is neither theoretically innocent nor politically vacuous.[28]

In the second place, and closely connected to these considerations, the humanities are by no means devoid of theoretical sensibilities. Consider literary theory, a field in which Marxist scholars like Eagleton, Jameson, and Raymond Williams – all of them well-known in human geography, too – have been particularly prominent. Indeed, Meinig writes very warmly of Williams and holds him up as an example of what the humanities might

[27] Pierce Lewis, "Beyond description," *Annals of the Association of American Geographers* 75 (1985) pp. 465–77; the quotation is from p. 468.

[28] E. P. Thompson, *The making of the English working class* (London: Penguin, 1968); *idem, Whigs and hunters: The origin of the Black Act* (London: Allen Lane, 1975). For Thompson's critique of Althusser see "The poverty of theory, or an orrery of errors," in his *The poverty of theory and other essays* (London: Merlin Press, 1978) pp. 193–397; the "dialogue" between theory and evidence is invoked on pp. 225, 235–37 and *passim*.

hope to be. Now Williams was not interested only in literary theory, of course, and in any event literary theory is not all of a piece: It also has its opponents. But that theoretical sensibility seems to me to be of vital importance not only to Williams's work as a whole – including the range and command of *The Long Revolution* and the probing lyricism of *The Country and the City*, both of which Meinig commends – but to any serious attempt at practical criticism. To think otherwise is to misunderstand the whole business of interpretation and to foreshorten what *happens* in the practices of·writing and reading. A human geography that is supposedly scrupulously attentive to other voices – which insists, with Meinig, that the connections between human geography and the humanities "must serve as an exchange rather than a one-sided exploitation"[29] – and that then refuses to listen to theoretical arguments developed *within* the humanities places itself in an impossible position. As Daniels observes:

From this viewpoint authors appear not to work, as they do and very humanly, with the possibilities and restraints of artistic form and language or, in a larger context, with those of the society in which they work. They become instead vehicles for transcendent truths.[30]

There is on occasion more than a hint of Leavis's rasping elitism in the conservative canon of humanistic geography, in marked contrast to Tuan's seemingly simple homilies and Ley's assault on the barriers between high and popular culture. In those moments I suspect Eagleton's description of the movement that grew up around *Scrutiny*, the journal of criticism launched by the Leavises in the 1930s, might apply equally well to the aspirations of that geographical project. *Scrutiny* was concerned, so Eagleton says, with the unique value of the individual and the creative realm of the interpersonal.

These values could be summarized as "Life," a word which *Scrutiny* made a virtue out of not being able to define. If you asked for some reasoned theoretical statement of their case, you had demonstrated that you were in the outer darkness: either you felt Life or you did not. Great literature was a literature reverently open to Life, and what Life was could be demonstrated by great literature. The case was circular, intuitive and proof against all argument.[31]

In much the same way, the more patrician forms of humanistic geography had no need of theory: it was unnecessary and even vulgar in its intervention between "experience" and "appreciation."

---

[29] Meinig, "Geography as art," p. 318.

[30] Daniels, "Arguments," p. 149.

[31] Terry Eagleton, *Literary theory: An introduction* (Oxford: Blackwell Publishers, 1983) p. 42; these characterizations repeat those set out in Eagleton's *Criticism and ideology: A study in Marxist literary theory* (London: Verso, 1978) pp. 13–16. But in the earlier account the class-location of the architects of *Scrutiny* is given a much more strident treatment.

Of course, literature does not have to be interpreted like that, any more than humanistic geography has to be cast in a conservative mold. I say this to underscore the importance of the continuing conversations between humanism and historical materialism, but also to recognize the radical edge to other versions of humanistic geography. Ley's own work on social movements and neighborhood struggles seems to me exemplary. It draws upon a series of theoretical traditions – one of its strengths is its suspicion of any single source of ideas – but it is nonetheless at once theoretically informed and politically engaged. Like many others who have urged a closer engagement with the contextualities of social life, and in particular with the fine-grained texture of the lifeworld, Ley's interests have moved in the direction of a "postmodernism of resistance."[32] The very idea is viewed with deep suspicion by many Marxists, including Harvey, but by no means all versions of historical materialism are antithetical to what is in fact an equally particular version of postmodernism. One does not have to turn to the contemporary writings of Soja or Jameson for support. More than 40 years ago Mills thought that the world was already entering "a postmodern period" in which the tension between rationalization and freedom had become almost palpable. And as he made plain, the social sciences, through their pervasive functionalism, were complicit in the trauma of this postmodern condition.[33] But he also indicated that its pathologies were nonetheless susceptible to theoretical interrogation. Whatever the nuances distinguishing Ley's critique of theoretical involution from Mills's attack on Grand Theory – the one outside and the other inside historical materialism – they are united in their opposition to what Mills called "abstracted empiricism." For as Eagleton remarks, and as both of these authors make clear in their own work, hostility to theory "usually means an opposition to other peoples' theories and an oblivion of one's own."[34]

In the third place, the experience of working with theory is itself mediated by the processes that many of these humanist writers esteem so highly. Harvey, who played such a prominent role in the reorientation and radicalization of human geography, freely admits that his thinking about capitalism and urbanism "has been as much influenced by Dickens, Balzac, Zola, Gissing, Dreiser, Pynchon, and a host of other [novelists]" as it has been by "dry-as-dust science." In his view, "reflection and speculation

---

[32] See, for example, David Ley, "Modernism, post-modernism and the struggle for place," in John Agnew and James Duncan (eds.), *The power of place: Bringing together geographical and sociological imaginations* (Boston: Unwin Hyman, 1989) pp. 44–65; *idem*, "Can there be a postmodernism of resistance in the urban landscape?," in Paul Knox (ed.), *The restless urban landscape* (Englewood Cliffs, N.J.: Prentice-Hall, 1993) pp. 255–78.

[33] Mills, *Sociological imagination*, pp. 181–88 and *passim*.

[34] *Ibid.*, pp. 60–86; Eagleton, *Literary theory*, p. viii.

prepare the way for theory construction at the same time as they define an arena of open and fluid evaluation of theoretical conclusions."[35] Expressed like that, one can surely recognize the creative, imaginative capacities *of theory* and acknowledge that these are not the unique preserve of a canonized "literature." I know that some commentators are dubious about Harvey's ability (or willingness) to translate these claims into practice. They doubt the openness and fluidity of his own work, and they regard his hostility to all forms of postmodernism and his marginalization of most forms of feminism as symptomatic of an objectivist, masculinist, and ultimately authoritarian mode of theorizing.[36] Now I have no doubt that it is wrong to absolutize theory, and such is not my intention: If it is not always abusive or intrusive, neither is it invariably enlarging or energizing. But theory, in the sense that I want to use the term, *is* a sort of moving self-reflexivity, "human activity bending back upon itself," always incomplete and constantly responding to the problems and predicaments of human existence and human practice. Again, it will not do to endorse the expansion of the interpretative field willy-nilly: Unmasking the pretensions of objectivism ought not to be a cover for the installation of relativism and much depends on the *politics* of theory.[37] This is, I think, precisely the point at issue between Harvey and his critics. All of them clearly agree on the connective imperative between the theoretical and the political and their disagreements turn on both axes at once: in particular, on the identification and articulation of what Harvey terms *significant* differences. The criteria involved are simultaneously theoretical and political.[38]

Political culture, spatial analytics, and the capitalist economy

These considerations do not settle matters, I realize, and I will want to return to the question of theory in human geography in due course. But if, as I have suggested, the theoretical and the political are connected, it follows that the dialogue between historical materialism and human geography in the late 1960s and 1970s cannot be disentangled from the political culture of which they were both a part. These were the years of the student protest movements in Western Europe and North America; of a growing concern over civil rights and social justice; and of demonstrations against

---

[35] David Harvey, *Consciousness and the urban experience* (Oxford: Blackwell Publishers, 1985) pp. xv–xvi.

[36] See, for example, Rosalyn Deutsche, "Boys town," *Environment and Planning D: Society and Space* 9 (1991) pp. 5–30; Doreen Massey, "Flexible sexism," *loc. cit.*, pp. 31–57.

[37] Terry Eagleton, "The significance of theory," in his *The significance of theory* (Oxford: Blackwell Publishers, 1990) pp. 24–38; the quotation is from p. 27. Bruce Robbins, "The politics of theory," *Social Text* 18 (1987) pp. 3–18; see especially pp. 8–9.

[38] See David Harvey, "Postmodern morality plays," *Antipode* 24 (1992) pp. 300–326.

the American offensive in Vietnam, Cambodia, and Laos.[39] These concerns were expressed in different ways in different places, of course, and there were important differences between Anglophone political cultures in Britain, Canada, and the United States and other political cultures in France, Francophone Canada, Germany, and Italy: But there were also affinities and solidarities between them. Other disciplines were radicalized by the internationalization of these events in varying degrees too, of course, and in geography as elsewhere the move to Marxism was also shaped by the intellectual momentum of the research programs that preceded it.

In Harvey's case, he was plainly disenchanted with spatial science – with its inability "to say anything really meaningful about events as they unfold around us" – but he also retained a strong commitment to a recognizably *scientific* geography that could analyze the *structure* of the *space-economy*.[40] This triple emphasis helps to account for his interest in Marx's own writings and what I take to be his proximity to a tradition of more or less classical Marxism, developed in Britain by Maurice Dobb and Eric Hobsbawm who were primarily interested in the economic trajectory of the capitalist mode of production, rather than the so-called "cultural materialism" of E. P. Thompson and Raymond Williams, which was to attract a later generation of Anglophone geographers. What this does not explain, however, is Harvey's indifference to the development of a Francophone school of structural Marxism, which was also exercised by the scientific status of Marx's work and which sought to theorize the contradictory coexistence of multiple structures within capitalism. Their conceptions of science were plainly different, but the explanation of this indifference may lie more deeply in the gaps between these conceptual innovations and both the historicism that was embedded within Harvey's work – that powerful "logic" imparting a direction to historical eventuation and threading its way through many of his postpositivist essays – and his characteristic gesture of bracketing the movements of political or cultural-ideological levels in order to theorize and analyze the movements of the economic level. For the economy

---

[39] It should not be forgotten that geography was also entered on the other side of the ledger: that its increasingly technocratic character was complicit in the instrumentalities that many radical movements were contesting; that it was shot through with assumptions about class, ethnicity, and gender that all but paralyzed the search for social justice; and that it was implicated in the planning and execution of American strategy in Southeast Asia, not least in the waging of ecological warfare against the North Vietnamese people. *Antipode: A journal of Radical Geography*, which began publishing in August 1969, was at the forefront of a critical response to these developments within the Anglophone discipline. See Richard Peet, "The development of radical geography in the United States" in his *Radical geography: Alternative viewpoints on contemporary social issues* (Chicago: Maaroufa Press, 1975) pp. 6–30.

[40] David Harvey, *Social justice and the city* (London: Edward Arnold, 1973) p. 128. See also *idem*, "From models to Marx: Notes on the project to 'remodel' geography," in Bill Macmillan (ed.), *Remodelling geography* (Oxford: Blackwell Publishers, 1989) pp. 211–16.

remained at the center of this newly radical geography, for Harvey and for many others, as had been the case in spatial science.

But it was now conceived in strikingly different terms. Political economy registered an important advance over the models of spatial science because it *socialized* those abstract geometries. Spatial structures were no longer generated by point-process models in mathematical spaces, and geography's grids were no longer indifferent to the particular things that were put there. Neither were they collapsed into two-dimensional patterns etched onto isotropic surfaces by the schedules of supply and demand, tracing out their endless, empty, and entirely fictional equilibrium. Instead, human geography analyzed landscapes of capital accumulation that were produced by social processes in material spaces, and it was increasingly responsive to the disjunctures and specificities of those topographies, to the intricate mosaic of combined and uneven development.

This was more than a theoretical achievement; it had important political implications. Once geography was freed from the naturalizing cast of spatial science and once the delinquent particularities of the world were loosed from its carceral language of "deviation" and "residual," it was possible to think of social practices in other ways and to imagine other scenarios. I have to say, however, that this has turned out to be largely a promissory note. Critical theory involves both an explanatory-diagnostic and an anticipatory-utopian moment.[41] But for the most part critical human geography has shown more of an interest in analysis than prescription: Witness the speed at which discussion of Harvey's *Social justice and the city* moved away from the first two words of the title, with their intimations of moral philosophy and political practice, and fastened on the last two.

Some commentators might choose to interpret this as the product of an analytical approach that concentrated on capital accumulation rather than collective (class) action; the articulations between the two have long provided both the central pivot and the central problem for historical materialism. But I am talking about the movement of critical responses to Harvey's project, commentaries on *Social justice and the city*, not continuations of it. And David Livingstone has suggested to me that this silence might have been the result of a more general lack of "apparatus" to grapple with moral questions. The recourse to functional language, inside and outside spatial science, effectively replaced what he calls "the language of strong evaluation" with a purely utilitarian calculus.[42] There is much in this, but the same dilemma confounds other traditions of critical inquiry hostile to functionalism and often particularly disposed to humanism.

---

[41] Seyla Benhabib, *Critique, norm and utopia: A study of the foundations of critical theory* (New York: Columbia University Press, 1986) p. 226.

[42] David Livingstone, personal communication.

I think, for example, of Buttimer's sustained and sensitive meditation on values in geography, which drew upon phenomenology and existentialism and which was soon followed by a programmatic agenda for a geography of the lifeworld. Yet even here the anticipatory-utopian moment has been persistently deferred, perhaps for the countervailing reason. Throughout Buttimer's work, there is a constant concern to bring intellectual knowledge into closer harmony with lived experience, but this sometimes seems to translate into a way of not only recovering but also *romanticizing* lived experience. When she distinguishes between a "representational space" and a "lived space," it is the latter, the lifeworld, that is valorized. The distinction is common enough: The former conveys the experience of space "through scientific, logical, mathematical categories"; its systems have "functional requirements" which "lay claim to" the time-space horizons of individuals within the lifeworld.[43] And yet these encroachments and impositions – the whole process of colonization, which (to move beyond Habermas once more) certainly does depend in part on representations of (abstract) space – cannot be analyzed within her humanistic framework precisely because it lacks an adequate conception of system or structure. This matters so much because the explanatory-diagnostic moment in critical theory is intended to expose the present in such a way that we can glimpse, however faintly, the outlines of a better future: It is supposed to transform our grasp of the situation in such a way that the prospects for a more humane society are opened up rather than closed off. If that moment is forever suspended in either the crystalline logic of structure or the webs of meaning spun within the lifeworld, the one constantly set against the other, then it is scarcely surprising that the anticipatory-utopian moment, like Althusser's lonely last instance, never comes.

It may be, of course, that thinkers on both sides of this dilemma are involved in an anticipatory politics in their teaching and in other practical ways that the written record makes it difficult to recover. If that is the case, however, it seems to me quite wrong to regard the one as public (because academic) and the other as private (because political). That would be to seek refuge behind the very divisions that critical theory seeks to dismantle. Political practice is not achieved or sanctioned by theoretical fiat, to be sure, but theory has consequences as well as commitments. And it seems to me that Harvey's unwavering insistence on the connective imperative between the theoretical and the practical – on exposing the intersections between different conceptions of rationality, justice, and spatiality as they are embedded in social practices, in the medium of social process – has done much to keep alive the possibility of (literally) making human geography in ways far removed from the instrumentalities of spatial science.

[43] Anne Buttimer, *Values in Geography* (Washington: Association of American Geographers, Commission on College Geography: 1974); *idem*, "Grasping the dynamism of the lifeworld," *Annals of the Association of American Geographers* 66 (1976) pp. 277–92.

I hope this will not be misunderstood. It is possible to represent some of those social processes in mathematical terms, and the development of analytical political economy and its translation into human geography shows that political economy cannot be treated as the innumerate's response to spatial science.[44] But there are at least three significant differences between their procedures. First, analytical political economy depends upon the translation of substantive processes – not abstract spatial patterns – into mathematical languages. Two of the principal authors of analytical political economy within geography express it this way:

[D]istances are not some physical constraint to which all realizations of a process are subject in identical ways, as in the laws of physics, because spatial structures are socially constructed and far more complex than the isotropic and stationary spaces that are generally relied on in [locational] analysis.[45]

Second, analytical political economy does not define processes of equilibration as "normal." These may be technically more tractable (and ideologically more acceptable) assumptions for spatial science, but instabilities and irregularities are written into the calculus of analytical political economy. When this calculus is translated into human geography, the results are even more radical. For equilibrium is thus confounded not only by the grid of social practices that is necessarily implicated in the dynamics of a capitalist economy (including, centrally, class struggle) *but also by the inscription of those practices and those dynamics in space.* From this perspective, "society" and "space" simply cannot be prised apart and neither can they be brought together within the equilibrium horizon of some coherent totality.[46] Third, analytical political economy is characteristically more sensitive to the limitations of its representations than conventional locational analysis:

We recognize that many aspects of society and economy are not subject to analytical treatment, and even those aspects that are may well be more sensitively

---

[44] There are other ways of showing so too: Many of the sharpest critics of spatial science had either been instrumental in its development or socialized into its paradigms as undergraduates and graduate students. If they criticize spatial science, it is not because they (we) cannot do it.

[45] Eric Sheppard and Trevor Barnes, *The capitalist space-economy: Geographical analysis after Ricardo, Marx and Sraffa* (London: Unwin Hyman, 1990) p. 301.

[46] It is the possibility of stabilizing those processes within a regime of accumulation that has attracted the attention of the so-called Regulation School. But its formulations have been criticized precisely because they seem to minimize the *contradictory* constitution of regimes of accumulation; ironically, this is in part the result of a failure to provide a sufficiently rigorous analysis of the modes of regulation that are supposed to be one of the distinctive conceptual contributions of the school. See, for example, Julie Graham, "Fordism/Post-Fordism, Marxism/Post-Marxism: The second cultural divide," *Rethinking Marxism* 4 (1991) pp. 39–58 and *idem*, "Post-Fordism as politics: The political consequences of narratives on the left," *Environment and Planning D: Society and Space* 10 (1992) pp. 393–410.

treated by nonanalytical methods. As such, we think it crucial to delimit carefully the limits of analytical methods.[47]

Nevertheless, the critique of spatial science cast a long shadow, and few of the early engagements between political economy and human geography were analytical in quite this sense. Part of the attraction of Marx's work for Harvey was undoubtedly its analytical rigor, and I think it fair to say that his early, exuberant explorations of historical materialism were not altogether free of a residual positivism. In *Social justice and the city* he located Marxism at the intersection of "positivism, materialism and phenomenology," but the last of these received little sustained attention. Even his closing essay in that volume drew upon the most positivist of all structuralisms, if I may so describe the work of the psychologist Jean Piaget, and it was plainly all of a piece with his previous interest in the construction of a cognitive-behavioral location theory.[48] Like many others, however, Harvey had become skeptical of mathematical modeling – of its analytical rather than dialectical form – and suspicious of the unidentified calculus of spatial analysis.[49] It was this spatial formalism and its implied claim for an autonomous spatial science that perturbed Harvey and others, and thus Marx's critique of commodity fetishism was extended to "spatial fetishism." Spatial science was charged with severing spatial structures from the social relations that were embedded within them. It was not difficult to prosecute the case – nor to secure a conviction – but the verdict provoked a widespread withdrawal from the analysis of spatial structure altogether. Most of the early engagements between political economy and human geography were distinctly one-sided. Historical materialism was projected more or less directly into human geography, so that studies of the uneven development of capitalism conducted under its sign were inevitably limited by the uneven development of Marxism itself and, in particular, by its emphasis on history rather than geography. For this reason when Harvey so carefully delineated *The limits to capital*, he was marking not only the bounding contours of capitalist development – the ways in which capitalism was constrained by the sedimented landscapes which were laid down through successive cycles of acccumulation – but also an important lacuna in Marx's masterwork.[50]

---

[47] Sheppard and Barnes, *Capitalist space economy*, p. 13.

[48] Harvey, *Social justice*, p. 129; Harvey's discussion of Piaget will be found in his "Conclusions and Reflections" (pp. 286–314).

[49] See, for example, Harvey's assault on analytical Marxism in his review of Jon Elster's *Making sense of Marx* (New York: Cambridge University Press, 1985) in *Political Theory* 14 (1986) pp. 686–90. There Harvey describes the "reduction of dialectical reasoning to ordinary analytical language" as "catastrophic."

[50] David Harvey, *Limits*; the imagery of sedimentation and cycles of accumulation is derived from Doreen Massey, *Spatial divisions of labour: Social structures and the geography of production* (London:

Yet one of the most important achievements of Harvey's own inquiry was to show how Marx's critique of capitalism assumed, tacitly but nonetheless centrally, the production of an urbanized space-economy. I cannot do justice to the richness of Harvey's rereading of Marx here, but I must open one window on his work. There are a number of fundamental tensions within the landscape of capitalism, so Harvey claims, and one of the most basic is between processes working toward agglomeration in *place* and processes working toward dispersal over *space* (figure 7). The rationalization of capitalist production is supposed to pull clusters of economic activities into a "structured coherence" *within* particular regions, but insofar as capitalism is inherently expansionary, capital accumulation must evidently also depend upon time-space coordination *between* regions. Harvey argues that this tension can be resolved only through "the urbanization of capital." For the circulation of capital through an urbanized space-economy brings different labor processes at different locations into a general social relation:

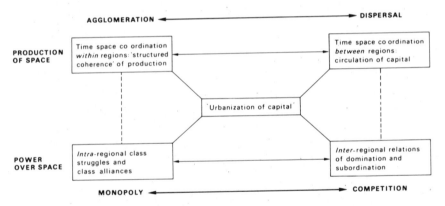

Figure 7 Tensions in Harvey's landscape of capitalism.

concrete labor processes acquire *abstract* qualities tied to value as socially necessary labor time. This is of immense importance because the notion of "socially necessary labour time" is absolutely central to Marx's labor theory of value. Without it, the whole structure of class exploitation through the appropriation of surplus value collapses. What Harvey is able to show, therefore, is that the mainspring of Marx's political economy, in contrast to the a-spatial calculus of neoclassical economics, depends upon the production of a differentiated and integrated space-economy.[51]

Macmillan, 1984). Both Harvey and Massey have made major contributions to the dialogue between Marxism and geography, but (through the 1980s at any rate) there was little obvious dialogue between *them*.

[51] Harvey, *Limits*, p. 375 and *passim*.

Even so, Harvey insists that this framework is chronically unstable, a sort of permanent holding operation, because the constant drive to reduce the turnover time of capital produces exactly that "annihilation of space by time" that Marx accentuated. Changes in transportation and communication systems, credit networks, and the like have all contributed to a remarkable increase in the speed of the global economy, and its accelerating rotations have progressively transformed the relative locations of existing production configurations. The deflation of old locational values has impelled changes in the geography of multinational investment, shifts in the international division of labor, and the emergence of new zonal and regional cities. These processes are not straightforward, Harvey points out, because a knife-edge has to be constantly negotiated "between preserving the values of past commitments made at a particular place and time, or devaluing them to open up fresh room for accumulation." The tension between the immobility of spatial structures and their capacity to stretch across ever wider spans of time and space is almost palpable.

The produced geographical landscape constituted by fixed and immobile capital is both the crowning glory of past capitalist development and a prison that inhibits the further progress of accumulation precisely because it creates spatial barriers where there were none before.[52]

The transcendence of these barriers to continued accumulation depends not only on the production of space, however, but also on power over space. The production of clusters of economic activities within regions is paralleled by the formation of equally unstable and place-specific "class alliances." These coalitions are territorially based social movements seeking to intervene in the competitive struggle between different regions within what is an increasingly complex and constantly changing hierarchy of domination and subordination. Class alliances are forged out of a mix of different interests, but their inherent instability is heightened by what Harvey presents as their double function. On the one hand, they have to protect those immobile regional infrastructures that provide for continued production and reproduction; on the other hand, they must secure their own upward spiral of accumulation by capturing new rounds of investment. Class alliances thus become caught between "the stagnant swamp of monopoly controls fashioned out of the geopolitics of domination" and the "fires of open and escalating competition with others."[53]

[52] David Harvey, "The geography of capitalist accumulation: Towards a reconstruction of the Marxian theory," in his *The urbanization of capital* (Oxford: Blackwell Publishers, 1985) pp. 32–61; the quotation is from p. 43.

[53] David Harvey, "The place of urban politics in the geography of uneven capitalist development," *loc. cit.*, pp. 125–64; the quotation is from p. 161.

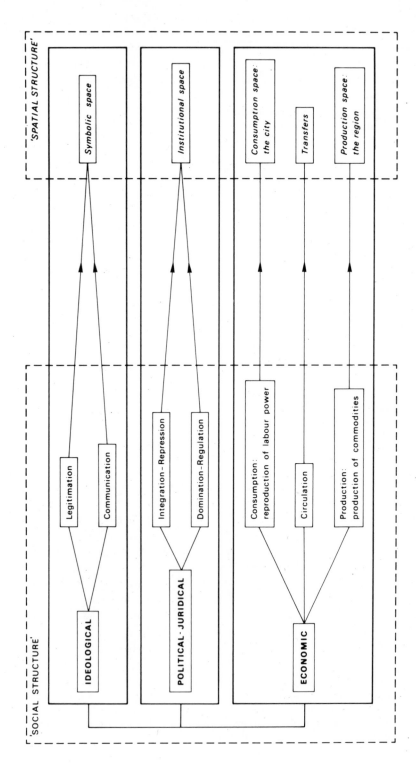

Figure 8 Social structure and spatial structure.

As I write these words, I am reminded again of how close they are to the spirit of Marx and of Harvey's seeming disinterest in much of post-classical Western Marxism. In failing to consider these contributions, it seems to me, Harvey runs the risk of ignoring the predicaments and problems that brought those later traditions of historical materialism into being. In his more recent work, it is true, Harvey has moved some way beyond his original concerns. His critique of *The condition of postmodernity* offers a more discriminating analysis than Marx's nineteenth-century problematic could reasonably be expected to provide, distinguishing between different regimes of accumulation within late twentieth-century capitalism, and illuminates a series of explicitly cultural landscapes through an historical geography of modern and postmodern representations of space. If this pushes back the perimeter of classical Marxism and elaborates a more complex geometry within it, however, the figure remains essentially the same: It is still centered on the spatial analytics of the capitalist economy.[54]

Reorientations: Western Marxism and human geography

Expressed like that, perhaps Harvey's indifference is not surprising. With the exception of lone figures like Benjamin and Lefebvre, the founders of Western Marxism maintained the same strategic silence about place and the production of space. Of the later generation, Althusser accentuated the "uneven development" of capitalism, and human geography could have learned much from the complex topography that he marked out. But his structural Marxism was concerned with the different *temporalities* of capital-ism, with the different and often discordant rhythms of its economic, political and cultural-ideological levels, not with what Lipietz and others subsequently identified as their *spatialities*. The most direct extension of Althusser's work into spatial analytics was provided by Castells, who projected the multiple levels of social structure onto the overlapping grids of spatial structure (figure 8); but I thought at the time (and still think) that the vocabulary of expression and projection was antithetical to the tenor of structural Marxism and that it is quite wrong to theorize social structure independently of spatial structure.[55] Poulantzas's brilliant sketches of what he called the "institutional materiality" of the state and its spatial

---

[54] Harvey, *Condition of postmodernity*. Many of Harvey's critics rail against his disinterest in work outside historical materialism, but while I sympathize with (some of) their concerns I find his disinterest in other traditions within historical materialism equally revealing. I discuss Harvey's project in detail in Chapter 6, this volume.

[55] See my *Ideology*, pp. 118–21. By including Althusser within Western Marxism, I follow Perry Anderson, *Considerations on Western Marxism* (London: New Left Books, 1976) and Martin Jay, "The topography of Western Marxism" in his *Marxism and totality: The adventures of a concept from Lukács to Habermas* (Cambridge: Polity, 1984) pp. 1–20. For the main texts referred to here, see Louis

matrices were much closer to the mark, but at the time they excited little attention in geography.[56] And apart from one or two philosophical forays in the direction of the Frankfurt School and its successors, mustering general support for an assault on the positivism that provided spatial science with its epistemological shield, the rest of Western Marxism received barely a passing glance.

Soja marks the 1980s as a decisive break: A period punctuated by a series of slips, hesitations and regressions but during which, nevertheless, the emphasis of classical Marxism on the making of history was joined by a new interest in the making of geography. Soja calls this a "provocative inversion," a reversal of the original mapping of historical materialism into geography, which set in motion a spatialization of critical theory whose intellectual momentum pushed far beyond Harvey's historico-geographical materialism. The "provocation" was supposedly double-sided. Not only was a spatial analytics incorporated within historical materialism as a central moment in its critique of capitalism, but many of these newer formulations were now directed toward a postclassical, Western Marxism that turned on more than the critique of political economy.[57] Soja makes so much of this, I take it, because the failings of classical Marxism were not confined to its prioritization of history over geography. Such failings also revolved around the articulations between capital accumulation and collective action – whose chronic tensions have so awkwardly skewered much of the preceding discussion – and they required some means other than the model of base and superstructure to articulate economy, politics, and culture. Although it is not easy to characterize Western Marxism with any precision, these twin problems provided it with much of its *raison d'être*, and while many areas of Western Marxism were given over to culture and cultural politics I do not think it can be adequately (or even accurately) described as "the Marxism of the superstructure" because so much of its effort was directed toward ways of transcending that classical model.[58]

And yet there is something strange about Soja's celebration of this inversion. As he himself acknowledges, many of the most recent chroniclers

Althusser and Etienne Balibar, *Reading capital* (London: New Left Books), first published in French in 1968; Alain Lipietz, *Le capital et son espace* (Paris: Maspero, 1977); Manuel Castells, *The urban question: A Marxist approach* (London: Edward Arnold, 1977), first published in French in 1972.

[56] See especially Nicos Poulantzas, *State, power, socialism* (London: Verso, 1978).

[57] Edward Soja, *Postmodern geographies: The reassertion of space in critical social theory* (London: Verso, 1989) pp. 39–41, 56–75.

[58] Cf. J. G. Merquior, *Western Marxism* (London: Paladin, 1986) p. 4: "What sets Western Marxism apart, beyond a mere shift from economics into culture, is the combination of a cultural thematics and the near absence of infrastructural weight in the explanation of cultural and ideological phenomena." Anderson, *Considerations op. cit.* also suggests that Western Marxism "came to concentrate overwhelmingly on the study of superstructures," but he provides a much more scrupulous survey of its genealogy and geography.

of Western Marxism, by no means unsympathetic to radical ideas and ideals, have described its achievements in considerably more muted terms; others have characterized it, more bluntly, as a failure. To take perhaps the most extreme example, Merquior portrays Western Marxism as a debilitating *Kulturkritik*, an apocalyptic indictment of capitalist modernity that derived from Hegel and Nietzsche as much as from Marx and which was paralyzed, from the very beginning, by the triumph of theory over analysis. I am not sure that the constellation of Hegel, Nietzsche, and Marx is always as sterile as Merquior makes out: Lefebvre's work suggests quite the opposite, and both Harvey and Soja draw upon it (though in different ways and to different ends).[59] But Merquior is unrelenting and cites Anderson's summary judgment on the project as a whole: "Method as impotence, art as consolation, pessimism as quiescence." Actually, Anderson is more subtle than Merquior allows. Western Marxism did produce a narrowing of focus, he agrees, but "in its own chosen fields this Marxism achieved a sophistication greater than that of any previous phase of historical materialism." Anderson's sophistication is Merquior's theoreticism, then, and it is not easy to adjudicate between them. But even Anderson has to concede that the high priests of Western Marxism spoke arcane languages that separated them not only from one another but also from ordinary people and everyday life. These are exactly the charges that Ley leveled against structural Marxism in geography (above, pp. 80–81), but they are given a greater force here by being directed against Western Marxism more generally and its apparent divorce from "political and economic realities" and from what both Anderson and Merquior call the *concreteness* of history.[60]

If one looks at the studies that Soja has in mind, however, and for that matter those that he has himself conducted, it becomes clear that most of them have been focused precisely on those politico-economic realities – on new configurations of state and economy, on local and global restructurings of contemporary capitalism – and that their emphasis on the multiple historical geographies of capitalism has endowed this newly spatialized critical theory with considerable concreteness. Certainly, many writers would now agree that the production of space has assumed a particular salience in the late twentieth-century world; that the emergent forms of high modernity, perhaps even of postmodernity, depend upon tense and turbulent landscapes of accumulation whose dynamics are so volatile and whose space-economies are so disjointed that one can glimpse within the dazzling sequences of deterritorialization and reterritorialization a new and intensified fluidity to the politico-economic structures of capitalism; that the hypermobility of finance capital and information cascading through the circuits

---

[59] See Chapter 6, this volume.

[60] Merquior, *Western Marxism*, pp. 10, 38–39, 186–87; Anderson, *Considerations*, pp. 92–94.

of this new world system, surging from one node to another in nano-seconds, is conjuring up unprecedented landscapes of power in which, as Castells put it, "space is dissolved into flows," "cities become shadows," and places are emptied of their local meanings; and that ever-extending areas of social life are being wired into a vast postmodern hyperspace, an electronic inscription of the cultural logic of late capitalism, whose putative abolition of distance renders us all but incapable of comprehending – of mapping – the decentered communication networks whose global webs enmesh our daily lives. In sum, and as Jameson has argued particularly forcefully, "a model of political culture appropriate to our own situation will necessarily have to raise spatial issues as its fundamental organizing concern."[61]

This image is a composite, of course, and there are important (and often sharp) disagreements between its authors. Even so, it seems clear that if this loosely collective project *is* part of Western Marxism, then it is being conducted at one or more removes from the original luminaries. One can perhaps glimpse the faint, dystopian outlines of Horkheimer and Adorno's dialectic of enlightenment or Marcuse's one-dimensional society, even Haber-mas's colonization of the lifeworld: but these are highly generalized influen-ces and affinities, which have not been developed in any systematic fashion. Jameson's writings have been directly informed by Althusser, and I take an immoderate pleasure in the interest shown in his account of post-modernism by some of the critics of structural Marxism, but Castells's more recent work has repudiated structural Marxism (and most other versions of Marxism) altogether.

Much the same difficulty arises in fixing the position of work that is ostensibly much closer to the central cultural thematic of Western Marxism. Soja himself displayed little interest in culture and its geographies, but the revival of cultural geography in a radically new form was one of the most striking developments of the 1980s. I will have more to say about this later, but one of its cardinal achievements has been a renewal of interest in the concept of landscape: a change that is built around a recognition *of* its conceptuality. To Hartshorne, the literal meaning of landscape as "the view of an area seen in perspective" was "of little importance in geo-graphy," and he complained that any more technical definition produced only a hodgepodge of different ideas. When Harvey rehabilitated the concept of landscape, he was not interested in optics or aesthetics either, and in his early work he made it virtually synonymous with the landscape

---

[61] This paragraph is a collage of many sources, including the work of Manuel Castells, Michael Dear, David Harvey, Fredric Jameson, Allen Scott, Michael Storper, and Sharon Zukin; the quotations are taken from Manuel Castells, *The city and the grassroots: A cross-cultural theory of urban social movements* (Berkeley: University of California Press, 1983) p. 314 and Fredric Jameson, *Postmodernism, or the cultural logic of late capitalism* (Durham: Duke University Press, 1991) p. 51.

of capital accumulation.[62] Against the grain of these readings, the new cultural geography has shown that the relations between landscape as "a way of seeing" and the geometries of linear perspective are of the utmost importance to geography and that the very idea of landscape is shot through with ambivalences, tensions, and grids of power that cannot be reduced to the marionette movements of the economy. As I show in Chapter 4, this impacts on Soja's own sketches of the landscape of late twentieth-century Los Angeles, and it also heightens the importance of a cultural materialism that has certain affinities with Western Marxism. When Daniels makes the illuminating suggestion that the concept of landscape might be seen as a "dialectical image," for example, shimmering with explosive contradictions which conjoin past and present in disturbing new constellations, he is evidently trading on the work of Adorno and Benjamin and deliberately sharing in and contributing to the generalized legacy of Western Marxism.[63] But the fact remains that the new cultural geography has invested that legacy far beyond the confines of continental European Marxism (the heartland of Western Marxism was a Western Europe supposedly immune to the peculiarities of the English), and that where this cultural geography draws upon historical materialism – as it often does – it usually looks to Anglophone art critics and historians like John Berger and Timothy Clark or cultural critics like Stuart Hall and Raymond Williams.[64]

To be sure, Soja recognizes the brittle tension between Western Marxism and human geography. "Just as contemporary Western Marxism seems to have exploded into a heterogeneous constellation of often cross-purposeful perspectives," so modern geography has "started to come apart at the seams." He describes this process of creative destruction as the "postmodernization of Marxist geography" and anticipates the construction of a fully postmodern geography that will "continue to draw inspiration from the emancipatory rationality of Western Marxism but can no longer be confined within its contours," and in which some of the key figures of Western Marxism – notably Gramsci, Lefebvre, and Sartre – now appear alongside other thinkers whose relation to historical materialism is considerably more complicated and contentious: Foucault, Giddens and, more recently, Baudrillard.[65] It is not

---

[62] Richard Hartshorne, *The nature of geography* (Lancaster, Pa: Association of American Geographers, 1939) p. 160; Harvey, "Geography of capitalist accumulation." His recent writings show much more interest in symbolic landscapes.

[63] On Soja and the landscapes of Los Angeles, see Chapter 4, this volume. On landscape more generally, see Stephen Daniels, "Marxism, culture and the duplicity of landscape," in Peet and Thrift (eds.), *New models*, vol. 2, pp. 196–220; the quotation is from p. 206. Other key contributions include Denis Cosgrove, *Social formation and symbolic landscape* (London: Croom Helm, 1984); *idem*, "Prospect, perspective and the evolution of the landscape idea," *Transactions of the Institute of British Geographers* 10 (1985) pp. 45–62.

[64] See Peter Jackson, *Maps of meaning: An introduction to cultural geography* (London: Unwin Hyman, 1989).

[65] Soja, *Postmodern geographies*, pp. 60, 72–73.

immediately clear what is "postmodern" about such a program, and others are markedly less enthusiastic about any postmodern turn. To conduct their own analytical conversation, for example, Pred and Watts invoke Benjamin's exploration of dialectical images and his critique of capitalism through the archaeology of the commodity to illuminate their vivid cartographies of commodity culture in Africa, Europe, and North America, and in doing so they retrieve the close links between Western Marxism and modernism.[66] My own view is that the relations between modernism and postmodernism are more complex than caricatures of opposition and reversal allow, and that both of them provide space for a critique of classical Marxism – of its historicism, reductionism, and essentialism – which is by no means incompatible with a renewed historical (or historico-geographical) materialism. But many writers have moved sufficiently far from the classical terrain to locate their projects within the horizon of a heterogeneous post-Marxism, and I now want to consider its developing topography in detail.

Mapping post-Marxism

Post-Marxism is a highly contested field and in figure 9 I have tried to fix some of its shifting contours as they traverse human geography: Other disciplines and other discourses would require other diagrams. This sketch is very much a map of the moment, too, and I would be surprised if its configurations remained stable for very long or if other (and better) maps of the same terrain could not be produced.[67] I distinguish two streams, each of which is in some sense a critical response to historical materialism – which is why I insist on its importance: unsurpassed, I believe, by any other discourse in the history of postwar geography – and each of which also maintains a tense relation to the other. It is this *agonistic* grid, with its matrix of critical responses and tense relations, that (I hope) converts the map into something more than a convenient classification by capturing the politico-intellectual dynamics that surge through and shape its continuing transformation. Before I set off with this rough map as my guide, however, I need to place some confidence limits around it.

---

[66] Allan Pred and Michael Watts, *Reworking modernity: Capitalisms and symbolic discontent* (New Brunswick, N.J.: Rutgers University Press, 1992); cf. Eugene Lunn, *Marxism and modernism: An historical study of Lukács, Brecht, Benjamin and Adorno* (Berkeley: University of California Press, 1982).

[67] Compare, for example, my rough sketches with Stuart Corbridge, "Marxism, post-Marxism and the geography of development," in Peet and Thrift (eds.) *New models*, vol. 1, pp. 225–54. In his review of that volume, Harvey wonders if Corbridge's post-Marxism amounts to much more than putting Marx's *Eighteenth Brumaire* back together with his *Capital*. My own discussion of post-Marxism is more catholic than Corbridge's, but even in his more precise sense the answer is surely that it does. For Harvey's review, see *Transactions of the Institute of British Geographers*, 15 (1990) pp. 376–78.

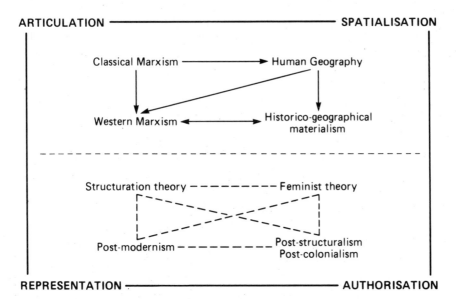

ARTICULATION ———————————————— SPATIALISATION

Figure 9   Marxism and post-Marxism.

First, I do not mean to imply that there is any simple progression as we move from the top of the diagram to the bottom. Marxism has not been left behind by the development of post-Marxism (or any of the other "posts" that have been spatch-cocked into it). Although events in Eastern Europe and the former Soviet Union have plainly and properly shattered its more rigid orthodoxies, historical materialism continues to develop, critically and for the most part constructively, in response to discourses that are in turn often still responding to its own formulations. One of the most significant interventions has been a renewed interest in a nonessentialist Marxism. In some versions the marks of Western Marxism are still visible, to be sure, most obviously in those that reach back to (and then go beyond) structural Marxism to reclaim and rework Althusser's concept of overdetermination – Hindess and Hirst yielding to Resnick and Wolff – and those that reinscribe Gramsci's concept of hegemony within a multiplication of political spaces and subject-positions: Laclau and Mouffe's project of a "radical democracy."[68] In other versions the latent positivism that Habermas (for one) saw in Marx's own writings has been replaced by

---

[68] For the former, see Stephen Resnick and Richard Wolff, *Knowledge and class: A Marxian critique of political economy* (Chicago: University of Chicago Press, 1987) and, in geography, Julie Graham, "Theory and essentialism in Marxist geography," *Antipode* 22 (1990) pp. 53–66 and *idem*, "Anti-essentialism and overdetermination," *loc. cit.*, 24 (1992) pp. 142–56. For the latter, see Ernesto Laclau and Chantal Mouffe, *Hegemony and socialist strategy: Towards a radical democratic politics* (London: Verso, 1985).

a theoretical realism that has its origins in the philosophy of science and includes work that is indifferent to and on occasion even antithetical to Marxism; but its appropriation by Bhaskar and Sayer in particular has done much to renew the scientificity of historical materialism (for those who consider it important) and to clarify the connections between diverse structures and the conduct of social life in ways that resist any slide into an essentialism of structures or actions.[69] These middling Marxisms, as I think of them, have attracted criticism from both flanks. On the one side, they are portrayed as arbitrary and politically disabling, dissolving the corpus of historical materialism in an acid-bath of relativism. On the other side they are seen as clinging to a foundationalism, deaf to the destabilizing interrogatories of deconstruction.[70] That the criticisms cancel one another out does not necessarily mean that these middling Marxisms have got things about right; but the passion with which the debates have been conducted surely implies that historical materialism is far from moribund. And for all the fireworks, these debates have imposed clarifications and raised further questions about the relations between capitalism, socialism, and democracy which make it impossible to think in terms of simple oppositions or indiscriminate condemnations. The rise of a series of new journals on the left in the course of the 1980s – *Capitalism, nature, socialism; Cultural critique; Inscriptions; New formations; Rethinking Marxism; Social text; Society and space; Theory, culture and society*, these are only a few – testifies not only to the continuing fertility of the marchlands between Marxism and post-Marxism but also to the increasingly subtle arguments and strategies being developed.

Second, and closely connected to these considerations, the discourses of post-Marxism share more than a critical orientation to Marxism. Those placed below the fold on figure 9 conceive of "theory" in a different way than those above it do. In general, they are much less architectonic in their inclinations and sensibilities: that is, they are much more suspicious of attempts to construct a single, centered conceptual system. This does not mean that post-Marxism marks a return to a radical empiricism, still less that it has succumbed to an extreme skepticism or a theoretical paralysis.[71]

[69] The most important figure is undoubtedly Roy Bhaskar: see in particular his *The possibility of naturalism: A philosophical critique of the contemporary human sciences* (Brighton: Harvester, 1979) and *Scientific realism and human emancipation* (London: Verso, 1986). In geography his ideas have been developed in detail by Andrew Sayer in his *Method in social science: A realist approach* (London: Hutchinson, 1984; Routledge, 1992, second edition).

[70] For more or less classical Marxist positions, see Norman Geras, *Discourses of extremity: Radical ethics and Post-Marxist extravagances* (London: Verso, 1990) and, in geography, Dick Peet, "Some critical questions for anti-essentialism," *Antipode* 24 (1992) pp. 113–30; for a deconstructive reading, see Donna Landry and Gerald MacLean, "Rereading Laclau and Mouffe," *Rethinking Marxism* 4 (4) (1991) pp. 41–60.

[71] For these concerns, see David Harvey and Allen Scott, "The practice of human geography: Theory and empirical specificity in the transition from Fordism to flexible accumulation," in Bill Macmillan (ed.), *Remodelling geography* (Oxford: Blackwell Publishers, 1989) pp. 217–29; Christopher

These anxieties are not idle fears, to be sure, and I understand their origins. But if post-Marxist discourses are suspicious of Grand Theory, this does not imply any disinterest *in large questions*. Attempts to think about human subjectivity and self-identity; about gender and sexuality; about "reason" and representation; about racism, colonialism and postcolonialism: attempts to think about all these things – and many more – in new and complex ways are hardly signs of a new parochialism. And to suggest that post-Marxism is somehow a-theoretical must seem almost comical to those of its critics who have lambasted what they see, contrarily, as its theoretical obsession: what Merquior diagnoses more specifically as griping "theorrhea."[72] Against both of these readings, I see post-Marxism as an attempt to undo theory's closures and to subvert its imperialisms *from within*. For this reason, I have some difficulty in accepting Skinner's sighting of a "return of Grand Theory" across the whole spectrum of the human and social sciences, if this is supposed to imply anything close to Mills's original observation, since in so many cases one of the most persistent claims of post-Marxist discourses is, precisely, that "concepts are not timeless entities with fixed meanings, but should rather be thought of as weapons (Heidegger's suggestion) or as tools (Wittgenstein's term), the understanding of which is always in part a matter of seeing who is wielding them and for what purpose." For these writers, the relations between the theoretical and the political are no longer to be adjudicated through a canonical grid.[73] Finally, and following from these considerations, if post-Marxist discourses are attentive to specificity and contextuality – and Skinner makes much of a new sensitivity to the importance of the local and the contingent – these cautions are not barriers to generalization (or political cooperation) but instead *conditions* of any attempt to reach beyond our own particularities, positionalities, and competences in ways that hold out some hope of avoiding the arbitrary and the authoritarian.

Third, the quadrants that frame figure 9 identify what I take to be some of the most important considerations in the conduct of any critical project. In their different ways, each of them reaffirms the deconstructive impulse

---

Norris, "Criticism, history and the politics of theory," in his *What's wrong with postmodernism: Critical theory and the ends of philosophy* (Baltimore: Johns Hopkins University Press, 1990) pp. 1–48.

[72] "More specifically" because he was referring to poststructuralism: J. G. Merquior, *From Prague to Paris: A critique of structuralist and post-structuralist thought* (London: Verso, 1986) p. 247, 253 and *passim*. He would presumably make the same diagnosis of Western Marxism.

[73] Quentin Skinner, "Introduction: The return of grand theory," in *idem* (ed.), *The return of Grand Theory in the human sciences* (Cambridge: Cambridge University Press, 1985) pp. 1–20; the quotation is from p. 13. I confess that I am extending Skinner's commentary here: He is in fact summarizing his discussion of Gadamer, Foucault, and Derrida (among others). If we turn to the writings of Althusser and Habermas, however, both of whom also figure in Skinner's collection, then the connections with Parsons's baroque constructions have more cogency and the claim for a return of Grand Theory has more force.

within the contemporary reconfiguration of human geography and the other humanities and social sciences. *Articulation* carries a double charge. It is intended to challenge the sedimented divisions between the economic, the political, the social and the cultural and to urge the necessity of bringing them into dialogue with one another; it is also meant to signal a resistance to attempts to close critical inquiry around the categories of either "structure" or "agency" and to indicate the importance of analyzing the various ways in which each is implicated in the constitution of the other. As I have said, these concerns mark the site where Western Marxism, among others, pitched its tent, but they also gesture toward a second quadrant. For it seems likely that the analysis of these articulations also requires an identification of the modalities through which time and space are bound into the constitution of social life.

Spatialization thus refers to those ways in which social life literally "takes place": to the opening and occupation of different sites of human action and to the differences and integrations that are socially inscribed through the production of place, space, and landscape. This is a redescription of the project of human geography, and deliberately so; but I have borrowed part of it from Schatzki – a philosopher who describes himself as an "honorary geographer" – because his Heideggerian phrasing points toward a third and a fourth quadrant, where versions of both postmodernism and poststructuralism are also indebted to Heidegger's writings.[74] Here too a double valency is implied – "space" is coded in both physical and social terms – and some writers have connected this double-coding to a chronic crisis of representation that has assumed a series of particularly acute forms in the course of the twentieth century.

Representation thus draws attention to the different ways in which the world is made present, re-presented, discursively constructed. I have no doubt that rendering this process problematic is one of the consequences of suspending the authority of Grand Theory, which has heightened a sense of uncertainty about the adequacy of established means of describing the world. But this implies no rejection of the theoretical as such: On the contrary, "elevated to a central concern of theoretical reflection, problems of description become problems of representation."[75] This was as much a concern of modernism as it has become of postmodernism, and here too the interventions of Western Marxism seem to me of considerable signi-

---

[74] Theodore Schatzki, "Spatial ontology and explanation," *Annals of the Association of American Geographers* 81 (1991) pp. 650–70.

[75] George Marcus and Michael J. Fischer, *Anthropology as cultural critique: An experimental moment in the human sciences* (Chicago: University of Chicago Press, 1986) pp. 8–9. Marcus and Fischer develop these claims in directions that intersect with both *articulation* (in their terms, the reconciliation of political economy with interpretative anthropology) and *spatialization* (the construction of multilocale ethnographies that register the ways in which the configurations of the world system break open the enclosures of local knowledge).

ficance in undercutting any simple opposition between the two; but in each case this is more than a theoretical matter because representation also has a related, more directly political meaning that has to do with giving voice to the concerns and situations of others. This process of "othering" is never neutral, however, and always works through grids of power. As Said reminds us, the "scrubbed, disinfected interlocutor is a laboratory creation"; it was only when subaltern figures made enough noise that they were admitted to the conversation, and the terms of their invitation, extended by those who had the power to do so (and, for that matter, the power not to do so), made sure that they were still trapped within the enclosures of colonialism.[76] These concerns strike at the very heart of Western Marxism too: Its characteristic Eurocentrism, carried in that seemingly (and self-consciously) innocent adjective that separated it not only from the orthodoxies of Eastern Europe and the Soviet Union but from the wider "non-West" is not a contingent but a constitutive orientation, imposed by what some of its most persistent critics see as its identity as a "white mythology."[77]

Authorization is an attempt to make that dimension more explicit by raising a series of questions about the inscription of subjectivity and the operation of power-knowledge: about the privilege of position, and about authorship and authority, representation and rights. When Spivak asks "Can the subaltern speak?" she is asking exactly those questions in the most compressed and the most agonizing of ways. One of the most productive characteristics of her own presentations is I think a steadfast refusal to substitute her marginality for her centrality. She constantly confronts the tensions between her privileged position as an intellectual now working in the United States and her subaltern status as a woman from the Indian subcontinent, and she strives to understand the constitution of her own subject-position: to "unlearn her privilege as her loss," as she puts it, without abandoning a strategic recourse to a high theory that includes not only feminism and poststructuralism but also historical materialism.[78]

These remarks do not anticipate any grand synthesis, which I frankly think is as unlikely as it is undesirable, but they do suggest that there is

---

[76] Edward Said, "Representing the colonized: Anthropology's interlocutors," *Critical Inquiry* 15 (1989) pp. 205–25; the quotation is from p. 210. Said also connects these concerns to *spatialization*: to the investments of cultural structures and geographical dispositions in the production of space.

[77] The term is Derrida's: "Metaphysics – the white mythology which reassembles and reflects the culture of the West: the white man takes his own mythology, Indo-European mythology, his own logos, that is, the mythos of his idiom, for the universal form of that he must still wish to call Reason." Jacques Derrida, *Margins of philosophy* (Chicago: University of Chicago Press, 1982) p. 213. For a particularly detailed discussion of its bearing on Western Marxism, see Robert Young, *White mythologies: Writing History and the West* (London: Routledge, 1990).

[78] Gayatri Chakravorty Spivak, "Can the subaltern speak?" in Cary Nelson and Lawrence Grossberg (eds.), *Marxism and the interpretation of culture* (Urbana: University of Illinois Press, 1988) pp. 271–313.

a fragile, often fractious but nonetheless persistent *communication* between historical materialism and post-Marxism. Although the messages delivered by so many of our contemporary "posts" speak to the abandonment of ideals of communicative rationality, which are supposed to be part of what Bernstein lists as "the now exhausted metaphysics of presence, logocentrism, phonocentrism, ethnocentrism, and phallocentrism which comprise the violent history of the West," my sympathies (as will become obvious in what follows) are much closer to Bernstein's own. In other words, I believe that the critical force of these messages – and that they *are* messages bears emphasis – is to open up the moral and political space of communicative reason to scrutiny – not rejection.[79] And since I have made so much of deconstruction in these pages, I should also say that this reading does not exclude Derrida. Bernstein shows that the constellation that can be formed through the conjunction of Derrida and Habermas bears directly on the vulnerable but, so it seems to me, nonetheless vital intersections between difference and dialogue.[80]

Signs: social theory and human geography

I have marked the first steps into the space of post-Marxism as "social theory" and since I am obviously using the term in a much more restricted sense than I have thus far I need to explain myself. The restriction is not one that I have imposed on the discussion – in human geography many of those who sought to move beyond Marxism identified their interest in this way – and in that sense it is perhaps not a restriction at all. The intention, I take it, was to signal a continuation of the project of a critical human geography ("social theory" inscribing the distance from spatial science) through an expansion of the discursive field to consider questions that lay beyond the horizons of political economy or historical materialism.

[79] Richard Bernstein, "The rage against reason," in his *New constellation*, pp. 31–56; the quotation is from p. 50. For a similar argument see also Steven Best and Douglas Kellner, "Towards the reconstruction of critical social theory," in their *Postmodern theory: Critical interrogations* (New York: Guilford Press, 1991) pp. 256–304 (though they have more faith in some grand synthesis than I do).

[80] Richard Bernstein, "An allegory of modernity/postmodernity: Habermas and Derrida," in his *New constellation*, pp. 199–229. Still more bluntly, Norris claims that deconstruction, properly understood, "belongs within that same 'philosophical discourse of modernity' that Habermas seeks to defend against its present-day detractors." See Christopher Norris, "Deconstruction, postmodernism and philosophy: Habermas on Derrida," in his *What's wrong with postmodernism*, pp. 49–76; the quotation is from p. 52. Since Bernstein constructs a tense *constellation* out of Habermas/Derrida – and out of concepts of modernity/postmodernity – I am not sure that he would endorse Norris's argument. However, taken together these two essays are particularly valuable in countering the more dismissive readings of Derrida.

Contemporary critiques of historical materialism

Figure 9 identifies two main streams. On the one side are those for whom political economy offered an analysis of the spatiality of capitalism that seemed to be too structural – too concerned with the logic of capital – and who sought to open a wider conceptual space for human agency. I have already said that this was not a new problem. The political project of historical materialism in both its classical and its modern forms had long been skewered by what Anderson calls a "permanent oscillation" between two principles of explanation: the structural logic of the mode of production and the conscious and collective agency of human subjects.[81] If that had been the only issue then the development of an analytical Marxism explicitly predicated on the actions of individuals might have seemed a sufficient response to those who wished to minimize the force of the structural turn. But their indictment turned out to have two counts. Political economy was accused not only of inflating "capital" but also of undervaluing "culture." Toward the end of the 1970s these double charges sparked a particularly vigorous debate in England among socialist historians, where Thompson's vicious attack on Althusser and what he claimed to be "the poverty of theory" – what he evidently meant was the poverty of somebody else's theory – was followed by a series of equally angry and even abusive exchanges in the pages of *History workshop journal*. If the tone and temper of these critical assaults is symptomatic of humanism, it is not difficult to sympathize with the protagonists of antihumanism, who were in any case raising questions about human agency and subjectivity whose cogency was not diminished by the abuse that greeted them. At the time, Anderson – whose political and historical writings had been influenced by Althusser and who had already crossed Thompson on more than one occasion – sought to moderate between the claims of structural Marxism and "humanist Marxism" in ways that seemed to me then, as they do now, extremely constructive.[82]

Anderson's interventions provided a brilliant contrast to the debate in human geography, where a series of strident objections to structural Marxism congealed into an opposition between Marxism and *anti*-Marxism. I do not want my dismay at the turn of events in my own discipline to be mistaken for intellectual fideism, but I do think that the local critique of structural Marxism was misconceived and doctrinaire. Although its authors sympathized with the "moral force" of Marx's critique of capitalism

---

[81] Perry Anderson, *In the tracks of historical materialism* (London: Verso, 1983) p. 34.

[82] E. P. Thompson, "Poverty of theory"; Perry Anderson, *Arguments within English Marxism* (London: Verso, 1980). His reading of Thompson's *Whigs and Hunters* – which I have long thought Thompson's most accomplished text – is particularly instructive.

and applauded modern Marxism's "demystification" of the inert categories of neoclassical economics, they also dismissed the concept of the mode of production as itself a mystification. Where, then, was the critical force of Marxism supposed to lie? The implication was that it had been blunted by the projection of Hegel directly into historical materialism, which had licensed the development of concepts at such high levels of abstraction that any attempt to challenge the primacy of structure and to open a space for human agency involved a cleansing immersion in "broadly Weberian streams of analysis."[83]

This seems to me a misreading of the work of Althusser and Balibar, whose project was conceived, in its very origins and for political as well as for intellectual reasons, as a *critique* of Hegelian Marxism. Such charges could more properly be laid – if they have to be laid at all – against Lefebvre, who played a prominent part in "Hegelianizing" French Marxism after World War II, but this would then complicate a humanism that aligns him with an existential Marxism without recognizing its proximity to Hegel: which was precisely Althusser's point.[84] The appeal to Weber is also a simplification. It is not necessary to accept those readings of his substantive work that identify important similarities between structural Marxism and a "Weberian structuralism" – which claim that there is "an ineluctable logic of structure" to Weber's account of the rise of the West – to realize that the contrary, voluntarist interpretations of his work are called into question by his own "developmental histories," which expose a definite logic to historical eventuation at a configurational level and which culminate in a critique of a generalized process of rationalization that issues in the "iron cage" of capitalism. If human agency is to be the measure of adequacy, at least Marx was optimistic about the capacity (and the consciousness) of human beings to break out of that cage, whereas Weber foreshadowed only "a polar night of icy darkness" as the ultimate fate of the modern world.[85] These twin simplifications are, I think, strategic. I do not doubt that much is to be learned from both Marx and Weber, as Habermas's theory of communicative action makes very clear, but dialogue becomes impossible when either side is silenced by caricature.

---

[83] James Duncan and David Ley, "Structural Marxism and human geography: A critical assessment," *Annals of the Association of American Geographers* 72 (1982) pp. 30–59. For a commentary and reply, see Vera Chouinard and Ruth Fincher, "A critique of 'Structural Marxism and human geography,'" *loc. cit.*, 73 (1983) pp. 137–46; Duncan and Ley, "Comment in reply," *loc. cit.*, pp. 146–50.

[84] See Chapter 6, this volume.

[85] For various structural inflections of Weber's project, see Lawrence Scaff, *Fleeing the iron cage: Culture, politics and modernity in the thought of Max Weber* (Berkeley: University of California Press, 1989) pp. 34–49 and *passim*; Wolfgang Schluchter, *The rise of western rationalism: Max Weber's developmental history* (Berkeley: University of California Press, 1981); Bryan Turner, "Logic and fate in Weber's sociology" and "Weber and structural Marxism," in his *For Weber: Essays on the sociology of fate* (London: Routledge, 1981) pp. 3–28 and 29–60; the quotation is from p. 10.

That is why I regard Anderson's commentary as so compelling. One of his central concerns was to register the sheer *complexity* of the terrain that could be mapped by the instruments of historical materialism and, in particular, to chart those cultural, social and political topographies that lay beyond the compass of political economy. In doing so, his *Arguments within English Marxism* effectively questioned the identities that had been placed at the center of the humanist critique in both historical materialism and human geography: ⟨structure: capital⟩ and ⟨agency : culture⟩.

It is against this background that we can most usefully foreground Giddens's attempt to construct what he calls a "post-Marxist" theory of structuration at about the same time and, significantly, in the same place. Anderson identifies a "topographical break" in the trajectory of Marxism in the 1970s, when the creative locus of historical materialism supposedly moved from French, German, and Italian Marxisms to other, Anglo-American traditions.[86] If that is so, however, Giddens's writings have to be hinged to both plates, since he draws so freely on "Continental" and American thinkers.[87] Indeed, in a subsequent commentary Anderson attributes the astonishing fecundity of British intellectual culture in the 1980s to a more or less complete rupture of its traditional insularity. By the end of that decade, he suggests, "there was probably no other country where influences from both sides of the Atlantic intermingled so freely."[88] Be that as it may, those influences were by no means confined to the trade winds of historical materialism, and the formulation of structuration theory was no exception. In his survey of *Central problems in social theory*, Giddens maintained that there are no simple dividing lines between Marxism and "bourgeois social theory" – in his view, it was no longer possible to remain true to the spirit of Marx by remaining true to the letter of Marx – and he later went still further and described structuration theory as a *critique* of historical materialism.[89] Even so, he has also repeatedly stressed that his project was never conceived as a dismissive rejection of Marx. Giddens may not be interested in reading Marx as the source of a transcendent and incontrovertible truth, nor as providing a continuous – or even discontinuous – progress toward a self-sufficient "science." But he does regard Marx's writings as the most

---

[86] Anderson, *Tracks*, pp. 23–24.

[87] The Continental influences are those most often identified in Giddens's work – including Althusser, Derrida, Foucault, Gadamer, Habermas, Heidegger, Ricoeur, Schutz – but for an exemplary discussion of the American connection see Alan Sica, "The California-Massachusetts strain in structuration theory," in Christopher Bryant and David Jary (eds.), *Giddens' theory of structuration: A critical appreciation* (London: Routledge, 1991) pp. 32–51.

[88] Perry Anderson, "A culture in contraflow – Part I," *New Left Review* 180 (1990) pp. 41–78; the quotation is from p. 48.

[89] The primary texts are: Anthony Giddens, *Central problems in social theory: Action, structure and contradiction in social analysis* (London: Macmillan, 1979); *idem, A contemporary critique of historical materialism.* Vol. 1: *Power, property and the state* (London: Macmillan, 1981) and Vol. 2: *The nation-state and violence* (Cambridge, England: Polity Press, 1985).

significant single fund of ideas on which to draw in order to illuminate the intersections between structure and agency: So much so, in fact, that it is by no means clear how far structuration theory has really moved from historical materialism.

In a general sense, one might say that Giddens's intention has been to think through the implications of Marx's deceptively simple claim – made in the *Eighteenth Brumaire* – that people make history, but not just as they please or in circumstances of their own choosing. But the affinities between the two are also precise and specific. Not only does Giddens reaffirm many of Marx's central characterizations of capitalism (including, on occasion, the labor theory of value), but some critics insist that many of his leading propositions that are supposed to strike out in radically different directions were anticipated in various ways by Marx and the development of other traditions of Western Marxism.[90] Giddens himself is unrepentant; his reformulation of social theory is not only a critique of historical materialism, he declares, but now also a "deconstruction" of historical materialism. And yet the connection remains, not least through Giddens's continuing critical interest in Habermas and his supposedly more affirmative "reconstruction" of historical materialism. Indeed, Giddens claims to have learned more from Habermas's writings than from those of any other contemporary social thinker, although he also says he disagrees "with very many, perhaps most, of Habermas's major conceptions."[91] Even so, there is no doubt that many of us who have explored the intersections between structuration theory and human geography have retained a much closer filiation with a somewhat diffuse Western Marxism than Giddens allows himself.

On the other side of figure 9 – and in some measure distanced from structuration theory for precisely the same reason – are those for whom political economy is a thoroughly masculinist discourse. The relations between feminism and historical materialism are complex, and I have no wish to minimize the connections between them. Many of the attempts to construct a feminist geography were (and are) indebted to an avowedly socialist feminism. But the development of such a geography has nonetheless had to confront the central objection that historical materialism, including historico-geographical materialism, is so preoccupied with questions of class that modalities of gender and sexuality have been marginalized. Susan Christopherson expresses this concern succinctly when she

[90] For a particularly constructive argument see Erik Olin Wright, "Models of historical trajectory: An assessment of Giddens's critique of Marxism," in David Held and John Thompson (eds.) *Social theory of the modern societies: Anthony Giddens and his critics* (Cambridge: Cambridge University Press, 1989) pp. 77–102.

[91] Anthony Giddens, "Labour and interaction," in John Thompson and David Held (eds.), *Habermas: Critical debates* (London: Macmillan, 1982) pp. 148–61; the quotation is from p. 155. For a comparison between Giddens and Habermas, see my "The crisis of modernity? Human geography and critical social theory," in Peet and Thrift (eds.) *New models*, vol. 2, pp. 348–85.

identifies the framework within which class has typically been theorized: "Capital is male." To say that historical materialism is a *critique* of capital is no answer because Christopherson's argument strikes deeper than the masculinity of the concept: It is the process of concept formation itself that she seeks to challenge. In common with several other feminist scholars, she argues that the marginalization (and, indeed, subordination) of gender is not simply an absence to be made good by incorporation – by theorizing capitalism and patriarchy together, for example – because such a maneuver almost always falls victim to the "seduction" of theory: to the characteristically immodest gesture of conceptual mastery enacted by high theory.[92]

This strategy of displacement (and, by implication, domination: like "mastery," a word whose etymology is deeply gendered) is not confined to historical materialism; as I have already indicated, structuration theory is open to the same objections. Most obviously, Giddens has proposed a typology of societies that is intended to clarify some of the distinctive features of modernity: and yet this schema turns on a distinction between "class-divided" and "class" societies that plainly confirms the centrality of class and fails to accord patriarchy the same constitutive systematicity within modernity. I suspect, too, that Giddens's conception of human subjectivity is not only insufficiently attentive to the process of gendering but also installs at its center a model of subject-formation drawn from a profoundly masculinist version of psychoanalytic theory. I know that some feminist critics would want to press these claims further and object to Giddens's style of theorizing too. In their view, its temper replicates a masculinist mode of knowing which is inclined to be abstract, authoritarian and univocal.[93]

For all the tensions between structuration theory and feminist theory, however, they both display the same double movement that characterized the engagement between historical materialism and human geography. In each case a form of social theory was projected into human geography as a way of meeting a series of supposedly "internal" concerns, and then, as a direct consequence of this conceptual mapping, the a-spatiality of the original formulation was itself called into question. It is those return movements that have widened the discourse of geography still further and impelled the continuing transdisciplinary exploration of "deep space."

---

[92] Susan Christopherson, "On being outside 'The Project,'" *Antipode* 21 (1989) pp. 83–89; see also Linda McDowell, "Multiple voices: Speaking from inside and outside 'The Project,'" *loc. cit.*, 24 (1992) pp. 56–72.

[93] The historical objection is obvious, but my understanding of these latter questions has been guided by Jessica Benjamin, *The bonds of love: Psychoanalysis, feminism and the problem of domination* (New York: Pantheon, 1988); Jane Flax, *Thinking fragments: Psychoanalysis, feminism and postmodernism in the contemporary West* (Berkeley: University of California Press, 1990); Ann Game, *Undoing the social: Towards a deconstructive sociology* (Toronto: University of Toronto Press, 1991).

Making geography: structuration theory and spatiality

In its early versions structuration theory was incorporated into human geography as a way of overcoming the dualism between structure and agency that had skewered the analysis of space and place. Whatever detailed criticisms might be made of structuration theory, and these are not inconsiderable, Giddens's general achievement was, I think, twofold. First, he showed how profound and how pervasive that dualism was by translating it into a philosophico-theoretical vocabulary that transcended the different language-games of the individual social sciences and humanities. Two anthropologies, two geographies, two histories, two sociologies: All of them constituted and seemingly confounded by a common problematic that had at its center what Abrams called an "estranged symbiosis" between structure and agency, whose quarrels were both "easily and endlessly formulated" and yet also "stupefyingly difficult to resolve."[94]

Second, Giddens proposed an explanation for the chronic failure of attempts to reconcile the two: the mistake, so he said, was to treat "structure" as establishing the parameters within which "agency" was able to exercise its own independent discretion. Giddens's central claim was that structure was instead implicated in every moment of action – that it was at once constraining and enabling – and that, conversely, structure was an "absent" order of differences, "present" only in the constituting moments of interaction through which it was itself reproduced or transformed (figure 10).[95]

This model replaces the traditional dualism at the heart of social theory with a duality, and it depends on three fundamental concepts: reflexivity, recursiveness, and regionalization. Giddens insists that people know a great deal about the social practices in which they are involved, although their stocks of knowledge are not always discursively formulated. Thus the production of social life is always and everywhere a skilled accomplishment ("reflexivity"); that structure is both the medium and the outcome of the social practices constituting social systems ("recursiveness"); and that social practices literally "take place" in time and space ("regionalization"). Taken together, these claims provide a social ontology that is, in effect, the nucleus of structuration theory. Arranged around it is a series of further claims about modernity that Giddens calls "conceptual salvoes" and which, on occasion, he separates from structuration theory; but I think this misrepresents the connections between these more substantive theses and the

---

[94] Philip Abrams, "History, sociology, historical sociology," *Past and Present* 87 (1980) pp. 3–16; the quotations are from p. 8.
[95] Giddens, *Central problems*, pp. 49–95.

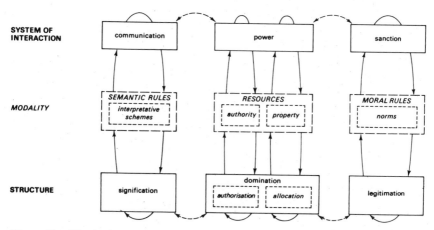

Figure 10 *The duality of structure [after Giddens].*

core propositions of his social ontology. For this reason, I prefer to read Giddens's project as a research program whose (uneven) development has been determined by both the coherence of its theoretical network and the concreteness of its substantive theses. Only in this way, I think, can we fix the bearing of structuration theory on "deep space."

There were at least two reasons why Giddens's program found a particularly receptive audience in human geography. In the first place, its philosophical and theoretical rigor dovetailed with the sensibilities that had been put in place by spatial science, by a number of post-positivist geographies, and by the development of historico-geographical materialism. There were other voices with other priorities and preferences, of course, and some of them undoubtedly preferred E. P. Thompson's way of working with the historical process of structuration (his term, too) over Giddens's more schematic formulations. This was often the response of humanistic geographers who were skeptical about the flight of yet another theoretical geography into abstract space, and who endorsed Thompson's vigorous affirmation of "experience" as a central (a moral and an existential) category of historical understanding and who admired his often moving recitations of what he called "the poetry of voluntarism." Whatever the parallels between structuration theory and socialist-humanist history. Thompson's indictment of the poverty of theory stood in sharp contrast to Giddens's apparent insistence on the power of theory.[96]

In some ways, perhaps, this was no more than a reflection of the distance that separated social theory and social history. Certainly, Thompson and

---

[96] See Anthony Giddens, "Out of the Orrery: E. P. Thompson on consciousness and history," in his *Social theory and modern sociology* (Cambridge, England: Polity Press, 1987) pp. 202–24.

Giddens understood different things by "theory"; Thompson stabled his *bête noire* close to the received model of the physical sciences, whereas Giddens was always scrupulously attentive to the limitations of naturalism and preferred to treat social theory as simply a "sensitizing device." But the development of structuration theory involved a sustained process of conceptual construction and clarification, whereas Thompson embedded his concepts in an empirical matrix and only pried them loose with the greatest reluctance. Giddens could insist that there are no logical or methodological differences between sociology and history – that both are implicated in the development of social theory – but many (perhaps most) sociologists and historians remained suspicious of any liaison, let alone union, between the two. It would, however, be wrong to conclude from all this that those geographers who were prepared to accept Giddens's parallel claim about their discipline – that there are no logical or methodological differences between sociology and human geography – were persuaded by purely theoretical, even theoreticist concerns.[97]

Consider, for example, Cole Harris, who would never regard himself as a theoretician and yet whose deep suspicion of the theoretical ambitions of spatial science in the 1970s developed into a keen interest in structuration theory in the 1990s. I say "developed" deliberately, because there was no radical break in his views: what had changed was the intellectual landscape within which he worked. Structuration theory was a different kind of theory, Harris argued, and one which offered a particularly suggestive set of ideas about the relations between power, space, and modernity that considerably enlarged the horizon of meaning within which his own, thoroughly concrete historical geographies of modernity could be written.[98] I would be inclined to go even further: The contemporary revival of historical sociology has been of great importance, but it has not dissolved the distinctions between sociology and history; whereas the project that Harris has in mind makes redundant any continued separation between historical geography and human geography.

This cannot all be laid at the door of structuration theory, of course, and in any case Giddens's original interest in the spatiality of social life was not prompted by the concerns or contributions of human geography. It derived from Heidegger's treatment of time-space as "presencing" and from concepts of "spacing through difference" developed in structuralism and post-structuralism. From these sources Giddens began to register the ontological claim that people make not only histories but also *geographies*:

---

[97] On the relations between sociology, history, and geography, see Anthony Giddens, *The constitution of society: Outline of the theory of structuration* (Cambridge: Polity Press, 1984) pp. 355–68.

[98] R. Cole Harris, "The historical mind and the practice of geography," in David Ley and Marwyn Samuels (eds.), *Humanistic geography: Prospects and problems* (Chicago: Marooufa Press, 1978) pp. 123–37; *idem*, "Power, modernity and historical geography," *Annals of the Association of American Geographers* 81 (1991) pp. 671–83.

that time-space relations are not incidental to the constitution of societies and the conduct of social life. The articulations of structure and agency could only be grasped, so he seemed to say, by explicating the ways in which time and space are "bound into" the conduct of social life. This was little more than a promissory note, and probably few geographers were convinced of its substantive purchase at the time. However, once these ideas were given a more concrete form it became clear that structuration theory also offered an analytical perspective on the geographical configurations of modernity.

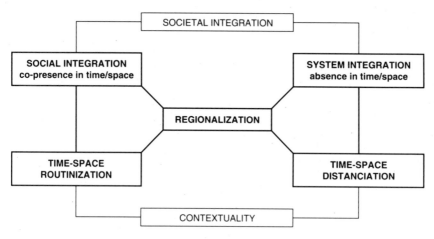

Figure 11   Time-space relations and structuration theory.

In the second place, therefore, Giddens extended his argument by drawing an old distinction in a way that spoke more directly to the substantive concerns of postpositivist human geography: between the spatialities of "social integration" and "system integration" (figure 11).[99] Social integration describes webs of face-to-face interaction that are embedded in place. In the course of their day-to-day lives, so Giddens argues, people move more or less routinely from one locale to another, tracing out paths

---

[99] A similar distinction can be found in Schutz and Luckmann's account of what they call the "spatial arrangement of the everyday lifeworld." They distinguish between the *primary zone* of operation – "the province of nonmediated action . . . the primary world within reach" – and the *secondary zone* of operation, "the world within potential reach," which "finds its limits in the prevailing technical conditions of a society." But they are concerned with the structures of the lifeworld and develop their arguments in the vocabulary of constitutive phenomenology, whereas Giddens objects to the persistent failure of approaches of this sort to recognize the centrality of power in the production and reproduction of social life. See Alfred Schutz and Thomas Luckmann, *The structures of the lifeworld* (London: Heinemann, 1974); Giddens never refers to this account, but he provides a general critique in his *New rules of sociological method: A positive critique of interpretative sociologies* (London: Hutchinson, 1976).

in time and space, and drawing upon elements of those different settings in the conduct of their affairs. This complex weaving of the fabric of everyday life was one of Goffman's primary concerns, but his microsociological studies are usually regarded as little more than illuminating vignettes of the minutiae of social life. Giddens sees a much deeper systematicity in those writings and describes Goffman as "the theorist of co-presence." But he also has two reservations about the way in which that project has been formulated.

First, although Goffman is sensitive to the significance of time and space for the flow of face-to-face interaction, Giddens argues that these sequences are captured more effectively in the traces of *time-space routinization* displayed on the screens of Hägerstrand's time-geography (figure 12). I have reservations about both time-geography and Giddens's critical appropriation of it, but I want to insist that its importance is more than methodological. To reduce time-geography to a series of templates for tracing trajectories

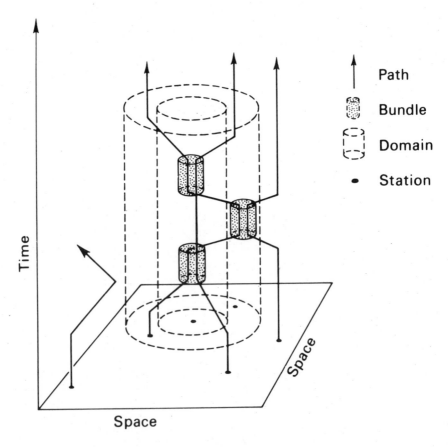

Figure 12   The web model of time-geography [after Hägerstrand].

in time and space is to treat it as another version of spatial science, whereas Hägerstrand's vision is profoundly ecological: It is a way of registering what he calls the "momentary thereness" of things in a constant and competitive flow of "interelated presences and absences." Its apparent simplicity is highly deceptive; Hägerstrand's scheme is not just a redescription of what we already know but rather a way of imaginatively reconfiguring our sense of what is going on: of accentuating the significance of "presencing," of the complex dialectic between presence and absence, in the time-space constitution of social life. Ironically, it is those "absences" that invite Giddens's second reservation about Goffman who, so he says, fails to explore the ways in which social integration is implicated in system integration: the ways in which face-to-face interaction is spliced into mediated processes of interaction with those who are absent in time or space. " 'Presence' – what the individual brings to and employs in any situation of conduct, whether there are others in that situation or not – is always mediated by what is absent."[100]

Yet Giddens has himself been accused of failing to give a satisfactory account of those "absences." His critics claim that he treats interdependence in the language of interaction and that for this reason has considerable difficulty in elucidating the dense, complex and multilayered connections between people who are not copresent in time and/or space. Kilminster puts this objection most forcefully when he asks: What is the connective tissue that binds people into multiple and overlapping networks that ripple beyond their immediate milieu? What makes it socially possible for people to enter into projects, break off and then take up different projects with other people in other places? These questions are directed at what Kilminster takes to be Giddens's persistent failure to specify the grid of *social* interdependencies put in place by structures of social relations within which the capacity of people to act – "to make a difference" – is distributed unevenly and asymmetrically. But these bonds are as much spatial as they are social, and Giddens's reading of time-geography seems inadequate to me precisely because he treats it as a useful notation rather than as a means of conceptualizing the *time-space* interdependencies that constitute what Hägerstrand calls the precarious "grain structure" of social life.[101]

Giddens attempts to mark the traces of "absence" in the conceptual register of structuration theory in quite another way: by identifying system

---

[100] Anthony Giddens, "Erving Goffman as systematic social theorist," in his *Social theory*, pp. 108–39; the quotation is from p. 137. On time-geography, see Torsten Hägerstrand, "Presence and absence: A look at conceptual choices and bodily necessities," *Regional studies* 18 (1984) pp. 373–80 and Giddens, *Constitution*, pp. 110–19.

[101] Richard Kilminster, "Structuration theory as a world-view," in Bryant and Jary (eds.), *Giddens' theory of structuration*, pp. 74–115; Derek Gregory, "Presences and absences: Time-space relations and structuration theory," in David Held and John B. Thompson (eds.), *Social theory of modern societies: Anthony Giddens and his critics* (Cambridge: Cambridge University Press, 1989) pp. 185–214.

integration with a process of *time-space distanciation*, which is supposed to establish the conditions that connect presence and absence. One of the most significant consequences of crossing the threshold of modernity, so he argues, is the generalized "disembedding" of spheres of social life from the immediacies of the here and now. This does not mean that social life is no longer anchored in place – it still "takes place" in the most obdurately physical of senses – but those mooring lines become "stretched" across variable spans of time and space. Giddens attaches particular importance to the *generalization* of this process because neither time-space distanciation nor disembedding are in themselves diagnostic of modernity. The empires of the ancient world were systems of extensive power, spanning considerable stretches of time and space, and although their capacity for intensive power was much less and the penetration of the day-to-day lives of their subject populations correspondingly restricted, the lifeworlds of local elites were usually bound into the assumptions and aspirations of a translocal ruling class.[102] But the generalization of distanciation/disembedding provides for the *globalization* of social life on a continuous and systematic basis that, precisely because it does reach into the most intimate areas of most people's day-to-day lives, is one of the most far-reaching consequences of modernity. It does not erase difference with planes of uniformity, however, because the process is a thoroughly dialectical one in which events at one pole often produce divergent or contradictory outcomes at another. Presence and absence are interleaved in a "local-global dialectic," whose restless animations are channeled through systems of mediated interaction that dissolve and recombine local networks of interpersonal relations across an increasingly global space.[103]

In this unelaborated form I suppose these propositions are not very novel. As Harvey makes clear, in the early twentieth century aesthetic modernism was preoccupied with the simultaneity of social life: Apollinaire, among others, thought that the modern poet's greatest challenge was "to express the sense of life being lived all over the globe at one and the same time."[104] Less expressive and more analytical in temper, modern human geography sought to index time-space distanciation through measures of

---

[102] These claims are derived from Ernest Gellner's model of an agroliterate polity set out in his *Nations and nationalism* (Oxford: Blackwell Publishers, 1983) and from the typology set out in Michael Mann, *The sources of social power.* Vol. 1: *A history of power from the beginning to AD 1760* (Cambridge: Cambridge University Press, 1986) pp. 6–10. All of this gives a rather different gloss to Giddens's account of these "class-divided" societies: Subordinate classes can now be seen to be divided by axes of literacy and property both *laterally* from one another and *vertically* from a transcendent and "universal" ruling class.

[103] Anthony Giddens, *The consequences of modernity* (Stanford: Stanford University Press, 1990) pp. 63–64 and *passim.*

[104] Harvey, *Postmodernity*, pp. 260–83; Margaret Davies, "*Modernité* and its techniques," in Monique Chefdor, Ricardo Quinones, and Albert Wachtel (eds.), *Modernism: Challenges and perspectives* (Urbana, Ill.: University of Illinois Press, 1986) pp. 146–58; the quotation is from pp. 149–50.

time-space convergence and its derivatives.[105] Most of those attempts were set within the framework of spatial science, and in consequence they brought out the spatial structures and variable geometries of distanciation much more clearly than did Giddens; but the encasements and entailments of the process for social life – for the constitution of "society" and "self" – plainly could not be elaborated within the problematic of an autonomous spatial science. And it is, I think, those analytical elaborations that make Giddens's discussion of time-space distanciation so interesting.

In one sense, of course, his account remains within the grammar of interaction that Kilminster criticizes. Thus Giddens draws attention to the development of successive systems of communication, including writing, printing, telecommunications, and electronic media, and to the development of successive systems of exchange, including money and networks of credit transfer. The modern world is supposedly shaped by the intimacy of the intersections between the two; most obviously through the ways in which electronic networks enable financial markets across the continents to implode into dizzying constellations of buying and selling. In another sense, however, I think the concept of time-space distanciation does represent an attempt to move beyond the grammar of interaction to identify "structural sets" whose interdependencies transcend the limitations of presence. This is most obvious in Giddens's original formulations, where time-space distanciation is made to depend on the mobilization of *authoritative* ("political") and *allocative* ("economic") resources that are embedded within structures of domination that interlace in different ways in different societies.[106] This marginalizes the importance of structures of legitimation and signification, but the emphasis on "structure" – for all its vagueness and imprecision[107] – does at least carry the implication of a connection beyond the contingent or discretionary.

In his later writings, however, Giddens describes modern time-space distanciation in terms of the development of "abstract systems" whose structural imbrications are much less clear. These include *symbolic tokens*, which are "media of exchange that have standard value and thus are interchangeable across a plurality of contexts," and *expert systems*, which

[105] The concept was originally proposed by Douglas Janelle in his "Central place development in a time-space framework," *Professional Geographer* 20 (1968) pp. 5–10 and "Spatial reorganization: A model and concept," *Annals of the Association of American Geographers* 59 (1969) pp. 48–364. It was subsequently extended by Ronald Abler, "Distance, intercommunications, and geography," *Proceedings of the Association of American Geographers* 3 (1971) pp. 1–4 and Pip Forer, "Space through time," in E. L. Cripps (ed.), *Space-time concepts in urban and regional models* (London: Pion, 1974) pp. 22–45. For Janelle's present views, see his "Global interdependence and its consequences," in Stanley Brunn and Thomas Leinbach (eds.), *Collapsing space and time: Geographic aspects of communication and information* (London: HarperCollins, 1991) pp. 49–81.

[106] Giddens, *Power property and the state*, pp. 51–56, 91–97, 113–21 and *passim*.

[107] Cf. John B. Thompson, "The theory of structuration," in Held and Thompson (eds.), *Social theory of modern societies*, pp. 56–76.

"bracket time and space through deploying modes of technical knowledge which have validity independent of the practitioners and clients who make use of them." Giddens argues that these abstract systems penetrate all aspects of day-to-day life and in doing so undermine local practices and local knowledges; they dissolve the ties that once held the conditions of daily life in place and recombine them across much larger tracts of space. Modernity is thus made inseparable from the operation of abstract systems that encompass social life and embed it in global webs of interdependence.[108]

Although these claims are not derived from the core model of structuration in any direct fashion, and Giddens makes no attempt to elucidate the connections, I think they can be given a structural inflection consonant with his original theses. Abstract systems are presumably a mix of rules and resources that are made available by structures and drawn upon by knowledgeable subjects. Those stocks of knowledge are, of course, variable, but the capacity to draw upon them is not confined to specialists and experts. Giddens is adamant that the involvements of abstract systems in the routines of day-to-day life are not tantamount to what Habermas sees as the "colonization" of the lifeworld by the instrumental rationalities of politico-administrative and economic systems. What such a perspective overlooks, Giddens contends, is the way in which "technical expertise is continuously reappropriated by lay agents as part of their routine dealings with abstract systems." So it is; but I think this foreshortens Habermas's argument. As a minimum, what Giddens is describing is the *rationalization* of the (posttraditional) lifeworld; and at the limit – where the incorporation of this technical knowledge becomes a matter of routine, of axiomatically drawing upon the rules and resources of abstract systems – this is surely exactly what Habermas has in mind in speaking of the *colonization* of the lifeworld. Of course, neither Giddens nor Habermas thinks that these processes are autonomic. The entanglements of abstract systems are often contested, both individually and collectively, and there are also discretionary spaces between lay and professional practices; but those "panoramas of choice," as Giddens calls them, are not without a certain structure of their own and (as he says himself) there are limits on how far anyone in the modern world can disengage from its abstract systems.[109]

The emphasis on knowledgeability nevertheless enables Giddens to grasp the experiential dimension of globalization by connecting absence in time and space to *trust* in abstract systems. Trust, which Giddens takes to mean reliability in the face of contingent outcomes, becomes a necessary condition

---

[108] These descriptions are drawn from Giddens, *Consequences*, pp. 21–29, 80, and *idem, Modernity and self-identity: Self and society in the late modern age* (Cambridge, England: Polity Press, 1991) pp. 18–19.

[109] Giddens, *Consequences*, p. 144; *idem, Modernity and self-identity*, pp. 137–39; Jürgen Habermas, *The theory of communicative action*. Vol. 2: *The critique of functionalist reason* (Cambridge, England: Polity Press, 1987).

of modern time-space distanciation. But it is also double-edged. Abstract systems create large areas of security for the conduct of social life; their incorporation into time-space routines provides the minimal continuity, coherence, and calculability necessary for the taken-for-granted, matter-of-factness of the social world. And the generalized guarantees that they provide, whatever confidence limits may be placed around them, encourage repetitive and renewed investments of trust. Yet these abstract systems also implicate day-to-day transactions in chains of interactions across time and space, the fleeting traces of webs of interdependence, which injects what might be called a "dialectic of vulnerability" into the logic of globalization. An unexpected event that cannot be contained will speed through the network and may impact on the system as a whole; conversely, the local fabric of everyday life is everywhere shot through with the implications of distant events.[110]

The reverse side of the local-global dialectic is thus what Harvey calls *time-space compression* (figure 13). He sees this as the product of the compulsion to "annihilate space by time" under capitalism, shaped by the rules of commodity production and capital accumulation, whereas Giddens is more distant – though by no means completely disconnected – from these conceptual mappings of capital circulation. Nevertheless, both writers accentuate the disorienting and sometimes even disturbing consequences

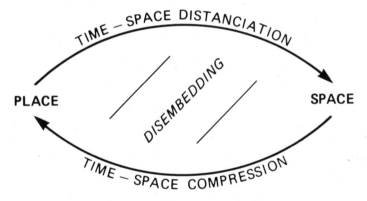

Figure 13   Time-space distanciation and time-space compression.

of contemporary changes in time-space dimensionality. For Harvey, "the foreboding generated out of the sense of social space imploding in on us" is wired to a crisis of identity: "to what space/place do we belong?"[111]

---

[110] Giddens, *Consequences*, pp. 29–36; *idem, Modernity and self-identity*, pp. 133–37.

[111] Harvey, *Condition of postmodernity*, p. 240 and *passim; idem,* "Between space and time: Reflections on the geographical imagination," *Annals of the Association of American Geographers* 80 (1990) pp. 418–34; the quotation is from p. 427. This does not mean that the question of identity *reduces* to a question of location.

Giddens develops his answer in a way that addresses the question of identity much more directly, but he too fastens on the tension between – in fact, the "separation" of – place and space. It is no longer possible to map the world in the old ways, so he seems to say, because we are caught up in a dizzying process of deterritorialization and reterritorialization – and it is on its perpetually shifting sands that we build our castles and shape our identities. Places are local condensations and distillations of tremulous global processes that travel through them and whose effects are reworked and inscribed within them. In the world of high modernity it has become virtually impossible to make sense of what happens in a place without looking beyond the local horizon. However intensely a sense of place is invoked, it is now always something of a holding operation – a strategic closure, a sort of "false summit" beyond which lies a series of higher elevations that impact on the apparent immediacies and intimacies of the present moment. In Giddens's technical vocabulary, places are "thoroughly penetrated by and shaped in terms of influences quite distant from them" and "the structures by means of which [they are] constituted are no longer locally organized." In short, he says, place has become *phantasmagoric*: a word that clearly resonates with the modern(ist) conception of landscape – what is seen/scene – as a dialectical image (above, p. 99).[112]

All of this is so unsettling, Giddens argues, that modernity has to be seen neither as a monster (Marx) nor as its iron cage (Weber) but instead as a *juggernaut* whose exhilarating, confounding, runaway momentum "derives ultimately from the dialectic of time and space."[113] This claim, made in less metaphysical terms, chimes with the arguments of other writers: Clifford's "traveling cultures," Harvey's "geographical imagination" and Jameson's "cognitive mapping." I use these as shorthand descriptions, nothing more, simply to indicate that the erraticism of high modernity has issued in a profound crisis of representation that turns on the possibility of comprehending, of somehow bringing into presence, these intricate, multiple and compound geographies that mix "presence" and "absence" in such new and volatile ways. In the face of such radically unstable configurations, which seem to defy every conventional cartographic possibility, some writers have turned to postmodernism for inspiration. In their eyes its celebration of the play of difference, of indeterminacy and contingency, seems peculiarly appropriate to the giddy, kaleidoscopic diversity of the late twentieth-century world. I am not sure that Clifford is quite as

[112] Giddens, *Consequences*, p. 18–19; *idem, Modernity and self-identity*, pp. 146–47. Zukin expresses the disassociation between place and space in more particular terms, as a separation of place and "market," but her identification of landscape as the major cultural product of our time, as a liminal space at once physically present and yet thoroughly mediated and so literally dis-placed, speaks directly to these concerns. See Sharon Zukin, *Landscapes of power: From Detroit to Disneyworld* (Berkeley: University of California Press, 1991).

[113] Giddens, *Consequences*, pp. 138–39, 153.

enthusiastic about this as some of his critics seem to think, but his essays do at least confront the limitations of an austere high modernism and challenge the boundaries it puts in place between avant-garde art and disciplinary science. To other writers, however, the postmodern turn is part of the problem not the solution. They see postmodernism as the cultural dominant of late twentieth-century capitalism and, like Harvey, argue that its inscriptions mark the turbulent landscapes of flexible accumulation in ways that require the renewal of historico-geographical materialism as a means of grounding the (post)modern production of space. A third response (there are no doubt others) is to capitalize on the critical tension between postmodernism and historical materialism as a means of reinvigorating Western Marxism – in Jameson's case, interestingly enough, through the pursuit of a politics of space – and thereby turning postmodernism against itself.[114]

Yet Giddens has made none of these moves, preferring instead to keep his distance from the postmodern. In saying this, I am aware that several commentators have connected structuration theory more or less directly to a postmodern sensibility. Giddens is one of Soja's postmodern geographers *malgré eux*, for example, and his own reluctance to endorse foundationalism coupled with his continuing interest in hermeneutics and in the contingency of historical eventuation suggests that there may well be an elective affinity (of sorts) between them – at least, for those who want to read structuration theory in this way. But the fact remains that Giddens has not pursued these affiliations, if such they are, in any systematic or even sympathetic fashion. Some critics suggest that there are good reasons for that.[115] Certainly, Giddens has shown remarkably little interest in tracing the contours of time-space distanciation in detail – its variable geographies that impact on different people and places in different ways – or in explicating the production of "postmodern hyperspace" at any more general level. He is, I think, right to be skeptical about the value of purely abstract models of spatial form, but quite wrong to ignore the concrete specificities of space, place, and landscape. He seems equally uninterested in the cultural politics of postmodernism. Not only does he insist that we live in a world of "high modernity" rather than postmodernity, but he also argues that much of what is mistaken for the postmodern concerns "the experience of living in a world in which presence and absence mingle in historically novel ways" which are the product of a *radicalization* rather than any

---

[114] For examples of the positions staked out in this paragraph, see James Clifford, *The predicament of culture: Twentieth-century ethnography, literature and art* (Cambridge: Harvard University Press, 1988); Harvey, *Condition of postmodernity*; Jameson, *Postmodernism*.

[115] Cf. Matthew Hannah and Ulf Strohmayer, "Ornamentalism: Geography and the labor of language in structuration theory," *Environment and Planning D: Society and Space* 9 (1991) pp. 309–27.

transcendence of modernity. Again, that may well be so. But Giddens treats postmodernism as a moment of "aesthetic reflection" upon modernity that properly refers to movements within literature and the other arts. The implication, presumably, is that it is of marginal interest to his own project.[116]

I think it a mistake to leave matters there, as if these developments had nothing to say to social theory or human geography. If structuration theory is to provide more than a series of abstract ideas, it is necessary to consider the textual strategies through which concrete processes of structuration might be set down on the page, and here both modernism and postmodernism underline the astonishingly conservative modes of representation used in most of the humanities and the social sciences. Instead of considering how to convey the duality of structure in the medium of print, however, Giddens recommends a "bracketing" of agency and structure that vitiates the original purpose of structuration theory; and instead of examining the attempts of experimental ethnographers and others to elucidate the contemporary interpenetrations of the local and the global, he is content to draw out the procedural rather than the representational implications of his project. The barriers between "art" and (social) science remain in place. My own view, in contrast, is that they should be dismantled. Whatever one makes of the postmodern, it seems to me that human geography has much to learn from both aesthetics and social theory. This is Harvey's conclusion too, but it can also be derived directly from the momentum of structuration theory which, so it seems to me, leads directly to the connections between power, space, and representation.

Unmastering geography: feminist theory and spatiality

Feminist geography turns on a crisis of representation too. Although this can hardly be explained by processes of time-space compression, it has important implications for the spatiality of social life. Feminist geography was initially concerned with challenging the orthodoxies of what might well be called a half-human geography, one which virtually ignored the other half of humankind, through the construction of a "geography of women" that drew attention to the specificity of women's experience of place and the contours of their activity spaces. Its proximity to historical materialism was indexed by its focus on geographies of waged and nonwaged work. These modalities of economic exploitation are immensely important, but

---

[116] Giddens, *Consequences*, pp. 45, 177. Giddens may not be interested in postmodern*ism*, but he does sketch out the contours of an immanent (but utopian) postmodern*ity*: a "post-scarcity" system; multilayered democratic participation; demilitarization; and the humanization of technology (pp. 163–73).

so too is a series of other questions, including (for example) male violence toward women, which have their own geographies but cannot be derived so directly from the Marxist tradition.

Although this projection of feminism into geography challenged the empirical bases, the vocabularies and, on occasion, even the logics of "malestream" geography – and in this sense it irrupted into the ways of thinking and, one hopes, the ways of being of men as well as women – it seems to have had little impact upon feminist theory itself. Part of the reason for this was perhaps the comparative simplicity of early feminist geography. If these first studies took little notice of more complex debates within other areas of feminism, however, I suspect that the strategy was, at least in part, a deliberate one. So many thematic and theoretical innovations within human geography had had their origins in the conceptual stratosphere that the introduction of ideas at a much more immediate and accessible level said much about the very different political and intellectual project represented by feminism. Feminist geography was conceived from the very beginning as a discourse *about* exclusion and not as a discourse *of* exclusion. That said, I know that many writers insist that men cannot do feminist work: that it is about the experience of women and that supposedly sympathetic men should have the good sense to realize that this is not available for new and more subtle forms of (intellectual) colonization. Suzanne Moore has described this as a kind of "gender tourism" in which "male theorists are able to take package trips into the world of femininity" secure in the knowledge that they have return tickets to the world of masculinity.[117]

This imposes limits on what I can say; one of its implications is perhaps that the proper subject of man is quite literally man. In one sense, of course, that is exactly what the dominant traditions of human geography and other disciplines have been studying for centuries – except that most men have written as though the particular, dominant masculinities in which they were positioned are all there is or could ever be. If in response the feminist critique requires an attempt to understand the world of masculinity – as a way of clarifying the modalities of patriarchal power – this is not to say that all constructions of masculinity are inherently repressive. But it does make it necessary to explore the multiple ways in which different constructions of masculinity and femininity are constituted over space.[118] Those last four words are not special pleading; subsequent developments

---

[117] Suzanne Moore, "Getting a bit of the other: The pimps of postmodernism," in Rowena Chapman and Jonathan Rutherford (eds.), *Male Order: Unwrapping masculinity* (London: Lawrence and Wishart, 1988) pp. 165–92; the quotation is from p. 167. See also the discussions in Alice Jardine and Paul Smith (eds.), *Men in feminism* (London: Routledge, 1987).

[118] Peter Jackson, "The cultural politics of masculinity," *Transactions of the Institute of British Geographers* 16 (1991) pp. 199–213.

have seen questions of spatiality placed at the heart of those feminist theories that articulate a distinctive "politics of location." This phrase is intended to have a much wider meaning than the conventionally geographical, to be sure, but it is also more than a literary trope.[119] When bell hooks chooses "the margin as a space of radical openness"; when she writes as an African-American woman about the effort "to change the way I speak and write, to incorporate in the manner of telling a sense of place, of not just who I am in the present but where I am coming from"; when she urges the effort "to create space where there is unlimited access to the pleasure and power of knowing": when she says all these things she means her spatial metaphors to be understood in the most insistently material of senses. She wants to open a "space of resistance" for the production of a counter-discourse *"that is not just found in words but in habits of being and the way one lives."*[120]

I should like to use this developing feminist critique to reflect further on the masculinities/spatialities of structuration theory. I take as my initial provocation an essay in which Gillian Rose explains that the connective imperative which structuration theory establishes between power and space – and in particular its assertion of power *over* space – suggests to her "that its space is masculine."

Only white heterosexual men can usually enjoy such a feeling of spatial freedom. Women know that spaces are not necessarily without constraint; sexual attacks warn them that their bodies are not meant to be in public spaces, and racist and homophobic violence delimits the spaces of black and gay communities. There's also a sense of powerful knowledge of space – the space of that kind of geography is constructed as absolutely knowable. It thinks spaces can always be known and mapped; that's what its transparency, its innocence, signifies; it's infinitely knowable.[121]

There is much in this passage with which I am in sympathy; and yet I am also troubled by its reading of time-geography. For Rose concedes that Hägerstrand's framework may reveal some of the time-space routines through which structures of patriarchy are produced and reproduced, but she also argues that its "theoretical space" remains one in which the liberties of a masculinist geography are celebrated. And I find myself trying to

---

[119] Elspeth Probyn, "Travels in the postmodern: Making sense of the local," in Linda Nicholson (ed.), *Feminism/Postmodernism* (London: Routledge, 1990) pp. 176–89.

[120] bell hooks, *Yearning: Race, gender and cultural politics* (Toronto: Between the lines, 1990) pp. 145–46, 149; my emphasis.

[121] Gillian Rose, "On being ambivalent: Women and feminisms in geography," in Chris Philo (ed.), *New words, new worlds: Reconceptualising social and cultural geography* (Aberystwyth, Wales: Social and Cultural Geography Study Group, Institute of British Geographers, 1991) pp. 156–63; the quotation is from p. 160.

clarify my unease, wondering exactly why I don't like this argument. My immediate answer is straightforward and, in its appeal to textual authority, thoroughly conventional. Rose draws on Gould's commentary on the conception of space embedded within time-geography to secure what she takes to be "its infinitude and unboundedness, its transparency." But she is able to do so, I want to say, only by misreading – in fact reversing – Gould's argument. The point of his essay turns on a distinction between, on the one side, the English *space* and the French *espace*, words which Gould suggests invoke a boundless expanse that he captures in the precious image of "the joyous unconstraining of the dancing child in a flower-spangled meadow"; and, on the other side, the German *Raum* and Hägerstrand's Swedish *rum*, which resonate with the compulsions and constraints of "the adult world where my actions constrain yours, where our interdependencies jostle and compete, demand and require." Gould concludes that the dominant spatialities of time-geography are those of competition and constraint, emphatically not "spatial freedom."[122] It is of course possible to question whether capability, coupling, and steering constraints are sufficiently discriminating concepts to allow the critical dissection of patriarchal power; equally, Gould's essay invokes a metaphoric of (lost) innocence that could be translated into a masculinist narrative: As childhood gives way to adulthood, so the feminine "naturally" yields to the masculine. Perhaps. And yet, does it follow from these possibilities (neither of which, to be fair, I have derived from Rose) that time-geography is an incorrigibly phallocentric discourse inscribing patriarchal power in space? If it registers the traces of patriarchal power, must that involve a *celebration* of "masculine space"?

Rose is surely (and sadly) right to insist that "women know that spaces are not necessarily without constraint"; so do people in other subjugated positions. Hägerstrand knows it too, but within his web model positionality is *carried within* the trajectory. A feminist time-geography would, I think, pay more attention to the ways in which positionality is *constituted through* the trajectory: to its embodiment and its construction through multiple sites. Such a view would treat not "place" but places and would describe them not only as meaningful sites within a symbolic landscape – a common omission from time-geography – but also as sites between and within which identities and subjectivities are negotiated. Several feminist geographers have already moved beyond pure time-space geometries to show how social practices are focused and guided through specific locales. Dyck's exceptionally important study of "mothering," for example, shows that the bundles of activities around which many mothers condense their daily routines are shaped by all sorts of constraints; but the configurations that

---

[122] *Ibid.*, p. 160; Peter Gould, "Space and *rum*: An English note on espacien and rumian meaning," *Geografiska Annaler* 63B (1981) pp. 1–3.

result are also enabling, even empowering because it is through those regular relationships with other women in a diversity of locales that conceptions of identity, of self-esteem and of mothering are constructed and negotiated. "Woman" and "motherhood" thus emerge as categories whose meanings are situationally defined, and the social construction of space is shown to be important not simply as a logistical exercise – time-budgets, time-space packing, and the like – but also as an essential component of the construction of social meaning.[123] For this reason I need to say that any absolutist critique of time-geography – even when it is conducted under the sign of a constructivism – is quite unhelpful in understanding the multiple, compound, and complex construction of human spatialities.

And yet I find that this won't do. It isn't so much that my countervailing argument is wrong (though it may well be) but that it is *beside the point*. I suspect that my more profound sense of unease, at once professional and personal, and which even now I find difficult to articulate, is that Rose's critique – that "initial provocation" – obliged me to consider the possibility of a feminist critique of a corpus of work I had never before thought of examining from such a perspective. Of course, I had registered several reservations about time-geography and structuration theory, but I still thought of them as essentially constructive, critical and, yes, "sensitizing" interventions. I had taken it for granted, I think, that the generality of the duality of structure and (in his later writings anyway) Giddens's careful use of nonexclusionary language had somehow ensured that the core model of structuration theory would be free from discriminatory assumptions. Rose's essay was then all the more disconcerting (and all the more effective) because it did not focus on the absences from structuration theory – Giddens is a past-master at showing how lacunae can be brought within the compass of his thought (and "mastered")[124] – but on its *presences*: on the deep gendering of his ontology of social life, which is put in place in part through his conception of space and spatiality.

I have tried to learn from this experience, and so I now want to return to structuration theory and attempt to disclose quite another sense in which I think it does tacitly reproduce a master-narrative of power over space. Giddens uses the concept of time-space distanciation to mark what he

[123] Isabel Dyck, "Space, time and renegotiating motherhood: An exploration of the domestic workplace," *Environment and Planning D: Society and Space* (1990) pp. 459–83. For a different treatment of the ties between identities and places, see Minnie Bruce Pratt, "Identity: Skin Blood Heart," in Elly Bulkin, Minnie Bruce Pratt, and Barbara Smith (eds.), *Yours in struggle: Three feminist perspectives on anti-semitism and racism* (Ithaca: Firebrand Books, 1984) pp. 10–63.

[124] His treatment of feminism is no exception; in response to feminist critiques, one has the impression that Giddens prefers to list what feminism can learn from structuration theory rather than listen to what structuration theory might learn from feminism. See, for example, the section on "Gender and social theory" (pp. 282–87) in his "Reply to my critics," in Held and Thompson (eds.), *Social theory of modern societies*, pp. 249–301.

variously calls the "bracketing" or even the "dissolution" of the restraints of time and space; as time-space is "bound in" to the conduct of social life so space as barrier and as void is transcended, incorporated, subjugated. And in describing these transformations as a more or less progressive movement, which I think *is* the implication and perhaps even the telos of his genealogy of modernity, Giddens repeats the characteristic movement of Western master-narratives more generally, which has been to recover and reincorporate what eludes them *as* lacunae, margins, "blank spaces" on the map.[125] Jardine argues that these spatial metaphors give themselves away: for "this other-than-themselves is almost always a 'space' of some kind (over which the narrative has lost control), and this space has been coded as *feminine*, as *woman*."[126] In fact, these are more than metaphors – or, rather, as hooks insists, metaphors are more than textual conceits. It is a commonplace of feminist history that "Nature" has been coded as feminine within the Western intellectual tradition; and if concepts of space can be derived directly from concepts of nature, as Smith implies, then it is scarcely surprising that socially produced space – spatiality – should have been coded in the same way: as a space to be mastered, domesticated and gendered.[127]

This is particularly intrusive in the sexualization of colonial landscapes, where two master tropes are characteristically deployed: on the one hand, a rich and fecund virgin land is supposedly available for fertilization; on the other hand, a libidinous and wild land has to be forcefully tamed and domesticated.[128] In the seventeenth century, for example, Jan van der Straet's engraving of "America" (figure 14) followed European graphic convention in showing the male figure of Amerigo Vespucci standing over and confronting the "new" continent depicted as a naked, reclining woman. De Certeau suggests that this iconography marks the site where the conqueror "will *write* the body of the other and inscribe upon it his own *history*." It establishes a threshold; after the encounter between Vespucci and "America" – the one surprised, even taken aback, and the other rising

---

[125] Giddens's claim that European voyages of discovery, exploration, and survey went hand in hand with advances in cartography to produce maps "in which perspective played little part" and through which the representation of space could proceed without reference to a privileged vantage point would be comical if the "empty space" that was supposedly constituted in this way were not also a space of violence, terror, and death: It was not empty but *emptied*. Giddens, *Consequences*, p. 19; cf. Michael Taussig, "Culture of terror, space of death," in his *Shamanism, colonialism and the wild man: A study in terror and healing* (Chicago: University of Chicago Press, 1987) pp. 3–36.

[126] Alice Jardine, *Gynesis: Configurations of woman and modernity* (Ithaca: Cornell University Press, 1985) p. 25.

[127] As Smith notes in *Uneven development*, p. 66: "[T]he production of space is a logical corollary of the production of nature." On the gendering of "nature," see Carol MacCormack and Marilyn Strathern (eds.), *Nature, culture and gender* (Cambridge: Cambridge University Press, 1980).

[128] Ella Shohat, "Imaging *terra incognita*: The disciplinary gaze of empire," *Public Culture* 3 (1991) pp. 41–70.

Figure 14 Jan van der Straet's "America."

invitingly from her hammock – this space, shown here as a wild and luxuriant "nature," will become a blank page on which, so de Certeau argues, the occidental and masculine "will-to-power" is to be written.[129] But it occurs to me that it is not only a "history" that is inaugurated in this way: so too is a "geography," inscribed through the very instruments of navigation and survey that, as the engraving shows, accompany and make possible Vespucci's presence on the shore of the "New World."[130] Its native inhabitants – its first nations – become not only "people without

---

[129] Michel de Certeau, *L'écriture de l'histoire* (Paris: Gallimard, 1975; 2nd ed., 1984) p. 3 and *passim*; see also Peter Hulme, *Colonial encounters: Europe and the native Caribbean 1492–1797* (London: Methuen, 1986).

[130] I am reminded of McClintock's reading of the map that accompanies Rider Haggard's *She*:

> The colonial map is a document that professes to convey the truth about a place in pure rational form, and promises at the same time that those with the technology to make such perfect representations are best entitled to possession.... Haggard's map is [thus] a fantastic conflation of the themes of colonial space and sexuality. The map abstracts the female body and reproduces it as a geometry of sexuality held captive under the technology of scientific form.

See Anne McClintock, "Maidens, maps and mines: The reinvention of patriarchy in colonial South Africa," *South Atlantic Quarterly* 87 (1988) pp. 147–92; the quotation is from pp. 151–52.

history," in Wolf's angry phrase, but also "people without geography." These are human beings whose relationship to their own land is denied – whose presence in Western narratives of exploration and discovery is made faint and marginalized – through a discourse which, in the course of the nineteenth century, typically "centers landscape, separates people from place, and effaces the speaking self." This geography "writes itself," so to speak, across a now-empty, made-empty space.[131] Similarly in colonial Africa, where "nature" and "space" were also made to stand in the closest of associations, Torgovnick argues that its supposedly "primitive landscape" was axiomatically identified with the female body:

> The masculine sexual metaphors of penetrating closed dark spaces no doubt help account for the West's attachment to the trope of center, heart or core of Africa. African landscape is to be entered, conquered; its riches are to be reaped, enjoyed. The phallic semiology accompanies the imperialist topoi, a conjunction based on the assumption that if explorers are . . . "manly," then what they explore must be female.[132]

These are arresting readings, but several qualifications are necessary. First, for much of this period most landscapes and spaces seem to have been coded as feminine within the Western political imaginary: *including those of Europe*. But I think it would be possible to show that in the course of the nineteenth century Western "nature" is made ever more elaborately feminine – particularly through constructions of landscape – whereas Western "space" is made over in the image of a masculine, phallocentric power that deals not so much with the rationalization *of* space as with a presumptive identity between rationality *and* space. Thus space is *itself* represented as the physical embodiment of (masculine) rationality whose structures are to be superimposed over "nonspace."[133] At the same time, the ur-history of Western space is suppressed and its old conceptions of space transposed to those non-Western "virgin lands" where space remains "other," "mysterious,"

---

[131] My reading is indebted to Mary-Louise Pratt, "Scratches on the face of the country; or, What Mr. Barrow saw in the land of the Bushmen," *Critical Inquiry* 12 (1985) pp. 119–43; the quotation is from p. 124. In fairness, I should emphasize that Pratt does not treat this as the only textual strategy used by colonial writers.

[132] Marianna Torgovnick, *Gone Primitive: Savage intellects, modern lives* (Chicago: University of Chicago Press, 1990) p. 61; the Comaroffs, *Revelation*, p. 116 make much the same point when they describe how the "erotic allure" of Africa was conveyed through "submerged sexual metaphors of encounter and conquest, of the voluptuous and accessible landscapes with their carefree sensuous inhabitants." These are both resolutely heterosexual readings, however, and more discriminating interpretations of the tropes employed by different explorers and surveyors are evidently necessary for any fuller account of the sexualization of colonial landscapes. I should also add that few women travelers treated these "other" countries as female, sexualized bodies: see Sara Mills, *Discourses of difference: An analysis of women's travel writing and colonialism* (London: Routledge, 1991) p. 82.

[133] For some particularly suggestive arguments about the constellation of visual-phallic-geometric power contained within modern, Western, "abstract" space, see Lefebvre, *Production of space*. A fuller argument would obviously have to consider the institutionalization of "public/masculine" and

"feminine," to be discovered and penetrated. There is, of course, a complex dialectic between the two spatialities – the relations between metropolitan and colonial modernism in fin-de-siècle France are now well-known, for example[134] – but the circle is closed when Freud borrows the metaphor of the "Dark Continent" from H. M. Stanley's account of his African adventures to conjure his own discovery of the "territory" of female sexuality. It is true, of course, that a more discriminating account would have to deconstruct such stark binaries; even within the West there were significant differences between the political cultures of (for example) Britain and France. But it is also true, I suspect, that these differences were articulated through an appeal to an ideological landscape – a political imaginary – in which "the West" was indeed given unitary force. Other discriminations are necessary too. Although the readings I have considered here accentuate the gendering and sexualization of space and spatiality, it is important not to absolutize or homogenize gender. Gender, sexuality, class, "race" and other ascriptive dynamics are all present in these texts and (in the colonial case) their intersections are overdetermined by what Chrisman calls "the dictates of a highly problematic imperialism." But I have made so much of these specific codings to resist the tendency to reduce such rationalities to the "logic" of capitalism: that logic is clearly inscribed in notations of class but it is also gendered.[135]

I cannot develop these skeletal suggestions in the depth or detail they require here, but it should be clear that this feminist project is a much more politicized one than the conjunction of structuration theory and postmodernism allows. In an essay that I have found particularly helpful in thinking through these questions, Bondi suggests that postmodernism characteristically elides "a dichotomy between ideas and reality": that it is preoccupied with (for example) "the coding of space as feminine" and fails to elucidate the existence and experience of "women in geographical space."[136] But I do not think that the arguments I have sketched in these paragraphs

---

"private/feminine" spaces and the ways in which this Western spatial imaginary and its subordinations were contested and subverted; see Gillian Rose, "The struggle for political democracy: Emancipation, gender and geography," *Environment and Planning D: Society and Space* 8 (1990) pp. 395–408. Rose traces this distinction back to seventeenth-century Europe, but it surely has a much earlier origin. The classical project of (patriarchal) "democracy" was embedded in and depended on the masculine/public spaces of the Hellenic city-state.

[134] Cf. Gwendoyln Wright, *The politics of design in French colonial urbanism* (Chicago: University of Chicago Press, 1991).

[135] Laura Chrisman, "The imperial unconscious? Representations of imperial discourse," *Critical Quarterly* 32 (1990) pp. 38–58; the quotation is from p. 42.

[136] Liz Bondi, "Feminism, postmodernism and geography: Space for women?" *Antipode* 22 (1990) pp. 156–67; the quotation is from p. 163. See also Liz Bondi and Mona Domosh, "Other figures in other places: On feminism, postmodernism and geography," *Environment and Planning D: Society and Space* 10 (1992) pp. 199–214.

amount to a textualism; as I read them, they insist that the texts that they examine carry within them meanings *that were embedded in thoughts and actions*; that they had real, practical, material consequences. And in doing so, they raise urgent questions about authority and authorization that draw upon the claims of poststructuralism and carry over into those of post-colonialism.

Traces: cultural studies and human geography

There is little doubt that the late 1980s and early 1990s have seen an explosion of interest in "culture" across the spectrum of the humanities and the social sciences, including cultural forms and practices that, in the West at any rate, were previously invisible to formal academic criticism. In human geography the emergence of a "new cultural geography" deliberately removed from (and in many cases in opposition to) the preoccupations of Sauer and the Berkeley School has contributed to a major refiguration of the discipline. The critique of Sauer is often misjudged, it seems to me, and I can think of few discussions of his work that avoid either uninteresting hagiography or unreflective dismissal; certainly the tradition he represents has been responsible for some of the most exacting scholarship in cultural geography. I also suspect that the reawakening of an interest in cultural ecology and environmental history may presage a return to some of Sauer's most pressing concerns (though one that is likely to be informed by more recent adventures in cultural theory and cultural politics). In the course of these more general realignments, as Huyssen shows, several "great divides" have been crossed, most notably those between historical and contemporary analysis and between high and popular culture. To understand how this has come about, another boundary has to be dissolved. For cultural criticism has unraveled the seam between text and context as a deliberate political gesture – as a move in a considered politics of interpretation – so that the renewed salience of culture in the critical vocabularies of the late twentieth century cannot be separated from the practical involvement of its "maps of meaning" in the diverse worlds they seek to represent.[137]

This makes the interest in culture more than merely responsive: It is also in many, crucially important ways *constitutive*. The diversity of those

---

[137] Andreas Huyssen, *After the Great Divide: Modernism, mass culture and postmodernism* (London: Macmillan, 1988). In geography this might be summarized as a move from the Berkeley School to the Birmingham School (the Centre for Contemporary Cultural Studies): see Peter Jackson, *Maps of meaning: An introduction to cultural geography* (London: Unwin Hyman, 1989), whose title is borrowed from the Centre's former director, Stuart Hall.

involvements needs to be underlined, because "culture" has become coded in multiple ways. On some readings, it marks the site of a reactionary politics, a "consumer culture" in which culture has itself become a product, and whose lubricious commodification of the aesthetic serves only to heighten the fetishisms of bourgeois ideology. Or, still more insidiously, it consecrates what Werckmeister calls a "citadel culture," a culture of permanent self-doubt and constantly deferred problems that acts as late twentieth-century society's "internalized, aesthetically sublimated reaction to the conflicts it sustains at its margins and beyond its frontiers in order to sustain itself."[138] On other, more affirmative readings culture becomes a site on which to regroup the critical faculties of both liberalism and the left, following the assaults of neoconservatism and its abusive intolerance of difference and dissent; culture cocking a collective snook at authoritarianism. All of these claims have merit, but none of them is exhaustive; the cultural turn includes all this and more. It inaugurates neither a culture of complicity, I think, nor a culture of consolation but instead maps a contested, negotiated domain of extraordinary and perhaps even unprecedented political significance.[139]

All of this (or anyway much of it) has been achieved under the banner of cultural studies. I have already trespassed on this field in the previous pages, but one of the implications of contemporary cultural studies seems to be that we are all trespassers now. It is impossible for me to survey its terrain with any precision. Cultural studies is made up of multiple discourses with different origins and different histories. Of these, two main sources are readily identified: on the one side, the study of literature, and in particular the disruption of the canonical status and textual unity of traditional literary studies; and on the other side, anthropology, and in particular its repatriation as a cultural critique that works through ethnographic sensibility rather than philosophical speculation.[140] It is tempting but I think ultimately tendentious to identify "literature" with postmodernism and "anthropology" with postcolonialism, because the independent integrity of these two streams is soon lost in a complex braiding of intellectual currents. The study of literature, most particularly comparative literature, is increasingly drawn to anthropology, to semiotics and structural anthropology and more recently to poststructuralism and postcolonialism, while anthropology often conceives of culture as text and borrows from literary theory and literary criticism to interrogate its own textual practices. Schematically:

---

[138] O. K. Werckmeister, *Citadel culture* (Chicago: University of Chicago Press, 1991) p. 19.

[139] Terry Eagleton, *The ideology of the aesthetic* (Oxford: Blackwell Publishers, 1990); Norris, "Criticism, history and the politics of theory," in his *What's wrong with postmodernism*, pp. 1–48.

[140] Antony Easthope, *Literary into cultural studies* (London: Routledge, 1991); George Marcus and Michael Fischer, *Anthropology as cultural critique: An experimental moment in the human sciences* (Chicago: University of Chicago Press, 1986).

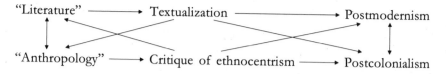

And, as I will show presently, both streams are also crisscrossed by ideas drawn from historical materialism and psychoanalysis.

Whatever theoretical and methodological concerns these interventions have in common, Hall attaches a particular significance to the importance of *cultural politics*. Although there may be a general reluctance to close off the field and police it, he insists that there is also "a determination to stake out some positions within it and argue for them." Openness does not slide into vacuousness. In this connection, Hall advances two leading claims. First, he argues for the critical importance of the linguistic turn in the humanities and the social sciences, and more particularly for its catholic interest in texts and their worldly affiliations: "the trace of those other formations, of the intertextuality of texts in their institutional positions, of texts as sources of power, of textuality as a site of representation and resistance." Second, he argues that cultural studies (in Britain at any rate, and perhaps elsewhere) developed as a critical project both profoundly influenced by Marxism and yet also against the grain of Marxism; against the grain of what he sees as its economism, reductionism, determinism, certainly, but for him most powerfully (given his own biography) against the grain of its "profound Eurocentrism."[141] These are probably both contentious claims, but I repeat them not because they provide a definitive view of cultural studies but because they seem to extend an open invitation to discuss postmodernism and postcolonialism in ways that intersect with the transdisciplinary exploration of deep space. Accordingly, I propose to situate both streams of thought in relation to the current displacement and dislocation of the imaginative geographies of the West.

That discussion is not without its frictions, especially along the lower diagonals of figure 9. The relations between feminism and postmodernism are characteristically tense, but for the most part they have been neither blindly enthusiastic nor bleakly dismissive.[142] In human geography this ambivalence has been captured with considerable perspicacity by Liz Bondi and Mona Domosh:

[141] Stuart Hall, "Cultural studies and its theoretical legacies," in Lawrence Grossberg, Cary Nelson, and Paula Treichler (eds.) *Cultural studies* (New York: Routledge, 1992) pp. 277–86; the quotations are from pp. 279 and 284.

[142] Nancy Fraser and Linda Nicholson, "Social criticism without philosophy: An encounter between feminism and postmodernism" and Seyla Benhabib, "Epistemologies of postmodernism: A rejoinder to Jean-François Lyotard," both in Linda Nicholson (ed.), *Feminism/Postmodernism* (London: Routledge, 1989) pp. 19–38 and pp. 107–30. The argument is rejoined in Seyla Benhabib,

[T]here is a tendency within postmodernism to indulge in a kind of "tourism," manifest in references to "woman" and the "feminine"; we have raised objections to what we see as an *appropriation* of women's voices and experiences by white, male theorists. But, as Haraway points out, there is a "very fine line between the appropriation of another's (never innocent) experience and the delicate construction of the just-barely-possible affinities, the just-barely-possible connections that might actually make a difference in local and global histories." This project of constructing affinities and connections remains an as yet latent possibility within postmodern geography.[143]

They draw a parallel between the cultural tourism of many colonial geographies – the textual appropriation of the "exotic" worlds of others – and the gender tourism of postmodern discourse, and they hold out the hope that in much the same way that the personal worlds of some colonial explorers were changed by their expeditions so too might those of the postmodern travelers. Along the other diagonal, and from a position far removed from these two writers and their sympathies with poststructuralism, Giddens is unequivocal. From the perspective of structuration theory, he declares bluntly that poststructuralism is a "dead tradition of thought" that has conspicuously failed to deliver the revolution in philosophy and social theory it once supposedly promised. This has not prevented him from appropriating some of the central ideas of Derrida and Foucault, however, and one critic argues that in doing so he has systematically and *necessarily* misrepresented them. "A serious reading of Derrida would find [Giddens] forced to respond to the charge that the theory of structuration is perfectly in harmony, and therefore (to put the matter in more disturbing terms) in league with the project of western reason." And although Giddens has not so far considered postcolonialism, his interest in non-Western societies seems to be essentially historical and at best typological: Anthropology is on one side of the Great Divide, (his) sociology on the other.[144]

Modernism, postmodernism, and cartographies of power

In order to maintain the momentum of my previous discussion, I will continue to talk about power, space, and representation and the discursive

"Feminism and postmodernism: An uneasy alliance," *Praxis international* 11 (1991) pp. 137–49; Judith Butler, "Contingent foundations: Feminism and the question of postmodernism," *loc. cit.*, pp. 150–65; Nancy Fraser, "False antithesis: A response to Seyla Benhabib and Judith Butler," *loc. cit.*, pp. 166–77.

[143] Bondi and Domosh, "Other figures in other places," p. 211; they are referring to Donna Haraway, "Reading Buchi Emeta: Contests for women's experience in Women's Studies," *Women: A cultural review* Vol. 1 (1990) pp. 240–55.

[144] Anthony Giddens, "Structuralism, post-structuralism and the production of culture," in *Social theory and modern sociology*, pp. 73–108; Roy Boyne, "Power-knowledge and social theory: The systematic misrepresentation of contemporary French social theory in the work of Anthony Giddens," in Bryant and Jary (eds.), *Giddens' theory of structuration*, pp. 52–73.

triangle they form around the question of authorization. I have already proposed that in the course of the nineteenth century dominant conceptions of space installed within the political imaginary of the West a presumptive identity between "rationality" and "space"; that the one was inscribed within the other. Since this claim is the base-line for what follows, I need to spell out some of its implications. I will use as my example what Rabinow calls "French modern." The specificity of this needs some emphasis, since the geographies of modernism are too often erased by casual generalization. But France did occupy a position of unusual prominence within the cultural landscape of late nineteenth- and early twentieth-century Europe. There, the equation between rationality and space that I have been seeking to establish seems to have provided a series of coordinates for those underlaboring discourses that occupied Rabinow's "middle ground between high culture and ordinary life" – a distinction that, as he says, is itself a typically modern one – and that was staked out and dug over by what he calls "technicians of general ideas." These were not the darlings of high culture, the painters, poets, and philosophers, nor the heroes and heroines of a rough-and-tumble popular culture. They were instead the mundane agents of what Foucault called "bio-power," an anonymous and dispersed system of power that in the course of the nineteenth century conjoined an anatomo-politics of the human body with a bio-politics of the human population.

As these phrases imply, one of the characteristic preoccupations of these new technicians was the transfer of concepts from the new science of physiology, defined by Bernard as the experimental study of the properties of organized matter, of the forms and conditions of existence of "life," to the realm of the social – particularly concepts of function, hierarchy, regulation, and norm. In Bernard's view, "life" was neither a given fact nor an essential property. It was, rather, an *effect* that was produced in nature and could be reproduced in experiment under specific environmental conditions. I have already indicated the importance of this metaphoric to Durkheim's sociology, but Rabinow identifies its more general translation into a search for "a spatial localization of function in society similar to that found in the body." This was, in part, a product of what he calls particular "practices of reason" – the rationalism of the problematic is surely obvious – but what is equally important is the interest in specifying the environmental conditions that make such a rational social organization possible; that establish the *space* within which, as Rabinow puts it, an individual's actions "were formed and normed."[145]

---

[145] This paragraph relies on Paul Rabinow, *French modern: Norms and forms of the social environment* (Cambridge: MIT Press, 1989); the quotation is from p. 10. My capsule summary of bio-power is derived from Michel Foucault, *The history of sexuality.* Vol. 1: *An introduction* (London: Penguin, 1981) pp. 135–45. For Bernard's physiology, I have relied on Paul Hirst, *Durkheim, Bernard and epistemology* (London: Routledge and Kegan Paul, 1975) pp. 58–77.

Rabinow's debt to Foucault is well-known and, like Foucault, he means that "space" to be understood both metaphorically and materially.

What characterized modernity, according to Foucault, was the advent of a normative age: the normalization of disciplines, the movement from discipline as blockade to discipline as mechanism, and correlatively the formation of a disciplinary society, which is characterized not by confinement, even if one continues to use the procedure, but rather by the constitution of a *space*: supple, interchangeable, without segregation, indefinitely redundant and without exterior.[146]

More specifically, Rabinow argues that toward the end of the nineteenth century there was in France (and elsewhere) a concerted effort to identify and bring into being "a common space regulated by scientifically derived social norms." Urbanism and the discourses of planning and administration had a central place in this problematic, and its "techno-cosmopolitanism" was given a clearer definition after World War I by a so-called "middling modernism" in which, so Rabinow argues, "society became its own referent, to be worked on by means of technical procedures which were becoming the authoritative arbiters of what counted as socially real." This was the space of administered welfare, the space of what Foucault would call "governmentality," and its systematic construction of "public spheres" was, as I have said, gendered. But this process also represented a move beyond the perimeter of the physiological paradigm, so that norms and forms were now "freed from previous historical/natural constraints, defined by their own operations and claiming universal status." And it was in that moment that rationality and space were finally fused in "an anonymous space of regulation and rationalization": in what Lefebvre called simply *abstract* space.[147]

These discourses belonged neither to high culture nor popular culture, but they were centrally present in what I propose to call "public culture" – a missing term that seems to me indispensable for any critical analysis of modernism or postmodernism – where they had the most material of presences in the day-to-day lives and arrangements of ordinary people. It is of course true that the certainties that they put in place were called into question in Europe and North America during the fin de siècle, but Kern

---

[146] François Ewald, "A power without an exterior," in Timothy Armstrong (ed.), *Michel Foucault: Philosopher* (London: Routledge, 1992) pp. 169–75; see also *idem*, "Norms, discipline and the law," in Robert Post (ed.), *Law and the order of culture* (Berkeley: University of California Press, 1991) pp. 138–61. Foucault is sufficiently interested in space and spatiality for Deleuze to describe his *Discipline and punish* as "a new cartography": Gilles Deleuze, *Foucault* (Minneapolis: University of Minnesota Press, 1988) pp. 23–43; see also John Rajchman, "Foucault's art of seeing" in his *Philosophical events: Essays of the 80s* (New York: Columbia University Press, 1991) pp. 68–102.

[147] Rabinow, *French modern*, pp. 12–13, 321–22; Lefebvre, *Production*, pp. 285–91. It was in this moment, too, that Vidal's *géographie humaine* was exhausted.

surely claims too much when he describes this as a period when a whole "culture of time and space" gave way. Many of those challenges were conspicuously avant-garde and had little immediate impact on public culture; many of them were still thoroughly rationalist in sensibility. One could go further and suggest that the public culture of governmentality was wired to the three elements which, after World War II, cut off the revolutionary vitality of those adversarial modernisms: that is to say, the rise of bourgeois democracy and the consolidation of the welfare state; the inauguration of Fordist regimes of accumulation and their associated modes of regulation; and the glacial stasis of the Cold War.[148]

But the modern equation between rationality and space has been thrown into much more complete disarray, at once intensive and extensive, in the late twentieth century. Whatever else postmodernism may be about, I propose to argue, it represents an attempt to come to terms with – to find the terms for – this bewilderment of the contemporary; what might be called, with apologies to Edward Said, the dis-orientation of Occidentalism. If in some ways postmodernism returns to many of the concerns of earlier modernisms, which I think it does, its vocabularies are neither marginalist nor rationalist. Many critics regard postmodernism as a cultural dominant (which is not to say that postmodernism is all there is); most agree that it calls into question the absolutisms of Reason. Of course, postmodernism is not all of a piece and one should perhaps be wary of extracting an "essential" postmodernism from such a resolutely antiessentialist constellation: though it is equally striking how often critiques of metanarrative conducted under its sign seem to advance a surreptitious counternarrative of their own. To say all of this is neither to decorate nor to domesticate postmodernism, to make it more complicated than it is or to confine it within the categories of yet another modernist imaginary, but to admit its implication in the ambivalent world it describes. What gives the debates surrounding postmodernism a particular inflection, however, is that a number of commentators have seen that involvement in spatial terms; as a spatialization which, in contrast to the coordinates of modern geometries, now interrupts and frustrates every immediate orientation. Postmodern theory, according to Jameson, "infers a certain supplement of spatiality in the contemporary period" and suggests that "even if everything is spatial, this postmodern reality here is somehow more spatial than everything else." Again, this is not a rhetorical gesture. Architecture – and I think the built environment more generally – is for Jameson, as it is for many other cultural critics, both a privileged aesthetic language and a paradigmatic site for the inscriptions of postmodern culture. Postmodernism has come to have its own, highly visible topoi: for Jameson, the Bonaventure Hotel,

[148] Stephen Kern, *The culture of time and space, 1880–1918* (Cambridge: Harvard University Press, 1983); Perry Anderson, "Modernity and revolution," *New Left Review* 144 (1984) pp. 96–113.

for others the West Edmonton Mall (supposedly a twentieth-century equivalent of Benjamin's nineteenth-century Arcades), and even Los Angeles itself, which two critics advertise in another parody of Benjamin as "the capital of the twentieth century."[149]

I want to distinguish two strategies of "cognitive mapping" that seek to make this postmodern disorientation intelligible: I shall refer to them as *textualities* and *spatialities*.[150] There are important connections between them, which help to account for the repeated invocation of Benjamin in the examples I have just given and, for that matter, in some of those I am about to consider. His Arcades Project used both strategies simultaneously as a way of situating modernism and modernity.[151] But there are also significant differences between them and so I will discuss each in turn.

Postmodern readings and textualities

"Textualities" is an awkward term, but I use it to cover the various modalities through which "text" has been extended beyond the spaces of the written page. If this is not sufficiently clear, I should perhaps emphasize that what follows is not a commentary on the textual strategies used in human geography and the other humanities and social sciences. There is undoubtedly a danger of obsessive introspection in the prominence accorded to those concerns at present, but this is such a contrast to the previous orthodoxy – which held that writing was merely a matter of record so that it was necessary to prize simplicity, clarity, and economy above all else, as though they were not, in their own way, culturally complex ideological constructions – that one is bound to welcome almost any critical reflection on the poetics and politics of writing. But what I have in mind here is a particular version of what Rorty calls "textualism," namely the textualization of landscape that has figured so prominently in the new cultural geography.

To contextualize this intellectual strategy, I should first say that there are several other ways in which textualism bears on the terrain of human geography. Most general of all is the attempt to read *space* (not "landscape") as a text. Here I am thinking of the work of Henrietta Moore, who has used Ricoeur's reflections on texts and textuality to read (or rather reread)

[149] Jameson, *Postmodernism*, pp. 37, 365; Rob Shields, "Social spatialization and the built environment: The West Edmonton Mall," *Environment and Planning D: Society & Space* 7 (1989) pp. 147–64; Edward Soja and Allen Scott, "Los Angeles: Capital of the twentieth century," *loc. cit.*, 4 (1986) pp. 249–54. Interestingly, all of these authors acknowledge their debt to Henri Lefebvre.

[150] I have borrowed "cognitive mapping" from Jameson, *Postmodernism*, pp. 51–54 and 409–18, who in turn borrowed it from Kevin Lynch's *The image of the city* (Cambridge: MIT Press, 1960) in order to triangulate the space between Lacan's symbolic, imaginary and real.

[151] See my Chapter 3, "City/commodity/culture."

social space. Although she regards space as a cultural representation, she makes it clear that "spatial order" (her term) is more than a physical reflection of social practice. Instead, she treats movements through and action in space as discursive practices – as simultaneous "speakings" and "readings" – in which meanings are embodied, inscribed, and reinscribed; and she cautions that those meanings, in all their plurality, typically escape the situated intentions of their originating authors through a process that Ricoeur called "distanciation." She is also aware of the limitations of Ricoeur's model of the text, however, and seeks to identify the specific conditions of meaning, the horizons of possibility, at once historical and social, within which particular modalities of power and practice come to prevail or yield. "A spatial text has no intrinsic essence," she remarks, "just as it has no inherent meaning; the truth of the text resides in practice." This inflection, which she shares with and in part derives from Bourdieu, gives her work a corporeality and materiality that is conspicuously and consciously absent from those textualizations less patient with the metaphysics of presence.[152]

In the middle ground is the textualization of landscape to be considered here.

Finally, there is the textualization of *architecture*, its inscription as what some commentators call "archetexture" in recognition of those for whom writing architecture has equal status to its production as building.[153] Here one thinks immediately of Peter Eisenman, whose work has been influenced by Baudrillard and Derrida. In his early writings Eisenman explored the possibility of a radical postmodernism that, unlike previous postmodernisms, would subvert rather than extend modernism. In fact, he argued that *all* postmedieval architecture in the West had been a variant of what, following Foucault, he called "classical" architecture. In contrast to Foucault's archaeology of other systems of thought, however, Eisenman maintained that architecture had remained an uninterrupted mode of representation from the fifteenth century to the present. On this reading, even modern and much postmodern architecture is "classical" architecture. All of them are variations on a single theme: They have all been predicated on three founding fictions – representation, reason and history – which acted as self-evident truths conferring legitimacy upon architecture. "Representation was to embody the idea of meaning; reason was to codify the idea of truth;

[152] Henrietta Moore, *Space, text and gender: An anthropological study of the Marakwet of Kenya* (Cambridge: Cambridge University Press, 1986) pp. 73–90; the quotation is from p. 89. Ricoeur's work provided the direct inspiration for Giddens's discussion of time-space distanciation too, but he is also acutely aware of the limitations of describing social life on the model of language. See John B. Thompson, *Studies in the theory of ideology* (Cambridge, England: Polity Press, 1984).

[153] For two very different commentaries, see Mary McLeod, "Architecture and politics in the Reagan era: From postmodernism to deconstructivism," *assemblage* 8 (1989) pp. 23–59 and Mark Taylor, "Nuclear architecture, or Fabulous architecture, or Tragic architecture, or Dionysian architecture, or Displaced architecture...," *loc. cit.* 11 (1990) pp. 8–21.

history was to recover the idea of the timeless from the idea of change."
Eisenman insisted that these assumptions produced merely "simulations"
(the term is Baudrillard's) precisely because they did not acknowledge their
fictive status; instead, they sought to simulate "reality" or "truth." In their
place, he proposed a radical break, a "not-classical architecture," so to
speak, which would inaugurate architecture as an independent and autonomous discourse "at the intersection of the *meaning-free*, the *arbitrary*, and
the *timeless* in the artificial." This would produce a dissimulation, which
could be read not as an image but as a *writing* and, most importantly, by
somebody bringing to the material "no preconceived knowledge of what
architecture *should* be": in other words, by a reader working outside the
classical fictions. This would involve a "pure reading of traces" in which

Knowing how to decode is no longer important; simply, language in this context
is no longer a code to assign meanings (and *this* means *that*). The activity of reading
is first and foremost in the recognition of something as a language (that *it is*).
Reading, in this sense, makes available a level of *indication* rather than a level of
meaning or expression.

In this way architecture would, for the first time, mark a "timeless" space
in the present "without a determining relation to an ideal future or to an
idealized past."[154] In his later writings – in every sense of that word –
Eisenman considers the possibility of introducing "textual time" into his
work to further subvert the modernist project: "to produce an architecture
which dislocates not only the memory of internal time but all the aspects
of presence, origin, scale and so forth." As I will show in due course, it
is precisely this effort that occasions Jameson's concern, and although he
does not cite Eisenman's project his characterization of postmodern architecture relies on a similar (though markedly critical) disjunctive reading of
the built environment. Indeed, in an open letter to Eisenman, Derrida uses
Benjamin's commentary on the baroque to put a series of questions about
the place of time, memory, or ruin in his architecture. Eisenman's purpose
is still to call into question what is conventionally assumed to be "natural"
or "authentic" by unmasking the conventions that masquerade as truths,
but he now seeks to suspend construction in what, following Derrida, one
might almost call an architecture of *différance*. The intention is to destabilize
the humanist imaginary through a displaced and dislocated architecture that

---

[154] This discussion is derived from Peter Eisenman, "The end of the classical: The end of the
beginning, the beginning of the end," *Perspecta* 21 (1984) pp. 155–72; the quotations are from
pp. 155–56, 166, 171–72. These may seem strange claims to advance but they are not exceptional.
Benhabib, "Epistemologies of postmodernism," uses Eisenman's essay as a summary statement of
"the conceptual self-understanding of postmodernism, not only in architecture but in contemporary
philosophy as well."

would exceed every form and escape all formation, as Taylor puts it, and yet that would also find its place in the *sublime*: a field that now has to be figured through the built environment (I assume) because it is no longer available in our late twentieth-century world as a response to "Nature."[155]

I should perhaps explain that the sublime is of immense importance to postmodern thought, and for that reason will reappear several times in the following discussion, because it marks that moment when we confront something that we are unable to represent as a purposive unity, something that exceeds our capacity for totalization, intuitively or conceptually, and when we are wrenched away from our tranquil contemplation of the world's seemingly obedient regularity. In the sublime, as Eagleton so marvelously puts it, "we are forcibly reminded of the limits of our dwarfish imaginations and admonished that the world as infinite totality is not ours to know."[156] This is the Kantian imaginary, of course, and Kant did not view the sublime as an occasion for despair. His philosophy turned on the essential claim that sublimity is a repudiation of sensuality *not reason* and that our common response to the sublime lies *within ourselves*: that it ushers in what Roberts calls "the material appearance of practical reason in the world."[157] This (modern) path leads back to a stilling of the waters, so to speak, a sort of domestication of dread, whereas the postmodern path leads to a perpetual guerilla warfare in which we "wage a war on totality" and become "witnesses to the unpresentable."[158]

It should now become apparent that I have not only arranged these paragraphs but also indicated these paths in order to illustrate the pivotal position occupied by those who conceive of landscape as text. On the one

---

[155] Peter Eisenman, "Architecture as a second language: The texts of between," *Threshold* 4 (1988) pp. 71–75; *idem*, "The authenticity of difference: Architecture and the crisis of reality," *Center: A Journal for Architecture in America* 4 (1988) pp. 50–57; Jacques Derrida, "A letter to Peter Eisenman," *assemblage* 12 (1990) pp. 71–73. I say "almost" an architecture of *différance* because, although he has collaborated closely with Derrida, Eisenman considers the use of deconstruction in architecture to be problematic. In architecture he argues that there is always a "third term," a condition that he calls "presentness," "that is neither absence nor presence, form nor function, neither the particular use of a sign nor the crude existence of reality, but rather an excessive condition between sign and the Heideggerian notion of being: the form and ordering of the discursive event that is architecture." See Peter Eisenman, "Post/El cards: A reply to Jacques Derrida," *assemblage* 12 (1990) pp. 14–17.

[156] Eagleton, *Ideology of the aesthetic*, p. 89.

[157] Julian Roberts, *German philosophy: An introduction* (Cambridge, England: Polity Press, 1988) p. 63.

[158] Jean-François Lyotard, *The postmodern condition: A report on knowledge* (Minneapolis: University of Minnesota Press, 1984) p. 82. Hence the claim that Kant is "the indispensable reference point for understanding postmodernism": Stanley Rosen, *Hermeneutics as politics* (Oxford: Oxford University Press, 1987) p. 20. Given the present interest in postmodern geographies, it is salutary to realize that traditionally the most influential discussions of Kant in the discipline have relied entirely on his lectures on physical geography; others have widened the debate to consider Kant's *Critique of pure reason* but their purpose has usually been to dissent from the prevailing orthodoxy and their

side they have much in common with those (I think modern) writers who
are drawn to Ricoeur; although they do not always mark the limitations of
his model of the text as carefully as one would wish, they are committed
to a depth hermeneutics, of a sort, which seeks to disclose orders of
meaning. On the other side they reach out, somewhat more cautiously,
toward those (postmodern) writers who use deconstruction and poststructuralism to set in motion a hermeneutics of suspicion and, on occasion,
to conjure with the sublime. If this middle position makes for ambivalence,
however, I also think it allows for domestication, and in what follows I
want to advance that contention by considering not postmodern landscapes
but rather postmodern *readings* of landscapes: that is to say, textualizations.

Rorty thinks there are close affinities between idealism and textualism,
between those nineteenth-century philosophers "who argued that nothing
existed but ideas" and those twentieth-century figures "who write as if
there were nothing but texts." Both groups are antagonistic to the received
model of the natural sciences and both of them turn to literature (and
particularly literary criticism) to advance an altogether different model of
intellectual activity in which language occupies a central place. This is
Rorty's characteristically blunt way of expressing things, of course, and
I am not sure that the linguistic turn is fairly summarized in this way. I
suspect that a series of finer distinctions is necessary to accommodate the
differences between (for example) Ricoeur, Foucault, and Derrida.[159] But it
is the case that many of those who have been drawn to this metaphoric
in human geography have been humanist geographers, of one sort or
another, and that they share a common suspicion of the certainties and
privileges of spatial science and often historico-geographical materialism as
well. Here I will consider the work of James Duncan as a way of bringing
these assumptions and dispositions into focus.

Landscapes of domestication: authority, aesthetics and the text

Traditionally, cultural landscapes were seen as reflections of the peoples
(and specifically the material cultures) that inhabited them: a palimpsest of
the past, a "medal struck in the likeness of a people," whose faint traces

---

arguments have invariably been framed by its closures. It is difficult to see how Kant's philosophy
can be properly understood without consideration of his other two critiques, and yet in geography
the only substantial discussions of the *Critique of judgement* (to which I refer here) are largely
unconcerned with Kant's aesthetics: Clarence Glacken, *Traces on the Rhodian shore* (Berkeley:
University of California Press, 1967) pp. 530–35 and J. A. May, *Kant's concept of geography* (Toronto:
Toronto University Press, 1970) pp. 136–47.

[159] Richard Rorty, "Nineteenth-century idealism and twentieth-century textualism," in his *The
consequences of pragmatism* (Minneapolis: University of Minnesota Press, 1982) pp. 139–59; some of
the distinctions I call for are provided by Alex Callinicos, *Against postmodernism: A Marxist critique*

could be used to mark cultural areas and follow diffusion tracks. But Duncan urges the need to recover layers of meaning lying beyond (or "beneath") those surface remains and relict features. "We need to 'fill in' much that is invisible," he writes, and "to read the subtexts that lie beyond the visible text." This is no casual analogy. Duncan argues that landscape "is one of the central elements in a cultural system, for as an ordered assemblage of objects, a text, it acts as a signifying system through which a social system is communicated, reproduced, experienced and explored." Although Duncan is concerned to distance himself from structural Marxism and poststructuralism, in some respects at any rate (which are all bound in to his continuing commitment to a humanism), his readings are by no means *un*-structured. On the contrary, there is a clear discipline at work and his writings convey a strong sense of uncovering a set of master codes. This is not the result of the peculiarities of what is, on his own admission, the highly (and unusually) textualized landscape of nineteenth-century Kandy that has been one of his main preoccupations; it is a generalizable claim that, although it is articulated *en arrière*, so to speak, is registered in advance.

Landscapes anywhere can be viewed as texts which are constitutive of discursive fields, and thus can be interpreted socio-semiotically in terms of their narrative structure, their synecdoches, and recurrence. This applies as much to late twentieth-century America as it does to early nineteenth-century Kandy. The way that these concepts are articulated with reference to different times and places will, of course, vary greatly. Nevertheless, the thrust of the interpretive method will be the same – to uncover the underlying multivocal codes which make landscapes cultural creations, to show the politics of design and interpretation, and to situate landscape at the heart of the study of social process.[160]

It is an important part of Duncan's project that these codes cannot be collapsed into a single key. Inspired in particular by Barthes, his intention is to treat landscape as an anonymous text and to identify the various textual communities that form around (and jostle around) its (intertextual) interpretation:

A text encourages the reader to carve it up, to rework it, to produce it. Although it cannot mean anything at all [i.e. there are *limits* to interpretative possibility], it is a space in which the reader as writer can wander, in which signifiers play, signifieds becoming signifiers in an endless process of deferment.[161]

(Cambridge, England: Polity Press, 1989) pp. 68–73 and Martin Jay, "The textual approach to intellectual history," *Strategies* 4/5 (1991) pp. 7–18.

[160] James Duncan, *The city as text: The politics of landscape interpretation in the Kandyan kingdom* (Cambridge: Cambridge University Press, 1990) p. 184.

[161] James Duncan and Nancy Duncan, "(Re)reading the landscape" *Environment and Planning D: Society and Space* 6 (1988) pp. 117–26; the quotation is from p. 119.

This is a highly abbreviated summary of a dense and complex account, but I think it is sufficient to reveal two meta-theoretical devices that are instrumental in what I take to be an attempt to domesticate postmodernism: namely interpretive stasis and interpretive privilege. I am aware that these will seem perverse claims to advance. After all, doesn't Duncan constantly accentuate the *contingency* of the given, its constitutive historicity, and seek (as he says) "to decipher society's signs and to reveal the *complexity and instability* beneath the apparent simplicity of the everyday cultural landscape"?[162] And isn't the whole tenor of his work to deny the authority of the author? A landscape may possess "a similar objective fixity to that of a written text" and a symbolic landscape may be created to convey a particular set of meanings, but Duncan insists that, like a literal text "it also becomes detached from the intentions of its original authors, and in terms of social and psychological impact and material consequences, *the various readings of landscapes matter more than any authorial intentions.*"[163] So how can such a textualization possibly contain the play of difference?

Perhaps the clearest way to explain myself is through a consideration of Geertz's cultural anthropology. I don't think much hangs on whether or not Geertz has moved toward postmodernism; some critics have read his recent writings in that way, certainly, but others continue to frame his work within the assumptions of a broadly Parsonian-Weberian tradition. What is of more moment is that Geertz has been of considerable importance to the cultural turn in human geography, and that he is also in many ways the historian's anthropologist: in much the same way, I suspect, that E. P. Thompson is probably the anthropologist's historian. I propose to pursue this comparison as a way of bringing my concerns into focus. Geertz and Thompson are very different in their political inclinations and in many other ways, of course, but both of them are centrally concerned with questions of meaning and with what Geertz often calls "self-fashioning." Yet reading culture-as-text through Geertz's magnifying glass – one which is absent from the desk of Thompson's historian – all too often imposes an uncanny sense of social life suspended like a butterfly between a pair of tweezers: delicate, beautiful, yet profoundly melancholy in its immobility. To put the matter more prosaically, Geertz's critics object that metaphorizing culture as text erases process. "A text is written; it is not writing," one of them argues, so that "to see culture as an ensemble of texts or an art form is to remove culture from the process of its creation." Another

---

[162] James and Nancy Duncan, "Ideology and bliss: Roland Barthes and the secret histories of landscape," in Trevor Barnes and James Duncan (eds.), *Writing worlds: Discourse, text and metaphor in the representation of landscape* (London: Routledge, 1992) pp. 18–37; the quotation is from p. 18 (my emphasis).

[163] Trevor Barnes and James Duncan, "Writing worlds," *loc. cit.*, pp. 1–17; the quotation is from p. 6.

considers Geertz's cultural analysis to be "as static as any structuralism" – a charge that is given a particular force here by the imaginative proximity of Thompson – and again this is attributed to his characteristic emphasis on "the culture, not the history; the text, not the process of textualizing."[164]

This strategy is not peculiar to Geertz, and in fact Mitchell invokes the "world-as-exhibition," the modern process of enframing and representing the world, to suggest that such a metaphoric derives from (and contributes to) a more general constellation of power-knowledge:

> However much the cultural text is said to "find articulation" in "particular performances," it is assumed to enjoy a separate nature as an unphysical "structure" or "frame of meaning." The distinction between particular practices and their structure or frame is problematic not simply because it may not be shared by non-western traditions [in other words, this model of text may itself be a peculiarly Western one] but because ... the apparent existence of such unphysical frameworks or structures is precisely the effect introduced by modern mechanisms of power and it is through this elusive yet powerful effect that modern systems of domination are maintained.[165]

Some of his most trenchant critics argue that in Geertz's case this sort of effect – and the interpretive stasis that it installs – is achieved in very large measure through a mannered preoccupation with the aesthetics of his own textualizations. If there is a cultural politics at work within Geertz's readings, then Schneider has identified its assumptions and consequences with an almost forensic precision:

> [Geertz's] treatment of the "text" [of the Balinese cockfight] seems an example less of thick description than of a particular aesthetic wherein the sublime possibilities always latent in complex cultural practices are activated for the edification to be had. Turning cockfights into stories the Balinese tell themselves about themselves extends the sphere of textuality beyond publicly coded documents, and thus into a realm where cracking the "code" is apt mainly to reveal the sensibility of the cryptographer.... The perception of some meaning that, though only vaguely traceable, appears to knit together disparate orders of experience indeed activates sensations of sublimity, yet the nagging question remains

---

[164] These remarks are taken from William Roseberry, "Balinese cockfights and the seduction of anthropology," in his *Anthropologies and histories: Essays in culture, history and political economy* (New Brunswick, N.J.: Rutgers University Press, 1989) pp. 17–29; the quotation is from p. 24; and Aletta Biersack, "Local knowledge, local history: Geertz and beyond," in Lynn Hunt (ed.), *The new cultural history* (Berkeley: University of California Press, 1989) pp. 72–96; the quotations are from p. 80.

[165] Timothy Mitchell, "Everyday metaphors of power," *Theory and society* 19 (1990) pp. 545–77; the quotation is from p. 561; see also his "The world as exhibition," *Comparative studies in society and history* 31 (1989) pp. 217–36. The ostensible "physicality" of the landscape in Duncan's work does not blunt Mitchell's point, because Duncan's attention is directed toward the "unphysical" signifying structures that underly cultural landscapes.

whether this occurs commonly in Bali or perhaps only in the anthropological analysis. . . .

Generally, infusions of the sublime simply make culture appear as our elite contemplatives would have it, as something rich enough to express a refined sensibility – and textuality has of late been the primary vehicle by which such a sensibility reads itself into the cultural materials at hand. The effect of such a bias is not very fruitfully discussed in the abstract, but it appears part of a broader mode of totalizing discourse through which authority is granted insight in the measure that it masters diversity and constructs coherence. . . . It seems clear that textualizing the object of our inquiry, so that each nonsemantic gesture is presumed to inscribe meanings, will occasionally impute an integrity to culture that it does not necessarily have.[166]

Another way of expressing this objection is to say that Geertz's textualizations are never dialogical; that there is only "an invisible voice of authority who declares what the you-transformed-to-a-they experience."[167] This is plain not only in the stylistic devices and embellishments with which Geertz furnishes his argument – the elaborately staged hesitations, the rhetorical flourishes, the stage-asides: it is not difficult to imagine him performing his compositions – but also in the formal structure of his essays. In "Deep play," for example, Geertz proceeds through what Watson describes as "a carefully managed series of disclosures" designed to show that "what appears to be the case" is merely "the superficial emanation of an underlying reality" to which Geertz, as both auditor and author, has privileged access. And in his company, hushed, we are allowed to eavesdrop – to "listen," as Thompson once instructed his apprentice audience – and be enlightened.[168]

Even if these objections can be sustained, however, the fact remains that Duncan writes of nineteenth-century Kandy, Geertz of nineteenth- and mid-twentieth-century Bali. Surely these hermetic worlds (at least as they are presented in these texts) are far removed from the multiple imbrications of places in the postmodern world? The British occupation of the Lankan coast is not only physically peripheral but also textually peripheral, a dull

---

[166] Mark Schneider, "Culture-as-text in the work of Clifford Geertz," *Theory and society* 16 (1987) pp. 809–39; the quotations are from pp. 820, 829. Schneider is here referring specifically to Geertz's "Deep play: Notes on the Balinese cockfight," in his *The interpretation of cultures* (New York: Basic Books, 1973) pp. 412–53.

[167] Vincent Crapanzo, "Hermes' dilemma: The masking of subversion in ethnographic description," in James Clifford and George Marcus (eds.), *Writing culture: The poetics and politics of ethnography* (Berkeley: University of California Press, 1986) pp. 51–76; the quotation is from p. 74.

[168] Graham Watson, "Definitive Geertz," *Ethnos* 54 (1989) pp. 23–30. The question of whether we are listening to Geertz or the Balinese – or, for that matter, to Thompson or the radical artisan – is by no means easy to answer; see Renato Rosaldo, "Celebrating Thompson's heroes: Social analysis in history and anthropology," in Harvey Kaye and Keith McClelland (eds.), *E. P. Thompson: Critical perspectives* (Cambridge, England: Polity Press, 1990) pp. 103–24.

and distant presence in the margins of Duncan's analysis; the Balinese enact their dramas on a stage from which Geertz has removed most of the distracting clutter of global geopolitics.[169] But to ask the question in these ways is to beg the question. Postmodernism turns not so much on the limits of local knowledge – which, as those last sentences imply, are scarcely confined to the end of the present century – as on the particularity of its response to difference. Ricoeur expressed the matter very well:

When we discover that there are several cultures instead of just one and consequently at the time when we acknowledge the end of a sort of cultural monopoly, be it illusory or real, we are threatened with the destruction of our own discovery. Suddenly it becomes possible that there are just others, that we ourselves are an "other" among others. All meaning and every goal having disappeared, it becomes possible to wander through civilizations as if through vestiges and ruins. The whole of [humankind] becomes an imaginary museum.[170]

What, then, have these textualizations to do with postmodernism and its response to the contemporary disorientation of the West? The answer is, in part, that the strategies I have identified here – the metaphor of text, the flattening of historicity, the deliberate aestheticization – are among those invoked by many postmodern writers as a way of rendering copresence. But the answer is also, and more importantly, that Duncan's cultural geography and Geertz's cultural anthropology are both moments of a cultural criticism that, whatever it may tell us about those people and those places, is also supposed to speak *to us* in *our* clattering variousness. "Although the cosmic symbolism of Kandy may seem radically 'other' to us," Duncan declares, "fundamentally" – *fundamentally* – "it is not different from any other landscape in that it constitutes a text concerning which there is a politics of reading." Within such an interpretive practice the point is not "either to become natives (a compromised word in any case) or to mimic them," Geertz informs us, but to *converse* with them: to enlarge the space of our own intelligibility.[171] The same is true, I think, of Thompson's desire to reach back into the past and shake Swift by the hand; in doing so, as Rosaldo puts it, "the past and the present become contemporaneous."

[169] Cf. Vincent Pecora, "The limits of local knowledge," in H. Aram Veeser (ed.), *The new historicism* (London: Routledge, 1989) pp. 243–76. In a separate essay, Duncan does sketch the symbolic landscapes of colonial and postcolonial Kandy, but in this (I think much more vigorous) narrative the metaphor of landscape-as-text is conspicuous by its absence; see James Duncan, "The power of place in Kandy, Sri Lanka, 1780–1980," in John Agnew and James Duncan (eds.), *The power of place: Bringing together geographical and sociological imaginations* (London: Unwin Hyman, 1989) pp. 185–201.

[170] Paul Ricoeur, *History and truth* (Evanston: Northwestern University Press, 1965) p. 278.

[171] Duncan, *City as text*, p. 7; Clifford Geertz, "Thick description: Toward an interpretive theory of culture," in his *Interpretation of cultures*, pp. 3–30; the quotation is from p. 13.

Similarly, in seeking to rescue the poor stockinger from the condescension of posterity, Thompson is unavoidably setting him to work in *our* present and on *our* behalf.[172]

All of these otherwise different hermeneutics thus circle around a single center: cultures and landscapes — cultural landscapes — are not just random noise, empty signifiers whirling and clashing, forever lost in a past that is always another country but can be made intelligible in ways that enlarge our *own* understandings in our *own* present. If the immodesty of interlocution is inavoidable, then the implications for postcolonialism are obviously extremely troubling. But on this (thoroughly humanist) reading, postmodernism's celebration of difference need not produce a babble of sound or a riot of language-games. Duncan and Geertz assure us, and in a conspicuously therapeutic gesture I think *re*-assure us, that it is possible to *make* sense of those texts, to *make* them mean. Their readings then become more than fictions in the sense of being "something made"; they are also fabulations: moral tales told, as Rorty would say, for our own edification. And their response to the sublime is a thoroughly Kantian one.[173]

If these seem unremarkable objections, as I expect they might, think for a moment of Foucault. The effect of having Duncan or Geertz read these cultural landscapes is to make them strangely familiar: so much so, in fact, that Smith sees in readings of this sort a desire "to claim both the empirical substance of a discourse's object and also the humane mystery and innocence — the cleanliness — of interpretive procedures." In such a project, he continues, one can glimpse the symptoms of a paranoia: "the paranoia of a humanism that wishes to maintain its rights on a reality that it will not yet recognize as its own offspring or construction." Yet Foucault has the quite contrary ability to deliberately make strange our otherwise familiar fictions of the past, and in doing so to raise urgent and unsettling questions about the construction of subjectivities and differences that escape these textualities.[174]

Postmodern cartographies and spatialities

Other versions of postmodern cartography, distant from these humanist conceits, are much more discomforting. Relph suggests that postmodern

---

[172] Thompson, *Poverty of theory*, p. 234; Rosaldo, "Celebrating Thompson's heroes," p. 117; E. P. Thompson, *The making of the English working class* (Harmondsworth, England: Penguin, 1968) p. 13.

[173] Geertz, "Thick description," p. 15. This is cultural criticism standing in for philosophy, of course, and in gesturing to Rorty I am simply following Geertz; see, for example, his *Local knowledge: Further essays in interpretive anthropology* (New York: Basic Books, 1983) pp. 222–25.

[174] Paul Smith, *Discerning the subject* (Minneapolis: University of Minnesota Press, 1988) p. 87. It is only fair to note that Smith is more skeptical of Foucault's project than I am.

spatiality may be captured in Foucault's concept of a heterotopia, which he interprets as "a space with a multitude of localities containing things so different that it is impossible to find a common logic for them, a space in which everything is somehow out of place." In fact, Foucault thought that heterotopias were marginal sites of modernity, constantly threatening to disrupt its closures and certainties, but Relph argues that postmodernity is a *generalization* of the heterotropic:

> Heterotopia is the geography that bears the stamp of our age and our thought – that is to say it is pluralistic, chaotic, designed in detail yet lacking universal foundations or principles, continually changing, linked by centerless flows of information; it is artificial, and marked by deep social inequalities. It renders doubtful most of the conventional ways of thinking about landscapes and geographical patterns. It is also a serious challenge for cartographers.[175]

This sense of discomfiture and disorientation is revealed most clearly in the work of those writers who simultaneously celebrate the pluralities of postmodernism – its multiple voices, myriad subjectivities, many logics – and yet also offer, from within that same intellectual culture, a critique of postmodernity and the postmodern city that sustains a congeries of public and popular cultures: Castells's city of shadows and Soja's Citadel-LA. To Michael Dear, for example, postmodernism is about complication, about enfranchising and empowering those outside the traditional centers of scholastic authority and legitimizing and liberating differences. This is no abstract intellectual culture; it is embedded in practices of everyday life, which through their openness to globalization have produced "an intense localization and fragmentation of social process." Traditional rationalities, and with them established authorities, have withered on the vine; these local autonomies have spawned countless formal and informal, political and para-political alliances located within the interstices of once mainstream constellations of power and knowledge. In late twentieth-century Los Angeles, which Dear (like Soja) thinks may be the quintessential postmodern city, "Angelenos daily invent their city." This may be a testament to human creativity but it is also a matter of practical necessity, a tactic of survival. Dear also portrays the postmodern city as the site of an intensification of

[175] Edward Relph, "Post-modern geographies," *Canadian Geographer* 35 (1991) pp. 98–105; the quotation is from p. 105. See also Michel Foucault, "Of other spaces," *Diacritics* 16 (1986) pp. 22–27. Foucault's discussion of heterotopias is more complicated than Relph allows. Although a heterotopia is indeed "capable of juxtaposing in a single real space ... several sites that are in themselves incompatible," most often in symbolic form, it is not always and everywhere a space of illusion. It may also be what Foucault calls a space of compensation: "a space that is other, another real space, as perfect, as meticulous, as well arranged as ours is messy, ill constructed and jumbled." This is not quite what Relph (or Soja in *Postmodern geographies*) has in mind, but the common thread running through these suggestions, as I have said, is that heterotopias are *counter-sites*.

polarization and privatism, what he once called – in imagery recalling Habermas rather than any postmodern theorist – "a mutant money machine" wired to "a political economy of social dislocation" and driven by "the twin engines of (state) penetration and (corporate) commodification."[176]

The appeal to political economy is not unusual in critical reflections on postmodernity, of course, but what interests me rather more (here) is the fact that so many of these commentaries also turn to psychoanalytic theory to offer a diagnosis of the postmodern. As I will show in a moment, it is variously described in terms of paranoia, schizophrenia, and psychasthenia. Critical theory is clearly no stranger to pschyoanalysis, but this postmodern turn is closely tied to the interrogation of rationality, subjectivity, and identity that takes place under its multiple signs of complication and fragmentation and, in particular, to the construction of new cultural spaces. Much of this discussion is characteristically (and probably appropriately) ambivalent. Gilles Deleuze and Félix Guattari, for example, do not accept the appropriation of their *Capitalism and Schizophrenia* by postmodernism – Guattari describes it as a new conservatism – and yet they are often treated as "exemplary representatives of postmodern positions" in their celebration of difference, deterritorialization, and multiplicity.[177] While Deleuze and Guattari enjoy a creative relationship with historical materialism, I prefer to consider postmodern spatiality through the work of Fredric Jameson because he entertains (in addition) a similarly productive relationship with postmodernism, neither opposition nor affirmation. He clearly finds much that is exciting, even exhilarating within postmodern culture and particularly within the visual arts. Postmodernism is essentially a visual culture, he says, and he likes much of the architecture, photography, film, and video produced under its sign. But his politico-intellectual position within what one might call "late Marxism" – the title of his own discussion of Adorno – means that he also has to turn postmodernism against itself, against the "late capitalism" within which he sees postmodernism inscribed as a cultural dominant and articulated as a cultural logic.[178]

---

[176] Michael Dear, "Postmodernism and planning," *Environment and Planning D: Society and Space* 4 (1986) pp. 367–84; the quotation is from p. 380; *idem*, "Taking Los Angeles seriously: Time and space in the postmodern city," *Architecture California* August 1991 pp. 36–42; the quotations are from pp. 38–41; *idem*, "The premature demise of postmodern urbanism," *Cultural anthropology* 6 (1991) pp. 535–48.

[177] Gilles Deleuze and Félix Guattari, *Capitalism and schizophrenia* was published in two volumes: *Anti-Oedipus* (Minneapolis: University of Minnesota Press, 1983) (originally published in France in 1972) and *A Thousand Plateaus* (Minneapolis: University of Minnesota Press, 1987) (originally published in France in 1980). The postmodern ascription of Deleuze and Guattari is taken from Best and Kellner, *Postmodern theory*, p. 76; see also Brian Massumi, *A user's guide to Capitalism and schizophrenia: Deviations from Deleuze and Guattari* (Cambridge: MIT Press, 1992).

[178] Jameson, *Postmodernism*, pp. 5–6, 46, 298–99; see also his *Late Marxism: Adorno or the persistence of the dialectic* (London: Verso, 1990), especially his closing essay "Adorno in the postmodern" (pp. 229–52).

This deconstructive doubling is symptomatic of Jameson's description of the postmodern condition as a *schizophrenia*. He goes to some considerable trouble to establish that he is only offering a description, not a diagnosis, but in doing so he sanitizes and desexualizes much of the analysis that follows. Although he draws on Lacan's account of paranoia, he pushes the trauma – the terror – of the original discussion to the margins of his own analysis. There are, I think, good reasons for his drawing back from the full implications of Lacan at this point; indeed, Deleuze and Guattari offer their reading of schizophrenia as a critique of Lacanian psychoanalytic theory. But Jameson's unwillingness to pursue the psychosexual implications of Lacan's account is troubling; the return of Lacan's "Name-of-the-Father" is then inscribed in the penetrations of late capitalism without comment.[179] I can only suppose that this happens because Jameson's primary purpose is to reflect on and indeed to recuperate the sense of historicity that has always been central to his own critical practice. He uses Lacan's work to treat schizophrenia as a *language disorder* so that, by analogical projection, he can represent postmodernism as the discursive space produced by the breakdown of the signifying chain into "a rubble of distinct and unrelated signifiers." If the experience of temporality is inscribed within language – if "it is because language has a past and a future, because the sentence moves in time, that we can have what seems to us a concrete or lived experience of time" – it follows that a breakdown in the movement from one signifier to another will also mark a breakdown of temporality itself.[180] This speaks directly to Eisenman's proposal for a postmodern architecture and its "timeless space" (above, p. 142), and in doing so identifies what I take to be Jameson's essential agony, for such a project interrupts – indeed ruptures – the historicity inscribed at the heart of historical materialism.

This postmodern rupture is supposed to have two moments. On the one side, and most obviously, Jameson identifies postmodernism with a new *synchronicity*. He believes that the postmodern has issued in "a culture increasingly dominated by space and spatial logic," one in which the past as referent finds itself gradually bracketed and then effaced altogether. This process of erasure installs a profound depthlessness, mirrored in the play of multiple surfaces within postmodern architecture and the other visual and textual arts. The image is real, and Jameson takes as his example the freestanding wall of Wells Fargo Court in downtown Los Angeles, a surface which seems to be unsupported by any volume.

---

[179] Jacqueline Rose, "*The man who mistook his wife for a hat* or *A wife is like an umbrella* – fantasies of the modern and postmodern," in Andrew Ross (ed.), *Universal abandon? The politics of postmodernism* (Minneapolis: University of Minnesota Press, 1988) pp. 237–50; Samuel Weber, *Institution and interpretation* (Minneapolis: University of Minnesota Press, 1987) pp. 54–58.

[180] Jameson, *Postmodernism*, p. 27; idem, "Postmodernism and consumer society," in Hal Foster (ed.), *The anti-aesthetic: Essays on postmodern culture* (Seattle: Bay Press, 1983) pp. 111–25. Jameson also invokes Lacan in his discussion of cognitive mapping; *Postmodernism*, pp. 51–54.

This great sheet of windows, with its gravity-defying two-dimensionality, momentarily transforms the solid ground on which we stand into the contents of a stereopticon, pasteboard shapes profiling themselves here and there around us.[181]

This is more than *trompe l'oeil*. Olalquiaga, following Jameson, diagnoses this postmodern disorientation as a generalized form of what Caillois calls *psychasthenia*: "a state in which the space defined by the coordinates of the organism's own body is confused with represented space." By extension, she suggests that postmodern urban culture resembles this condition "when it enables a ubiquitous feeling of being in all places while not really being anywhere": the experiential response to what Sorkin calls "a universal particular, a generic urbanism inflected only by appliqué," in which space has been departicularized and postmodern spatialities are reproduced in endlessly repetitive a-geographic simulacra.[182] For Olalquiaga, as for Jameson, architecture is the canonical text and the privileged site of the postmodern:

Casting a hologram-like aesthetic, contemporary architecture displays an urban continuum where buildings are seen to disappear behind reflections of the sky or merge into one another, as in the downtown areas of most cosmopolitan cities and in the trademark midtown landscape of New York City. Any sense of freedom gained by the absence of clearly marked boundaries, however, is soon lost to the reproduction ad infinitum of space – a hall of mirrors in which passersby are dizzied into oblivion. Instead of establishing coordinates from a fixed reference point, contemporary architecture fills the referential crash with repetition, substituting for location an obsessive duplication of the same scenario.[183]

The epitome of this sense of bewilderment and disorientation is, of course, Jameson's own celebrated odyssey through (but never quite out of) the Bonaventure Hotel in downtown Los Angeles (figure 15).[184]

---

[181] *Ibid.*, p. 13.

[182] Celeste Olalquiaga, *Megalopolis: Contemporary cultural sensibilities* (Minneapolis: University of Minnesota Press, 1992) pp. 1–5, 17–18; Roger Caillois, "Mimicry and legendary psychasthenia," *October* 31 (1984) pp. 16–32; Michael Sorkin, "Introduction" in idem, *Variations on a theme park*, p. xiii.

[183] Olalquiaga, *Megalopolis*, p. 2.

[184] Jameson, "Cultural logic"; see also Donald Preziosi, "*La vi(ll)e en rose:* Reading Jameson mapping space," *Strategies* 1 (1988) pp. 82–99, which suggests that Jameson's disorientation is a generic characterization of his work: "a case of the analyst losing his place amidst the scenographies generated by his analysand" (p. 83). For a deliberate echo of Jameson, see Jean Baudrillard's account of the Bonaventure in his *America* (London: Verso, 1988) p. 97. But Baudrillard treats the Bonaventure as the epitome of radical modernism, and other critics agree that it is late modern – not postmodern. Many protagonists of postmodernism have seized on this as a rod to beat Jameson, but in doing so they have failed to consider how Jameson conceives of the relation between modernism and postmodernism. More importantly, they have fallen into the trap of misplaced concreteness by ignoring the possibility that Jameson is not (and knows he is not) describing a postmodern building – he concedes that it is in many ways uncharacteristic of postmodern architecture – but rather, as I will seek to show in a moment, evoking a postmodern *response*.

Figure 15 *The Bonaventure Hotel, Los Angeles [photograph: Michael Dear, by permission].*

On the other side, and more covertly, Jameson envisages postmodernism as a new *constellation* in something very like the original Benjaminian sense: a moment of explosive arrest, as it were, in which the possibilities of the present are thrown into frozen relief.

[Postmodernism] suddenly releases this present of time from all the activities and intentionalities that might focus it and make it a space of praxis; thereby isolated, that present suddenly engulfs the subject with undescribable vividness, a materiality of perception properly overwhelming, which effectively dramatizes the power of the material – or better still, the literal – signifier in isolation.

Jameson argues that this "schizophrenic disjunction or *écriture*, when it becomes generalized as a cultural style, ceases to entertain a necessary relationship to the morbid content we associate with terms like schizophrenia and becomes available for more joyous intensities."[185] There is then the possibility of a radical political response to late capitalism *within the postmodern*, which I imagine is why Jameson thinks it possible to turn postmodernism against itself – why he refuses the clinical force of Lacan's original account – and why the postmodern effacement of the divide between high and popular culture is so important for his vision of a radical democratic politics.

Like Deleuze and Guattari, Jameson insists that he is not using the term in any clinical sense. While this may be a clever way to construct the argument, there is also something cruel – at the very least insensitive – about analogizing schizophrenia like this. All the same, if this reading invokes a redemptive moment in the postmodern, as I think it is intended to do, then it owes as much to Baudrillard as it does to Benjamin. For taken together, Jameson argues that these two moments are punctuations in a postmodern *hyperspace* that dramatically exceeds our existing capacities for representation and figuration. It is no longer possible to appeal to the icons of previous modernisms – "not the turbine, nor even Sheeler's grain elevators or smokestacks, not the baroque elaboration of pipes and conveyor belts, nor even the streamlined profile of the railroad train" – because this new space is overwhelmingly a space of fiber optics and video screens, calling up, dissolving, and (re)producing a world of traces in nanoseconds. Postmodern architecture assumes its significance precisely because it offers

---

[185] Jameson, *Postmodernism*, pp. 27–29. Cf. Deleuze and Guattari, *Capitalism and schizophrenia*, who treat "the schizophrenic process" of deterritorialization – freeing the flows of desire within the social machine – as a process of liberation. "For Deleuze and Guattari, the schizophrenic process is the basis for a postmodern emancipation, which is to say, an emancipation from the normalized subjectivities of modernity, and they see the schizo-subject as the real subversive force in capitalism, 'its inherent tendency brought to fulfillment, its surplus product, its proletariat, its exterminating angel.'" See Best and Kellner, *Postmodern theory*, p. 90; Deleuze and Guattari, *Anti-Oedipus*, p. 35. But I doubt that their "angel" is Benjamin's (or rather Klee's) *Angelus Novus*, the Angel of History, whose face is turned toward the past, staring at "one single catastrophe that keeps piling wreckage on wreckage and hurls it in front of his feet"; see Gershom Scholem, "Walter Benjamin and his Angel," in Gary Smith (ed.), *On Walter Benjamin: Critical essays and reflections* (Cambridge: MIT Press, 1988) pp. 51–89. Deleuze and Guattari seem to me to have a conception of historicity that is radically different from both Benjamin and Jameson.

a visible approximation of this new spatiality. Many other cultural critics have used architecture in this way, of course, both to celebrate and to demolish the postmodern, and whether it can properly service all the critical demands that are placed upon it is problematic. Postmodern sensibilities in architecture do not mimic those in other spheres, and in many cases the recourse to architecture seems to involve what Rose criticizes as "an unexamined conceptuality without the labor of the concept." But Jameson's argument is an exception, it seems to me, because it turns so fully on the claim that architecture provides a "representational shorthand" capable of evoking – experientially as much as physically – what he calls a *technological sublime*. In the Bonaventure Hotel, for example, one enters – both synchronically and (so to speak) synecdochically – a vast, decentered hyperspace, one that involves

the suppression of distance (in the sense of Benjamin's aura) and the relentless saturation of any remaining voids and empty places, to the point where the postmodern body . . . is now exposed to a perceptual barrage of immediacy from which all sheltering layers and intervening mediations have been removed.[186]

And it is for *this* reason that "Benjamin's account of Baudelaire, and of the emergence of modernism from a new experience of city technology which transcends all the older habits of bodily perception, is [for Jameson] both singularly relevant and singularly antiquated."[187] It is relevant for its sense of disorientation and relentless determination to press against the limits of the discursive; but antiquated because it is simply unable to figure what has become, in Baudrillard's eyes too "a new form of schizophrenia," a space of "immanent promiscuity" and "continual connection," saturated with "the unclean promiscuity of everything which touches, invests and penetrates without resistance, with no halo of private protection, not even [one's] own body." Within the postmodern aesthetic, Jameson argues, "the representation of space has come to be felt as incompatible with the representation of the body: a kind of aesthetic division of labour far more pronounced than in any of the earlier generic conceptions of landscape."[188]

Bodies and spaces

This state of affairs was prefigured by Lefebvre, whose work has been important to both Baudrillard and Jameson, but who was more emphatic about the implicit gendering of that "body" and more circumspect about

---

[186] *Ibid.*, pp. 37–38, 413; cf. Gillian Rose, "Architecture to philosophy – the postmodern complicity," *Theory, culture and society* 5 (1988) pp. 357–71.

[187] Jameson, *Postmodernism*, p. 45.

[188] Jean Baudrillard, "The ecstasy of communication," in Foster (ed.), *Anti-aesthetic*, pp. 126–34; Jameson, *Postmodernism*, p. 34.

the seemingly casual sexualization of its "promiscuities" and "penetrations" than either of them. Lefebvre identified from within the genealogy of modernity the immanent production of a supremely abstract, hyper-rationalized space that was fast becoming decorporealized through a tripartite constellation of geometric-visual-phallic power. He also foreshadowed the emergence of a pervasive "urban society" in which cities as discrete physical forms would lose all definition and in which, without a radical urban revolution, abstract space would be hypostatized.[189] In one of his recent essays on "exopolis" – "the city without" – Soja suggests that this process of de-definition is already far advanced in southern California, which he represents as the generic postmodern landscape, the simulacra of all our futures. In his eyes it is also, for the most part, a landscape without figures. The anonymous voices of advertisers and PR agents, the knowing voices of fashionable critics and commentators, and the disembodied voice of former president Richard Nixon saturate this hyperspace so completely that one wonders whether the emphasis should not fall on the *hype*. From Soja's perspective, these various disorientations and decorporealizations are no longer confined to discrete heterotopia, as they were within modernity; in their place, and as Relph suggests, postmodernity has *itself* become heterotropic.

There has been a second wave that has carried hyperreality out of the localized enclosures and tightly bounded rationality of the old theme parks and into the geographies and biographies of everyday life, into the very fabric and fabrication of exopolis. Today the simulations of Disneyland seem almost folkloric, crusty incunabula of a passing era. The rest of Orange County is leaving these absolutely fake cities behind, creating new magical enclosures for the absolutely fantastic reproduction of the totally real.

Something new is being born here, something that slips free of our old categories and stereotypes, resists conventional modes of explanation and befuddles long-established strategies for political reaction. The exopolis demands more serious attention, for it is fast becoming the nexus of contemporary life – the only remaining primitive society, Jean Baudrillard calls it: a primitive society of the future.[190]

In much the same way, Zukin suggests that these new cultural landscapes mark the sites of emergence of a now pervasive condition of liminality. She represents postmodernity as a landscape of indeterminacy,

open to everyone's experience yet not easily understood without a guide. Defining the symbolic geography of a city or region, liminal spaces cross and combine the

---

[189] See Chapter 6, this volume.

[190] Edward Soja, "Inside exopolis: Scenes from Orange County," in Sorkin (ed.), *Variations on a theme park*, pp. 94–122; the quotation is from p. 101.

influence of major institutions: public and private, culture and economy, market and place. As the social meaning of such places is renegotiated by structural change and individual action, liminal space becomes a metaphor for the extensive reordering by which markets, in our time, encroach upon place.[191]

I juxtapose these two readings for several reasons. They both raise the same question: What possible critical response can there be to such configurations? In their not so different ways, both Soja and Zukin are calling for new cognitive maps, what Soja calls a "different cartography of power." This new world – always in the New World – seemingly "slips free of our old categories" and is apparently "not easily understood without a guide." If these phrases are not a restatement of the privileges of the universal intellectual, then what politico-intellectual response can they offer beyond an excavation of the so-called "industrial base," the productive periphery that "undergirds every exopolis," and a renewed and presumably collective resistance to the encroachments of the market? It seems clear that an unreconstructed historical materialism can offer no solution. Both Soja and Zukin insist on the spatiality of these transitions in ways that transcend the classical problematic and also what I take to be Jameson's brilliantly imaginative reinstatement of it.

For all his interest in the spatiality of postmodern culture, Jameson seems strangely reluctant to pursue the implications of his own inquiries. At one point, apparently following a suggestion of Baudrillard's, he worries that the strategy of cognitive mapping might respatialize "something we were supposed to think of in a different manner altogether." His original intention, so he now says, was to displace the geographical figure and transcend the limits of mapping by introducing "cognitive mapping" as a code word for a new kind of class consciousness.[192] Not surprisingly, neither Soja nor Zukin seems to have got the joke; but in their odysseys through postmodern spaces and over postmodern landscapes they have also – and less accountably – lost sight of Lefebvre's defiant insistence on the body as the site of resistance. "The body, at the very heart of space and of the discourse of power, is irreducible and subversive."[193] I think Lefebvre's proposition is worth exploring, in spirit at least, because it suggests a reformulation of our critical response: How is it possible to conceive of the relations between bodies and spatialities within these new configurations? Olalquiaga sketches the beginnings of an answer when she sees in these surfaces of emergence a substitution of intertextuality for indexicality that dissolves one's sense of bodily identity and biography:

---

[191] Zukin, *Landscapes of power*, p. 269.

[192] Jameson, *Postmodernism*, pp. 416–18.

[193] Henri Lefebvre, *The survival of capitalism: Reproduction of the relations of production* (London: Allison and Busby, 1976) p. 89.

Bodies are becoming like cities, their temporal coordinates transformed into spatial ones. In a poetic condensation, history has been replaced by geography, stories by maps, memories by scenarios. We no longer perceive ourselves as continuity but as location. ... It is no longer possible to be rooted in history. Instead, we are connected to the topography of computer screens and video monitors. These give us the language and images that we require to reach others and see ourselves.[194]

In such circumstances, if we even approximately recognize ourselves in the description, a passionate appeal to historicity is unlikely to be enough. Olalquiaga presents two options: either contemporary identity "can opt for a psychasthenic dissolution into space," affixing itself to any scenario by a change of costume, or "it can profit from the crossing of boundaries, turning the psychasthenic process around before its final thrust into emptiness, benefiting from its expanded boundaries."[195]

Let me pursue that second option – which, in different forms, recurs in a number of texts that straddle the seam between feminism, postmodernism, and poststructuralism[196] – through a consideration of the work of Donna Haraway.

Landscapes of technoculture: guerillas, cyborgs, and insurgency

Haraway's "cyborg manifesto" is written across that same topography of computer screens and video monitors and, like Olalquiaga, she thinks that the technologization of everyday life is not inherently conservative: that it is possible to enter into something approaching Jameson's technological sublime, to trace its monsters and grotesques through their lairs, and to turn technology around so that conventional hierarchies and enclosures are called into question and new, affirmative cultural practices activated.[197] This is a captivating vision, and Haraway's original essay is by turns dazzling and disturbing, playful and passionate, but above all (I think) hopeful. Her images are arresting, and necessarily so: "If we are imprisoned by language," she declares in what must surely be a deliberate reference to Jameson, "escape from that prison-house requires language poets, a kind of cultural

---

[194] Olalquiaga, *Megalopolis*, p. 93.

[195] *Ibid.*, p. 17.

[196] See, for example, Rosalyn Diprose and Robyn Ferrell (eds.), *Cartographies: Poststructuralism and the mapping of bodies and spaces* (Sydney: Allen and Unwin, 1991).

[197] The composite in this paragraph and the next is drawn from Donna Haraway, "A cyborg manifesto: Science, technology and socialist-feminism in the late twentieth century" and "The biopolitics of postmodern bodies: Constitutions of self in immune system discourse," both in her *Simians, cyborgs and women: The reinvention of nature* (London: Routledge, 1991) pp. 149–81 and pp. 203–30; and from *idem*, "The promises of monsters: A regenerative politics for inappropriate/d others," in Grossberg, Nelson and Treichler (eds.) *Cultural studies*, pp. 292–337.

restriction enzyme to cut the code."[198] This sensitivity to the creative capacities of language and the poetic ability to imaginatively "undo" the present means that I must stay close to her own words in presenting her argument. But I will stay close for another reason too, for in many ways Haraway is a more confident guide than Jameson. His bewilderment in the Bonaventure was not entirely gestural, I think, whereas Haraway seems able to find her way through this strange landscape with great facility, moving like a guerilla rather than a surveyor within the hyperspaces of its new technoculture. As her preferred escape route from the prison-house of language implies, her cultural politics confronts science directly (not obliquely), as both culture and colonization. In one sense, no doubt, this is unexceptional:

Science remains an implicit genre of Western exploration and travel literature.... Science as heroic quest and as erotic technique applied to the body of nature are utterly conventional figures.[199]

So they are. But "nature" is no longer conventional: or, rather, it is no longer conventional in quite that (or only that) way. In our late twentieth-century world "nature and culture are reworked," Haraway argues, and "the one can no longer be the resource for appropriation or incorporation by the other." Instead, "nature for us is made": What we call "Nature" is "a co-construction among humans and nonhumans."[200] She is very well aware that this dislocates our conventional categories and disrupts our usual distinctions, but that is precisely the point. The world has turned, and to indicate the new (techno-) culture of nature, if I may so call it, Haraway provides a summary listing, of which I provide an extract here:

Representation	Simulation
Reproduction	Replication·
Physiology	Communications engineering
Depth	Surface
Nature/culture	Fields of emergence

It is easy to see from this – the hypostatization of expert systems? – how the time-space coordinates of French modern and of the Western political imaginary, marshalled in the column on the left, have given way to the

---

[198] *Idem, Simians*, p. 245n; the allusion is to Fredric Jameson, *The prison-house of language: A critical account of structuralism and Russian formalism* (Princeton: Princeton University Press, 1972).

[199] Haraway, *Simians*, p. 205; see also pp. 221–23.

[200] *Ibid.*, p. 151; *idem*, "Monsters," p. 297. See also Alexander Wilson, *The culture of nature: North American landscape from Disney to the Exxon Valdez* (Toronto: Between the Lines, 1991) and Paul Rabinow, "Artificiality and enlightenment: From sociobiology to biosociality," in Jonathan Crary and Sanford Kwinter (eds.), *Incorporations [Zone, 6]* (New York: Zone, 1992) pp. 234–52.

new grids of postmodernism set out in the column on the right.[201] But I find Haraway's work interesting for several other reasons too, some of which I'm not quite sure about but one of which has to do with the way in which she moves between the conventional enframing of science-as-exhibition to a tactic of inscription that interrupts and dissolves that quite extraordinary epistemological gesture. Given the preceding discussion, I have most particularly in mind her view of bodies as "maps of power and identity." In the manifesto she presents the cyborg as a cybernetic organism, a creature simultaneously machine and organism, populating worlds ambiguously "natural" and "cultural." As such, it is at once a creature of social reality – "a kind of disassembled and reassembled, postmodern collective and personal self" – and also a creature of fiction: "a fiction mapping our bodily and social reality."[202]

But it is a deeply serious fiction; Haraway regards her imaginative explorations as forms of political-scientific-fictional analysis. From one perspective, she admits, "a cyborg world is about the final imposition of a grid of control on the planet, about the final abstraction embodied in a Star Wars apocalypse waged in the name of defense, about the final appropriation of women's bodies in a masculinist orgy of war"; it is the terminal stage of biopolitics whose possibility has not, I think, been sensibly diminished by the continued imperialisms of the so-called "new world order." Yet from another perspective Haraway holds out the (utopian) prospect that "a cyborg world might be about lived social and bodily realities in which people are not afraid of their joint kinship with animals and machines, not afraid of permanently partial perspectives and contradictory standpoints."[203] This is a promissory note, Haraway concedes, but the technology of the self that it invokes is not entirely implausible. She is particularly encouraged by the experience of "those [women of color] refused stable identity in the social categories of race, sex or class" who, both in spite of *and* by virtue of those refusals, have been able to fashion a political identity that is personally empowering, "a kind of postmodernist identity [constructed] out of otherness, difference and specificity." In order to convey the profusion of spaces and identities that such a vision makes possible, Haraway invokes the image of "the integrated circuit," which is meant to conjure up those polymorphous global networks, structured through the social relations of science and technology, which disperse and interface social locations in myriad ways.[204] This *is* a cognitive mapping, I suggest, but one which, *contra* Jameson, is clearly *not* a synonym for class consciousness. It is within this circuit (yet against its polarities) that Haraway wants boundaries in the personal body and the body politic to be dissolved

---

[201] Haraway, *Simians*, pp. 161–62, 209–10. It should be noted that Haraway is uneasy in talking about the modern and the postmodern, and in her later essay insists that the cyborg is neither modern nor postmodern but rather *a*-modern: "Monsters," pp. 299, 329–30n.

[202] *Idem, Simians*, pp. 149–50.   [203] *Ibid.*, p. 154.   [204] *Ibid.*, pp. 161–73.

so that new, affirmative cyborg identities can be created. To describe what she has in mind, she borrows the word "inappropriate/d" from Trin Minh-ha. It means, she says, "not to fit into the *taxon*, to be dislocated from the available maps specifying kinds of actors and kinds of narratives, not to be originally fixed by difference." In Haraway's view, this implies a model of *diffraction*, a mapping not of replication, reflection, or reproduction – the usual codings – but of *interference*, in which boundaries take provisional, never fixed, never finished shape and in which all sorts of fusions become possible, in which all sorts of monsters are loosed.[205]

This model of diffraction suggests another parallel to me, but one which is not altogether estranged from Trinh Minh-ha's postcolonial problematic. For the refusal to be appropriate(d) is also a refusal to "know one's place," at least as it is assigned from a superordinate position, and de Certeau has described the practices that are embedded within such a refusal as a *tactic* that is typically "marginalized by the Western form of rationality" and its constitutive strategies of power-knowledge.

[A tactic is] a calculated action determined by the absence of a proper locus. No delimitation of an exteriority, then, provides it with the condition necessary for autonomy. The space of a tactic is the space of the other. Thus it must play on and with a terrain imposed on it and organized by the law of a foreign power. It does not have the means to keep to itself, at a distance, in a position of withdrawal, foresight and self-collection: it is a maneuver "within the enemy's field of vision," as von Bülow put it, and within enemy territory. It does not therefore have the options of planning general strategy and viewing the adversary as a whole within a dist[inct], visible and objectifiable space. It operates in isolated actions, blow by blow. It takes advantage of "opportunities" and depends on them, being without any base where it could stockpile its winnings, build up its own position, and plan raids. . . . [It has] a mobility that must accept the chance offerings of the moment, and seize on the wing the possibilities that offer themselves at any given moment.[206]

I will want to return to these ideas – and in particular the distinction between a strategy and a tactic – in due course, but for now I hope that this metaphoric of insurgency will clarify my earlier description of Haraway as a guerilla rather than a surveyor.

Haraway is not the first person to write about the radical possibilities of technology, nor even the first to write about the radical possibilities of this technology; but she may well be the first to write about them in quite this way. There are, I realize, distant parallels in human geography. But

---

[205] *Idem*, "Monsters," pp. 299–300; Trinh T. Minh-ha, *Woman, native, other: Writing postcoloniality and feminism* (Bloomington: Indiana University Press, 1989). How different this is from the taxonomies and maps of natural history and its successor projects (above, p. 23–26 and *passim*)!

[206] Michel de Certeau, *The practice of everyday life* (Berkeley: University of California Press, 1984) pp. 36–37, 50.

when one compares Haraway's cyborg world with Harvey's reading of *Blade Runner* (a film that she also invokes, though to rather different ends), or her account of what she also calls "the production of nature" with Smith's much earlier but still highly provocative discussion of uneven development, it becomes clear that her critical appropriations of poststructuralism, and particularly the work of Derrida and Foucault, place her at a considerable distance from the classical canon of historical materialism.[207] That in itself is important; yet it is also where I am most hesitant about the implications of her presentation (and for reasons other than politico-intellectual fideism). Part of my unease arises from the sheer suasive force of her insistence on the constructedness and conventionality of concepts of "nature" or "the body": her brilliant demonstration that these are fictions in exactly the sense that Geertz and others use the term, that they are indeed "something made," and that their seemingly incontrovertible boundedness, far from being grounded by modern science has in fact been *subverted* by biotechnology. These are powerful claims that need close attention; but my fear is that, through their fictional tropes, they license too expansive a sense of political possibility: a worry that Susan Bordo identifies very clearly when she talks of deconstruction slipping into its own fantasy of "escape from human locatedness – by supposing that the critic can become wholly protean by adopting endlessly shifting, seemingly inexhaustible vantage points, none of which are 'owned' by either the critic or the other." She brings these concerns to bear directly on Haraway's text by suggesting that "the spirit of epistemological *jouissance*" which is suggested by the images of cyborg and trickster and the metaphors of dance and so forth unfortunately obscures "the located, limited, inescapably partial and always personally invested nature of human 'story making.'"

To deny the unity and stability of identity is one thing. The epistemological fantasy of becoming multiplicity – the dream of limitless multiple embodiments – is another. What sort of body is it that is free to change its shape and location at will, that can become anyone and travel anywhere?[208]

Bordo's point is that there is still a materiality involved in the relations between bodies and spaces that slides away from Haraway's screens. This overlaps with my second source of unease, which arises from her failure to capture the geographies of the cyborg world. The society-technology interface that she describes is constituted differently in different places; the subject-positions that might become available are not freely transposable around the integrated circuit; the idealized locations that she

---

[207] Harvey, *Condition of postmodernity*, pp. 308–14; Smith, *Uneven development*, pp. 32–65.
[208] Susan Bordo, "Feminism, postmodernism and gender scepticism," in Nicholson (ed.), *Feminism/postmodernism*, pp. 133–56; the quotations are from pp. 142–45.

identifies are not each implicated within the other: There are variable and varying degrees of implication. It is surely not necessary to elaborate these concerns to see that they all circle around a single troubling question: For all the appeals to Spivak, Trinh Minh-ha and others, is the cyborg myth a First World fantasy? The electronic and biotechnological freedoms that Haraway anticipates are withheld from many people in many places, and the high technology that she invokes in her deconstruction is disproportionately concentrated in the North/First World. Haraway knows all this, of course, and her brilliant reconstruction of *Primate Visions* provides a vivid record of the complicity of science and technology in the cultural-intellectual colonizations and appropriations of the West. But her manifesto, written out of and against that constellation, cannot escape its dispositions by a wilful act of reversal, however defiant or spirited. I share the desire to dissolve "the 'West' and its highest product – the one who is not animal, barbarian, or woman; man, that is, the author of a cosmos called [H]istory." I also think (and hope) that "as orientalism is deconstructed politically and semiotically, the identities of the occident destabilize."[209] But I am not so sure that the cyborg manifesto, for all its utopian passion, manages to transgress the powers of orientalism. Neither, I suspect, is Haraway. She now wonders whether there could be "a family of figures who would populate our imagination of these postcolonial, postmodern worlds that would not be quite as imperializing in terms of a single figuration of identity?"[210]

Reverse mappings: postcolonialism and time-space disorientation

Which brings me to another reason for juxtaposing Soja and Zukin (above, p. 159). Soja borrows Baudrillard's description of late twentieth-century America as "the only remaining primitive society" – in Baudrillard's view, "the same mythical and analytic excitement that made us look toward those earlier societies today impels us to look in the direction of America" – and Zukin takes the concept of liminality from Turner's seminal studies of African tribal cultures in order to suggest the pervasive ambivalence of postmodernity in the West.[211] These are not uncontroversial descriptions,

[209] Donna Haraway, *Primate visions: Gender, race and nature in the world of modern science* (New York: Routledge, 1989); *idem*, "Manifesto," p. 156.

[210] Constance Penley and Andrew Ross, "Cyborgs at large: Interview with Donna Haraway," *Social text* 25–26 (1991) pp. 8–23; the quotation is from p. 18.

[211] Jean Baudrillard, *America* (London: Verso, 1988) pp. 7 and 29; Victor Turner, *The ritual process: Structure and anti-structure* (London: Routledge and Kegan Paul, 1969). Although Turner constructed the concept of liminality from his studies of the Ndembu, he did not confine it to African tribal culture but also used it to reflect on (for example) medieval Europe and the counterculture of the West in the 1960s.

but I want to suggest that these "gifts" – or, more exactly, the system of exchange and appropriation in which they are embedded – should receive careful scrutiny. Let me explain what I have in mind. In a vigorous discussion of what he calls "the Jameson raid," Young suggests that postmodernism's "loss of the old imperial maps," of which both Soja and Zukin make so much, might be the condition not just of late capitalism, as Jameson argues, "but also of the loss of Eurocentrism, the loss of 'History' as such." In a glorious irony, the history (or rather History) that was denied to colonized peoples in the very act of colonization – those "people without history" – is now finally taken from the colonizers too. "In Jameson's terms, postmodernism would then be Orientalism's dialectical reversal: a state of dis-orientation."[212] I have already indicated my basic agreement with this position, which I think is also Haraway's; but I also think these cartographic metaphors should be taken literally. For, as I now want to suggest, it is also the West's *geography* (or rather Geography) that has been seized, interrupted, and confounded.

If this claim can be sustained, then a critical reading of Western responses to what, following Harvey and Giddens, one might call time-space disorientation will evidently require an examination of postcolonialism. A concern with the figures of place and displacement is a leitmotif of all postcolonial literatures, but I should say at once that I did not come to this somewhat diffuse body of work through these issues. My own understanding of the postcolonial critique remains deeply partial in ways that I need to declare in advance. My intellectual interest was aroused most immediately by Robert Young's *White mythologies: Writing history and the West*, a brilliant book in so many ways and one that has already informed much of my argument; but also one that – despite its subtitle – conspicuously fails to engage with the writings of historians. Young explains that his primary interest is in the sphere of literary theory and cultural criticism, and in the way in which poststructuralism has reformulated the question of "history" to confront the Eurocentrism of classical (and indeed Western) Marxism: perhaps even, through Derrida's deconstruction, to set in train a process of "decolonization" of European thought.[213] But turning a deaf ear to discussions beyond those rather formal boundaries seriously weakens his reading of Spivak in particular (whose work I shall have occasion to discuss in some detail). It is true that Spivak translated Derrida's *Of grammatology* into English, but she has a much wider appreciation of the contemporary crisis of historicity than such a project might seem to suggest.

---

[212] Young, *White mythologies*, p. 117; see also Dipesh Chakrabarty, "The death of history? Historical consciousness and the culture of late capitalism," *Public culture* 4 (1992) pp. 47–65.

[213] Young, *White mythologies*, pp. vi–vii, 18–19 and *passim*. See also Derek Attridge, Geoff Bennington, and Robert Young (eds.), *Poststructuralism and the question of history* (Cambridge: Cambridge University Press, 1987).

Although she is not an historian herself, some of Spivak's most important interventions in politico-intellectual debates – interventions forming if not the culmination then at least the temporary terminus of Young's own argument[214] – simply cannot be understood without an engagement with the work of Indian historians.

I will attend to the ambiguities of that phrase – "Indian historians" – in a moment. But I must also note that my own work has, for the most part, been concerned with the historical geography of Europe. While I have tried to remedy my ignorance of the rest of the world (most of the world) by reading the work of historians, geographers, and others working in those fields, I have not worked in them myself. That is not disabling, I hope, but it is limiting. I should say too that I began with the work of one particular group of historians, whose work goes under the collective title of *Subaltern Studies*, and their reconstructions soon made it clear that postcolonial discourse is itself what Bhabha would call a "hybrid."[215] Those "Indian historians" work on India; they are by no means all working in India and neither are they all Indian. Far from turning their backs on the West, their writings emerge at a conjunction of multiple, often competing· discourses – including Marxism, deconstruction, and, on occasion, feminism – and they are given a common focus through a deliberately loose, sometimes agonistic articulation of a particular conception of Indian history. In their texts, "India" becomes a discursive site into which concepts and constructions are shipped and unpacked, in which others are created and refined, and in which all of them are reworked and then, evidently, exported. Hybridity, then; but the subaltern historians would be the first to urge caution in transferring their ideas to other settings and situations. It should not be necessary to underscore the complex (and continuing) historical geographies of colonialism. Living in Canada, where the "two (European) solitudes" are at last slowly attending to the rights of native peoples, I have been made aware of the general importance of many of these ideas, but aware too, that they do not come ready-made. In short, it is necessary to avoid turning subaltern studies into a paradigm for post-colonialism. Bhabha sharpens the point when he suggests that precisely because postcolonialism has no single paradigm or text this allows for a particularly creative and intimately political form of *communality*. It is that possibility which will guide my own discussion of post-colonialism.[216]

---

[214] My hesitation here is the result of Young's insistence that the readings he offers in *White mythologies* do not constitute "a succession of analyses that move toward the present, demonstrating the gradual production of an adequate theory" (p. vi) and my own, contrary sense of a powerful linearity underwriting his own narrative. Spivak is the subject of the final chapter.

[215] Homi Bhabha, "Signs taken for wonders: Questions of ambivalence and authority under a tree outside Delhi, May 1817," *Critical inquiry* 12 (1985) pp. 144–65.

[216] Homi Bhabha (with David Bennett and Terry Collits), "The postcolonial critic," *Arena* 96 (1991) pp. 47–63. I should explain that I do not discuss Bhabha's own work for related reasons.

Strategies of dispossession

But I will return to history; I must begin with geography. One important concern of postcolonialism, though by no means the only one, is an attempt to bring into focus the dispossession that the West visited upon colonial societies through a series of intrinsically *spatial strategies*. Those last two words have full and equal force. As Said makes very clear, it is impossible to conceive of colonialism or imperialism "without important philosophical and imaginative processes at work in the production as well as the acquisition, subordination, and settlement of space."[217] That much should be obvious from geography's own complicity in colonialism, but what has been acknowledged much less often, at least until recently, is the connective imperative between those colonial projects, the production of space and various modalities of power-knowledge. Said's own critique of Orientalism is of immense importance here, but it does not provide the conceptual apparatus necessary for a systematic exposition of those connections (and neither was it intended to). But de Certeau, who like Said is indebted to Foucault, offers one possible basis for such an analysis when he proposes the concept of a "strategy" as a way of describing the means through which a constellation of power-knowledge is inscribed in its own – its "proper" – place:

Strategies are actions which, thanks to the establishment of a place of power (the property of a proper), elaborate theoretical places (systems and totalizing discourses) capable of articulating an ensemble of physical places in which forces are distributed. They combine these three types of places and seek to master each by means of the others. They thus privilege spatial relationships. At the very least they attempt to reduce temporal relations to spatial ones through the analytical attribution of a proper place to each particular element and through the combinatory organization of the movements specific to units or groups of units.[218]

This might seem forbiddingly abstract, but when de Certeau summarizes the "successive shapes" of a strategy and indicates some of their filiations with a particular historical (and, by implication, geographical) configuration

---

Like Young, *White mythologies*, p. 146, I am concerned at his apparently relentless setting aside of historicity. His successive claims to describe the conditions of colonial discourse

> seem always offered as static concepts, curiously anthropomorphized so that they possess their own desire, with no reference to the historical provenance of the theoretical material from which they are drawn, or to the theoretical narrative of Bhabha's own work, or that of the cultures to which they are addressed. On each occasion Bhabha seems to imply through this timeless characterization that the concept in question constitutes the condition of colonial discourse itself and would hold good for all historical periods and contexts.

[217] Said, "Representing the colonized," p. 218.

[218] de Certeau, *Practices of everyday life*, p. 38.

of rationality – a particular constellation of power, knowledge, and spatiality – the implications of his argument for studies of colonialism and post-colonialism become considerably clearer. He makes three suggestions: that strategies mark the triumph of space over time (the production of a "proper place" – one's own/owned space); that strategies typically involve the mastery of places through sight (through a system of surveillance); and that strategies are revealed in the power "to transform the uncertainties of history into readable spaces."[219]

I realize that these are still very general characterizations, and it is important not to allow this to obscure the heterogeneity and uncertainty of the process: what de Certeau calls "the sly multiplicity of strategies." Colonialism took characteristic, even hegemonic forms, but these cannot be reduced to devising plans in one set of places and imposing them on another set of places. Colonial projects were shaped by the diverse and dependent settings in which and through which they worked. For this reason I want to make these claims more concrete by identifying three of the principal modalities through which a particular version of the ration-ality/spatiality couple was imposed upon colonial societies. It will become obvious, I think, that all three owe something to Foucault's "ascending" analytics of power, in which its "capillary points" become invested and colonized by ever more general mechanisms and forms of global domina-tion. But it is important not to lose sight of the persistent – and often overshadowing – presence of other modalities of power which, in their "descending" force, could be terrifying in their savagery and exemplary in their execution. Some of these were sovereign in the juridical-political sense that Foucalt originally described, but others operated so far outside even colonial legal formularies that reports of their effects were greeted with incomprehension and even disbelief in Europe.[220]

I begin with *dispossession through othering* because of the extraordinary importance of Edward Said's *Orientalism* for the development of the postcolonial critique, but this should not be taken to imply that the process of othering is somehow coterminous with nineteenth-century European colonialism: It is not.[221] Said's central concern, however, and the founding

---

[219] *Ibid.*, p. 36.

[220] Michel Foucault, "Two lectures" in his *Power/knowledge: Selected interviews and other writings 1972–1977* (Brighton, England: Harvester, 1980) pp. 78–180. Compare Bernard Cohen, "Repres-enting authority in Victorian India," in Eric Hobsbawm and Terence Ranger (eds.), *The invention of tradition* (Cambridge: Cambridge University Press, 1983) pp. 165–209 and Taussig, *Shamanism, colonialism and the wild man.*

[221] See, for example, the attempt at an archaeology of anthropology which, following in Foucault's footsteps, claims to identify distinctive Renaissance, classical, and modern constructions of alterity: Bernard McGrane, *Beyond anthropology: Society and its other* (New York: Columbia University Press, 1989). For an attempt to put nineteenth-century Orientalism in its historical context, see Tierry Hentsch, *Imagining the Middle East* (Montreal: Black Rose, 1992).

anguish of his investigation, is undoubtedly the production of "the Orient" by Britain and France in the course of the nineteenth century. He constantly accentuates the operative agency of Western systems of power-knowledge: "Orientalism is – and does not simply represent – a considerable dimension of modern politico-intellectual culture, and as such has less to do with the Orient than it does with 'our' world." Here too it is necessary to emphasize – rather more insistently than Said does himself – the *heterogeneity* of these systems of power-knowledge: Orientalism did not speak in a single voice. But Said never allows that voice to be other than a monologue: "What gave the Oriental's world its intelligibility and identity was not the result of his [sic] own efforts but rather the whole complex series of knowledgeable manipulations by which the Orient was identified by the West." And again:

It is Europe that articulates the Orient; this articulation is the prerogative, not of a puppet master, but of a genuine creator, whose life-giving power represents, animates, constitutes the otherwise silent and dangerous space beyond familiar boundaries.[222]

The violence of representation is equaled only by the apparent passivity of its subject, the Other through which the Western Self comes to know itself. Said traces its marks in the inscriptions of what he calls an "imaginative geography," through which the Orient was presented as a *tableau vivant* for inspection by the distanced and detached Western audience that became virtually its *raison d'être*. Following Bachelard, Said claims that

Space acquires emotional and even rational sense by a kind of poetic process, whereby the vacant or anonymous reaches of distance are converted into meaning for us here.... [T]here is no doubt that imaginative geography and history help the mind to intensify its own sense of itself by dramatizing the distance and difference between what is close to it and what is far away.[223]

Within that space – "there" not "here" – knowledge consisted in enframing and envisioning, in making seen/scene, and the spectre of Foucault is conjured up in several further passages where Said identifies a series of spaces of constructed visibility:

The Orient was viewed as if framed by the classroom, the criminal court, the prison, the illustrated manual. Orientalism, then, is knowledge of the Orient that

---

[222] These descriptions are taken from Said, *Orientalism*, pp. 12, 40, 57. For a fuller discussion of the heterogeneity of Orientalism, see Lisa Lowe, *Critical terrains: French and British orientalisms* (Ithaca: Cornell University Press, 1992). Lowe's account is much closer to Foucault than Said's own, and she draws attention to the unevenness or, more technically, the *heterotopicality* of Orientalism: its contradictory and unstable inscriptions across multiple sites and positions.

[223] *Ibid.*, p. 55 (my emphasis); cf. Gaston Bachelard, *The poetics of space* (Boston: Beacon Press, 1969). For the *tableau vivant*, see Said, *Orientalism*, pp. 103, 126–27, 158.

places things Oriental in class, court, prison or manual for scrutiny, study, judgement, discipline or governing.[224]

The "irrational" Orient was brought within the technologies of Western rationality and constructed as "a sort of imaginary museum without walls" – what Mitchell later called the world-as-exhibition – but not, Said insists, *en arrière*: Orientalism was an active process of othering, the exhibiting of "the" Oriental in a profoundly worldly set of texts which, in a quite fundamental sense, made colonization and dispossession possible. "To say simply that Orientalism was a rationalization of colonial rule is to ignore the extent to which colonial rule was justified in advance by Orientalism, rather than after the fact."[225]

Overlapping with this spatial strategy, but installed on the ground – and installed *as* the ground – was *dispossession through naming*, which can be exemplified through Paul Carter's explorations of *The road to Botany Bay*. Said describes this as a compelling work of great intellectual power that uses the colonization of Australia to advance a thesis about imperialism that has considerable relevance to other regions of the world. Carter writes against the grain of what he calls "imperial history" (including historical geography), a discourse that has much in common with Orientalism in the privileges it accords to vision, sustained by the illusion of the past as theater and, more exactly, by "the unquestioned convention of the all-seeing spectator," and in its assimilation of particularities and specificities to the logic of a supposedly universal reason. But the diagnostic of imperial history is its unexamined assumption that the land was already there, so to speak, a vacant space awaiting its colonizers: that it was a stage on which history could immediately unfold, its movements and positions blocked out.

Imperial space . . . with its ideal, neutral observer and its unified, placeless Euclidean passivity, was a means of foundation, a metaphorical way of transforming the present into a future enclosure, a visible stage, an orderly cause-and-effect pageant.[226]

Against this view, Carter sketches a "spatial history" that seeks to show how the landscape of Australia had to be brought within the horizon of European intelligibility through the multiple practices of naming in order for colonization and dispossession to be set in contingent motion. The

---

[224] *Ibid.*, p. 41; I have borrowed "spaces of constructed visibility" from Rajchman, "Foucault's art of seeing."

[225] *Ibid.*, pp. 399, 166; Timothy Mitchell, *Colonizing Egypt* (Cambridge: Cambridge University Press, 1989). Cf. Ronald Inden, *Imagining India* (Oxford: Blackwell Publishers, 1990).

[226] Paul Carter, *The road to Botany Bay: An essay in spatial history* (London: Faber, 1987) pp. xv, 20, 304. For all his criticisms of historical geography, Carter nevertheless draws on the work of R. L. Heathcote, D. N. Jeans, Donald Meinig, J. M. Powell, and Dan Stanislawski.

very act of naming was a way of bringing the landscape into textual presence, of bringing it within the compass of a European rationality that made it at once familiar to its colonizers and alien to its native inhabitants. Its practices created a network of places within which events could unfold in time, in which history could begin to take place once it had *taken* place: "differences that made a difference." Thus "space itself was a text that had to be written before it could be interpreted."[227] What Carter is arguing, I think, is that the system of articulable difference that is language was an instrument of *physical* colonization; that there is an important sense in which the landscape had to be differentiated through naming in order to be brought into an existence that was meaningful for the colonizers and within which they could frame their own actions:

Possession of the country depended on demonstrating the efficacy of the English language there. It depended, to some extent, on civilizing the landscape, bringing it into orderly being. More fundamentally still, the landscape had to be taught to speak.[228]

But this was not an abstract exercise, everywhere driven by the relentless Cartesian logic of surveyors and scientists. It was often much less formal, deeply implicated in and constituted through everyday practice. If language was an instrument of physical colonization, then by the same token traveling was not simply (Carter actually says "not primarily") a physical activity: "It was an epistemological strategy, a mode of knowing."

The historical space of the white settlers emerged through the medium of language. But the language that brought it into cultural circulation was not the language of the dictionary: on the contrary, it was the language of naming, the language of traveling. What was named was not something out there; rather, it represented a mental orientation, an intention to travel. Naming words were forms of spatial punctuation, transforming space into an object of knowledge, something that could be explored and read.[229]

These formulations owe something to the later Wittgenstein, I suppose, in their insistence on the embeddedness of language-games. In early colonial Australia those attachments were evidently as precarious as they were provisional, however, and these hesitations and doubling-backs are mirrored in the structure of Carter's own text, most obviously in its fragmentary, essay form.[230] But Carter is also indebted to Merleau-Ponty, Deleuze and

---

[227] Carter, *Botany Bay*, pp. 41, 48. Like Said, Carter claims that Bachelard was an important source for these reflections, but this remark also echoes one of Lefebvre's most telling claims: "Space had to be produced before it could be read." See Lefebvre, *Production*, pp. 142–43.

[228] Carter, *Botany Bay*, pp. 58–59.

[229] *Ibid.*, p. 67.

[230] See Carter's conversation with David Malouf, "Spatial history," *Textual practice* 3 (1989) pp. 173–83.

Guattari, and Derrida – not the likeliest of intellectual companions – for his attempt to disclose the imbrications of intentionality and desire in the cultural construction of the colonial landscape. "White invasion was a form of spatial writing that erased the earlier meaning," he remarks at one point, and its erasure marked the site where translation was impossible, "the absence of a shared intentional space in which translation could occur."[231] The possibility of an aboriginal spatiality was effaced, faint "scratches on the face of the country," as Pratt puts it in a different context. Its oral traditions were consigned beyond the perimeter of possibility for a colonial spatiality that was articulated through writing, and converted into an absence that made the construction of European cultural space both possible and comprehensible (which turned out to mean the same thing).[232]

Finally, also sharing in Said's problematic but much more concerned with the concrete inscriptions of disciplinary power in colonial space, is *dispossession through spatializing.* I have already drawn attention to Timothy Mitchell's *Colonizing Egypt*, in which he documents the production of disciplinary power through the imposition of new spatialities within the landscape of nineteenth- and early twentieth-century Egypt. Following, as he says himself, paths opened up by Heidegger, Derrida, and Foucault, Mitchell identifies a long list of colonial inscriptions, including the construction of barracks, training camps and military schools along the Nile and across the Delta in order to produce a new drilled and disciplined infantry corps; the reconstruction of villages as "model villages," whose uniform plans were repeated across what became "an ordered countryside of containers and contained"; the development of a comprehensive system of regulated, inspected education through new primary, preparatory, and final schools; and the reorganization of urban space, opening blind alleys and blazing through great boulevards to radiate from the buildings of the colonial administration at their center. In these various ways, Mitchell observes, "the space, the minds, and the bodies all materialized at the same moment, in a common economy of order and discipline."[233] That "common economy" was, I think, a strategy in exactly de Certeau's sense, a "strategy of arranging, of ordering everything up so as to reveal a preexistent plan, a political authority, a 'meaning,' a truth." For in the colonial order, Mitchell explains,

[T]he effect created of a framework would always appear as though it were a "conceptual structure," as we say. It would appear, that is to say, as an order of

---

[231] Carter, *Botany Bay*, pp. 163, 329.

[232] Mary Louise Pratt, *Imperial eyes: Travel writing and transculturation* (London: Routledge, 1992) pp. 58–67. For a much more unyielding insistence on the power of European writing over oral cultures – of *logos* over *mythos* – see Tzvetan Todorov, *The conquest of America: The question of the other* (New York: Harper and Row, 1984).

[233] Mitchell, *Colonizing Egypt*, pp. 38, 45, 68.

meaning or truth existing somehow before and behind what would now be thought of as mere "things in themselves." Political authority itself would now more and more reside in this effect of a prior, ordering truth.... Such political authority presides, as what is seemingly prior and superior. And yet it presides without ever being quite present. In the white mythology, it is that which stands apart from the world itself, as the meaning that things themselves represent. This political method is the essence of the modern state, of the world-as-exhibition. The certainty of the political order is to be everywhere on exhibit, yet nowhere quite accessible, never quite touchable.[234]

The structure of colonial power was a virtual order of differences, as Giddens might say, present only in its instantiations and institutions, and hence (in one sense) transcendent: and, by implication, triumphant.

I think that these three texts can all be read as studies of colonialism from a postcolonial perspective. By this I mean not only that they were all written after the main period of colonial occupation and dispossession, and after most subjugated peoples have won independence (of one sort or another) from their colonial rulers. I also mean that they are critiques of the strategies through which colonial constellations of power-knowledge were inscribed in space. "Critique" here has its proper, moral and political charge; these are all histories-for, studies that recognize that the legacy of colonialism is a powerful and continuing presence in all these societies, to such a degree that the postcolonial critique depends upon and is articulated through an urgent sense of historicity fully commensurate with – and indeed inseparable from – its analytics of space and spatiality. This distinguishes potscolonialism from the assertions of those postmodernisms (and postmodern geographies) that celebrate, in various ways, the prioritization of space over time.[235] But the postcolonial insistence on historiography is, at least in principle, far removed from the assumptions and intentions of imperial histories, and it is in that act of removal that the question of authorization lies.

Authority, authorization, and the West

And yet the legacy of colonialism is a powerful and continuing presence not only in "those" societies but in "our" societies too. I want to worry away at this relation, and with it the question of authorization, by registering the persistent presence of the West within these politico-intellectual projects. It is surely striking that the three case studies I have summarized in the preceding paragraphs all use European high theory to unmask (some of) the intrusions and brutalities of European systems of power-knowledge.

---

[234] *Ibid.*, pp. 178–79. The reference to "white mythology" is, again, to Derrida (see above, p. 105, fn. 77).

[235] See Chakrabarty, "Death of history?"

They appeal to Heidegger and Benjamin, to Deleuze and Guattari, to Foucault and Derrida; indeed, Young borrows the title of his entire discussion of "white mythologies" from Derrida. Had they appealed to Habermas's theory of communicative action, I would have been less surprised at the centrality accorded to the European voice. All the time that Habermas seeks to subvert the colonization of the lifeworld in one register, he is busily installing it in another through the privilege he systematically accords to the West. His Eurocentrism remains undiminished by the pathologies of the Enlightenment project, whose deformations he apparently registers in the West alone, and in a rhetorical flourish of imperial proportions he demands to know

Who else but Europe could draw from *its own* traditions the insight, the energy, the courage of vision – everything that would be necessary to strip from the... premises of a blind compulsion to system maintenance and system expansion their power to shape our mentality [?][236]

But Said, Mitchell, and Carter rely on sources that are, for the most part, inimical to Habermas's concerns, ones to which he has been profoundly hostile: and yet they still install Europe at the center of their accounts. In all three studies the primary purpose, so it seems to me, is to understand *Europe* (or the West more generally) and its time-space inscriptions.

But that understanding installs two silences. In the first place, I have suggested that a profound gendering of colonial/colonized space was one of the basic dimensions of the Western political imaginary in the nineteenth century. But neither Said nor Carter nor Mitchell pays much attention to the gendering of the strategies that they describe in such detail. Although Said concedes that Orientalism "was an exclusively male province," when he turns to Flaubert's constant associations between the Orient and female sexuality – the motifs of sexual promise, untiring sensuality, unlimited desire that recur throughout the discourse of Orientalism – he announces, unaccountably, that these constructions are "not the province of my analysis here." But it is *precisely* his province, and in failing to consider the ways in which this "feminized" Orient was displayed before the masculinist gaze of Western power-knowledge, both "there" and "here," Said becomes complicit in its contagions.[237] Similarly, Carter writes about the discursive

---

[236] Jürgen Habermas, *The philosophical discourse of modernity* (Cambridge, England: Polity Press, 1987) p. 367; this was originally published in German in 1985.

[237] Said, *Orientalism*, pp. 186–88, 207; cf. Rana Kabbani, *Europe's myths of Orient: Devise and rule* (Bloomington: Indiana University Press, 1986). Said subsequently suggested that Orientalism was "a praxis of the same sort...as...patriarchy in metropolitan societies," but this does little to recover their imbrications. See Edward Said, "Orientalism reconsidered," *Cultural Critique* 1 (1985) pp. 89–107; the quotation is from p. 103. For a fuller discussion of Said's lack of address to gender, see Mills, *Discourses of difference*, pp. 57–63.

formation of the Australian landscape without ever considering the ways in which feminization was written into its prospectus (and reproduced in its cultural geographies) through significations of male fear and desire. Indeed, few of his essays register the presence of women at all, until the discussion of "Intimate Charm," where he uses the image of the candle flame flickering in the parlor window to illuminate what he takes to be the narrow circle of their "private" sphere:

The light is tended by women, the same women whom it shelters and protects and whose outlines it throws invitingly against the curtains. The flame manifests the double aspect of the home: as seducer and seduced, as mistress and servant.[238]

The scene is plainly described from the perspective of the returning male traveler, and Carter places himself in the same position. He offers his reading without any critical comment: and it is, in its way, as imperial as the most imperial history. Women become silhouettes, shadows cast by the flame, objects of male desire and domination, comforters and seductresses, but their figures are never allowed to move beyond these "intimate spaces" into those *other* spaces from which a properly "public" history is to be made. Equally, Mitchell's brilliant exposé of the optics of colonialism ignores the ways in which the Orientalist gaze was not only gendered but was itself refracted through the complex visual codes of Arab society. Patriarchy was patently not a Western invention, and neither did it assume the same forms everywhere. The construction of colonial and Egyptian spatialities were shaped by assumptions about the powers and proprieties of gender, and the images registered through the Orientalist, masculinist gaze were re-worked and recirculated, in various ways, through metropolitan societies (and their world exhibitions) to reconfigure the relations between Orientalism and patriarchy.[239]

These omissions matter for more than historical reasons, because in failing to disclose the patriarchal and phallocratic inscriptions of colonial power – in attentuating some of its violence[240] – they serve to obscure the presence of those same modalities within "our" present. Mieke Bal puts this very well, I think, when she remarks:

[238] Carter, *Botany Bay*, p. 26; cf. Kay Schaffer, *Women and the bush: Forces of desire in the Australian cultural tradition* (Cambridge: Cambridge University Press, 1988).

[239] Janet Abu-Lughod, "The Islamic city: Historic myth, Islamic essence and contemporary relevance," *International journal of Middle East studies* 19 (1987) pp. 155–76; Zeynep Çelik and Leila Kinney, "Ethnography and exhibitionism at the *Expositions Universelles*," *assemblage* 13 (1990) pp. 35–59.

[240] I insist on the "some": It is not my intention to reduce colonial power to sexism. I also want to emphasize that Western women negotiated a number of different positions within the complex enclosures of colonial power and that many of them were directly involved in drawing some of its contours; see, for example, Margaret Strobel, *European women and the second British Empire* (Bloomington: Indiana University Press, 1991) and Nupur Chaudhuri and Margaret Strobel (eds.), *Western women and imperialism: Complicity and resistance* (Bloomington: Indiana University Press, 1992).

Focusing attention on the presence of the colonial imagination in today's post-colonial society is not a gesture of ahistoricism – on the contrary. Problematizing historical distance and analyzing the way streams of the past still infuse the present make historical inquiry meaningful.[241]

In doing so, they also call into question the deceptive autonomy of the "post" in postcolonialism. Bal's point is that it is not enough to draw attention to the racism and sexism of colonial discourse, to convert the architects of Orientalism into the collective Other of the contemporary critic, because "an unproblematic emphasis on the difference of the colonial past is a sure way to keep it alive in an unacknowledged present."[242]

In the second place, and closely connected to these considerations, all three studies have difficulty in creating a space for non-European agency. A number of critics have identified an incoherence in Said's construction of Orientalism which, at first, seems to speak directly to this silence. As he says himself,

Unlike Michel Foucault, to whose work I am greatly indebted, I do believe in the determining imprint of individual authors upon the otherwise anonymous collective body of texts constituting a discursive formation like Orientalism.[243]

His critics want to know how it is possible to restate the values of liberal humanism on the one side and appeal to Foucault's antihumanism on the other. In practice, so they claim, Said offers such a totalizing account of Orientalism that authors end up trapped within its enclosures, a maneuver that marginalizes those scholars "whose involvement with the foreign traditions they studied evolved into a deep personal and dialogical quest for comprehension."[244] But notice where the anxiety lies. The problem of agency is represented as a problem *for the authors*:

On the one hand, the argument of Orientalism amounts to a passionate statement of the values of liberal humanism; by exposing the consequences of Orientalism, the critic is able to build new and less oppressive visions of the oriental Other. On the other hand, however, Orientalism undermines the very foundations on which (western) humanism was built, namely the power of the enlightened self to speak the truth of the Other in the name of science.[245]

---

[241] Mieke Bal, "The politics of citation," *diacritics* 21 (1991) pp. 25–45; the quotation is from p. 34.

[242] *Ibid.*, p. 44.

[243] Said, *Orientalism*, p. 23.

[244] Clifford, *Predicament of culture*, p. 261.

[245] Felix Driver, "Geography's empire: Histories of geographical knowledge," *Environment and Planning D: Society and Space* 10 (1992) pp. 23–40; the quotation is from p. 32. See also Clifford, *Predicament of culture*, p. 264.

These are real concerns and I have no wish to minimize them – much of the postcolonial critique turns on them in one form or another – and yet, as Said subsequently insisted, any challenge to Orientalism must involve "a challenge to the muteness [it] imposed upon the Orient as object."[246] Even that may be too little; colonial knowledge was not always and everywhere a purely European construction. The entanglements were by no means equal, but subjugated peoples often possessed a striking capacity for challenging and changing European inscriptions. If this is suppressed, then one conjures the ghostly presence of an "essential" Other beneath the skein of colonial meanings that supposedly shrouds it. From there it is only a short step to the belief that, as two historians of India put it,

Once these alien representations had been stripped away...the authentic (and "other") India would emerge: an India that had remained in some way untouched by wider historical changes.[247]

History then becomes a refuge rather than a resource. The burden of recognizing the historicity of colonial and postcolonial societies – and of dispensing with the grotesque formularies of "people without history" – can only be taken up by comprehending the entanglements of colonizer and colonized: not as a celebration of Occidental reason, and not as an Orientalist veneer to be planed away, but stubbornly, awkwardly, indelibly *there* (and here). It is an historical geography in which we are, all of us, still involved, and it brings its attendant horrors and responsibilities, which cannot be annulled or evaded; they have played too great a part in making us who we are. Said knows all this, of course, but his own involvement with the Palestinian peoples and their cause shows not so much the power of his account of Orientalism but, paradoxically, its weakness. For against so many odds, including the imposition of Orientalism in new, ever more sophisticated forms, the Palestinian people have consistently maintained their claims to their history, which is at the same time their geography.

Carter confronts the same dilemma, and he reflects on its implications with eloquence and passion. He recognizes the immense difficulty of finding terms within the lexicon of imperial history adequate to the aboriginal presence. "A world view which cannot entertain the logic of other worlds," as he says, inevitably confines aboriginal people to the marchlands beyond its horizon as "figures of unreason," "credulous victims of superstition" and "children of nature."[248] And yet:

---

[246] Said, "Orientalism reconsidered," p. 17.

[247] Rosalind O'Hanlon and David Washbrook, "Histories in transition: Approaches to the study of colonialism and culture in India," *History Workshop Journal* 32 (1991) pp. 110–27; the quotation is from p. 116.

[248] Carter, *Botany Bay*, pp. 320–21, 332, 335.

Despite their marginal place in the hierarchy of knowledge, the Aborigines are, in another sense, an obvious presence throughout the same explorers' and settlers' narratives. The Aborigines may be historically enigmatic, but *spatially* they come into prominent view.[249]

So they do, but always through the eyes of the Europeans. Spatially, Carter insists, "the Aborigines informed the Whites at every turn"; they enter'd a country already replete with directions, one where "the very horizons had been channeled and grooved by aboriginal journeys."

But, as we review the overwhelming evidence for white dependence on the Aborigines' cooperation, it soon becomes clear that the bare fact that the two peoples frequently traveled together along the same tracks, that they admired and contested the same country, does not of itself throw light on the Aborigines' experience. To conceptualize the historical space of the Aborigines in terms of tracks, journeys and regions is already to appropriate it to the symbolic language of white history.[250]

The problem is not so much that Carter constructs spatial history through modes of writing that have little or no purchase on oral cultures; it is, rather, that spatial history is constituted as the dual of imperial history and hence remains (reversed) within its enclosures. Carter knows that it is not enough to deconstruct the devices of imperial history, and he is equally skeptical about the possibility of recuperating the spatialities of aboriginal culture through (postmodern?) experimentation. While these projects may involve a highly sophisticated understanding of "reading," he argues that they remain resolutely empirical in their assumptions about spatiality. In their stead, Carter offers the prospect of a poetics of space:

It might begin in the recognition of the suppressed spatiality of our own historical consciousness.... A history of space which revealed the everyday world in which we live as the continuous intentional reenactment of our spatial history might not say a word about "The Aborigines." But by recovering the intentional nature of our grasp on the world, it might evoke their historical experience without appropriating it to white ends.[251]

To this, Carter seems to suggest, his spatial history is only a prolegomenon – and not, I have to say, an especially likely one. It is a vision that, through the intensity of its focus on intentionality, looks very much like a rational reconstruction (and the shadow of imperial history immediately falls across the page); a vision that effaces the inscriptions of power in the jostling,

---

[249] *Ibid.*, p. 335.
[250] *Ibid.*, p. 337.
[251] *Ibid.*, p. 350. The echo of Bachelard is no longer modulated by Foucault.

colliding construction of these different spatialities; and a vision that fixes its gaze so firmly on the West that it can only catch glimpses of other human beings through a glass, darkly.

The same vanishing point determines the perspective of *Colonizing Egypt*. Mitchell begins with a marvelous account of the Arab delegation to the International Congress of Orientalists in the summer of 1889 as they pause in Paris, a city that Said describes as the capital of the Orientalist world, in order to visit the World Exhibition. Like many other travelers from Egypt, they were simply incredulous at the way of seeing that was naturalized in the exhibition and reproduced endlessly throughout what was, for Benjamin, the capital of the nineteenth century – the so-called "European century" – itself. But when Mitchell travels back to Egypt, midway through his opening chapter, those same travelers disappear: They fold their tents and dissolve into the landscape from whence they came. They rarely reappear.

In their place, we witness Egypt only through the eyes of European travelers, register only their astonishments at quite other ways of ordering up the world, and never discover what their subjects made of being subjected to the new modalities of colonial power. The focus of Mitchell's inquiry is unwavering: It is "the peculiar methods of order and truth *that characterize the modern West*."[252] In his view, Orientalism was not just a particular instance of the violence of representation "but something essential to the peculiar nature of the modern world." His title can then be read as a play on words. As I have previously suggested, "colonizing Egypt" can signal not only the European colonization of Egypt – Egypt as Object – but also, and perhaps more significantly, the constitution of a constellation of power-knowledge that was made visible in Egypt but which, through the figure of Egypt as Subject, was also steadily colonizing the West itself. These moves may even be doubled by the very process of reading that Mitchell follows. In deconstructing one (dominant) constellation of colonial power-knowledge, so one critic argues, he tacitly installs another.

Mitchell's persistent line of argument is that the Western metaphysic includes a conception of order (inside-outside, frame-content, real-symbolic, signified-signifier, and so on) which renders other forms invisible or illegible to the Western observer, who then concludes that he [the male pronoun is, of course, entirely appropriate] is confronting a "disorder." [Mitchell] then tries to delineate alternative conceptions of order outside the Western metaphysic. But does he? Order, it may be argued, is not given in a particular situation, but *read* into that situation. . . . The contrast, then, is not between European conceptions vs. Egyptian, but between two Western academic traditions: the structural-functionalist and the [poststructuralist] inspired by Derrida and ultimately Heidegger.[253]

---

[252] Mitchell, *Colonizing Egypt*, p. ix (my emphasis).

[253] Sami Zubaida, "Exhibitions of power," *Economy and Society* 19 (1990) pp. 359–75; the quotation is from p. 364.

In short, Mitchell's attempt "to dismantle Western metaphysics is fought on Western discursive grounds."[254]

Critical theory and symbolic capital

I hope that this persistent focus on the West helps to explain why I regard so many versions of postcolonialism as therapeutic discourses (for particular constellations of power-knowledge within the Western academy at any rate) as they seek to understand, in displaced form, the time-space disorientations of the Western political imaginary. It is, I think, significant that the contemporary circulation of ideas within the international academy is dominated by a flow of "theory" *from* the West and a return cargo of empirical materials *to* the West.[255] Some critics would want to press this further. The three texts I have discussed all draw upon poststructuralism, in different ways and to different degrees, and Dhareshwar suggests that this Occidental focus is not accidental:

Poststructuralist theory seemed to promise a language to demarcate and deconstruct the hegemonic discourse of the West; and yet it has been unable to conceive of alterity as anything other than its own reflection.[256]

"Theory," and most particularly the intellectual field of cultural studies, has become a form of symbolic capital, Dhareshwar wryly observes, and its acquisition has become crucial "in the symbolic struggle to map the world."[257] Like other global cartographic projects, its coordinates and conventions have usually been established by, centered on, and ratified through the offices of the West. Its imperial meridians have not, of course, been confined to poststructuralism, and commentators have warned of the consequences of conducting a "cognitive mapping" of the postcolonial world under the sign of postmodernism too:

In the name of postcolonialism the types and categories of the metropolis are once again being projected onto the periphery, only this time it is postmodern rather than Enlightenment verities that are assumed to be valid for all peoples,

---

[254] *Ibid.*, p. 364. Mitchell's work has attracted considerable critical interest, and other thoughtful responses include Barbara Harlow, "Travel documents: Redelegating representation," *Social Text* 27 (1991) pp. 72–87 and Charles Hirschkind, " 'Egypt at the Exhibition': Reflections on the optics of colonialism," *Critique of Anthropology* 11 (1991) pp. 279–300.

[255] Cf. David Slater, "On the borders of social theory: Learning from other regions," *Environment and Planning D: Society and Space* 10 (1992) pp. 307–28.

[256] Vivek Dhareshwar, "The predicament of theory," in Martin Kresiwirth and Mark Cheetham (eds.), *Theory between the disciplines: Authority/vision/politics* (Ann Arbor: University of Michigan Press, 1990) pp. 231–50; the quotation is from p. 240.

[257] *Ibid.*, p. 234.

periods and places. Postmodernism seizes on aspects of the neocolonial experience – hybridity, simulation, marginality, rootlessness – and universalizes them. The entire planet is crawling with composites, mimics, borderlines and nomads. Local distinctions are squeezed between the colonialism of the past and the cosmopolitanism of the future as postmodernism becomes a transnational and transhistorical formulary actually oblivious to the heterogeneity to which it appeals for its authority. What the Enlightenment was to colonialism, the postmodern is to neocolonialism.[258]

These are strong claims. There is undoubtedly a danger of perpetrating a new Orientalism, but I do not think that postmodernism, poststructuralism and postcolonialism can be folded so directly and indiscriminately into one another. Neither do I think that these anxieties about "theory" warrant a new empiricism (nor do those who voice them). If the refinement of theoretical sensibilities has achieved anything, it is surely to lay to rest the tired old claim that different cultures can communicate through an unmediated process of translation. It was, after all, just such a belief that underwrote much of Orientalism.[259] My own view, at present, is that one ought not to react to the "European" in European high theory (or the "Western" in Western Marxism) as the marks of Cain. But since they *are* signs of privilege, since they are the traces of both knowledge *and* power, they need to be examined: neither accepted without comment or dismissed out of hand. For these theoretical ideas are invested with their origins, scored by their tracks, and so their genealogies need to be interrogated, their politico-intellectual baggage declared and their closures prised open. None of this implies embargoes, to continue the metaphor, but it may well require import duties. For traveling theories are not innocent gifts:

Theory is no longer theoretical when it loses sight of its conditional nature, takes no risk in speculation, and circulates as a form of administrative inquisition.[260]

Vigilance, then, but not proscription. If theories are indeed, in some part, modalities of power – if they have the imaginative capacity to re-present and reconfigure the world like *this* rather than like *that* – there is a clear responsibility to call those powers to account and to map their spheres of influence: to spell out what it is that they do not only *here*, in their sites of construction, but also *there*, in their sites of reconstruction. The result of a traveling critique of this sort may well be equivocal, but where such ambivalence is achieved through a strategy of supplementarity – through *dis*-closing what theory closes off – then, I think, the certainties of theory

---

[258] Stan Anson, "The postcolonial fiction," *Arena* 96 (1991) pp. 64–66; the quotation is from pp. 64–65.

[259] Tejaswini Niranjana, *Siting translation: History, poststructuralism and the colonial context* (Berkeley: University of California Press, 1992).

[260] Trinh T. Minh-ha, *Woman, native, other*, p. 42.

can be capsized, its self-confidence interrupted, and its conditional nature reasserted. For this reason, and with these qualifications, Spivak seems to me to be right in saying that "only the elite playing at self-marginalization can afford the impossible luxury of turning their backs on those resources."[261]

She is also right, I think, to return to the question of liberal humanism. In her view, postmodernism "manages the crisis of postmodernity" just as "*the quick claims of postcolonial liberals* are attempts to manage the crisis of postcoloniality."[262] That emphasis is mine, and in what follows I want to try to explain it. In my own discipline there have been several (I think mistaken) attempts to reinstate liberal humanism under the sign of post-modernism – I am thinking of those that limit postmodernism to a critique of instrumental reason and a heightened sensitivity to difference – whereas poststructuralism, and those versions of the postcolonial critique that have been informed by it, have a critical and much more complex relation to that tradition. "Is theory, even in [this] new role, still the 'psyche' of the West," Dhareshwar asks, "the mirror in which the Western subject beholds himself, even if only to mourn his own disappearance?"[263] I can best respond to this question, and also make these comments more concrete, by considering Spivak and her reading of *Subaltern Studies*.

Spivak and Subaltern Studies

I need to begin with the Indian historians. As I understand it, Indian historiography has been dominated by two (not one) "imperial histories." On the one side is the modern, secular history that the British brought to the subcontinent, through which India is ushered from brigandage and feudalism into capitalist modernity under the tutelage of the Raj. On the other side is a nationalist historiography, which casts a native Indian elite in an heroic role, wresting the state apparatus from the imperialists and completing the political trajectory inaugurated by the British. For all the oppositions between the two, however, they subscribe to the same meta-narrative:

If the British emplotted Indian pasts in terms of a movement from "despotism" to a British-inspired "rule of law" (instituted by the colonial state), nationalist history-writing portrayed all antiimperialist struggles in India as steps toward a

---

[261] Gayatri Chakravorty Spivak, "Can the subaltern speak? Speculations of widow sacrifice," *Wedge* 7/8 (1985) pp. 120–30; the quotation is from p. 121. The remark does not appear in the revised and extended version of this essay, to which I shall refer from here on: *idem*, "Can the subaltern speak?" in Nelson and Grossberg (eds.), *Marxism and the interpretation of culture*, pp. 271–313.

[262] Gayatri Chakravorty Spivak (with Nicos Papastergiadis), "Identity and alterity: An interview," *Arena* 97 (1991) pp. 65–76; the quotation is from p. 66.

[263] Dhareshwar, "Predicament of theory," p. 232; the male pronoun is deliberate.

sovereign national state, a state that would one day stand on the very foundations that the imperialists themselves had laid.[264]

This abiding historicism is one of the legacies of Orientalism, in which, as Said puts it, "the one human history uniting humanity either culminated in or was observed from the vantage point of Europe, or the West."[265] This singular history installed at its narrative center the icon of Western humanism: the self-conscious subject capable of making and mastering history through Reason (and the gender of that second verb is not incidental to the project). The *Subaltern Studies* group seeks to disrupt these proceedings, to displace their discursive fields, by bringing into presence a composite agency which those twin imperial histories are constitutively incapable of registering: the "subaltern."[266]

This figure functions as a strategic fiction, I think, since the subaltern classes differ from region to region, and India is effectively decomposed into regional constellations of class-divided societies. These formulations consciously owe something to the work of Antonio Gramsci, one of the most influential architects of Western Marxism: his discussions of hegemony, of uneven development and of the plurality of political struggles provide an important conceptual grid for insurgent (postcolonial) historiography. In seeking to recover the subaltern, therefore, the group has been obliged to consider an alternative metanarrative, that of historical materialism. But it has done so from a markedly less hostile perspective:

"Subalternity" – the composite culture of resistance to and acceptance of domination and hierarchy – is a characteristic of class relations in [Indian] society, where the veneer of bourgeois equality barely masks the violent, feudal nature of much of our systems of power and authority.

The persistence of these relationships in the face of industrialization and capitalism cannot be explained by a theory that seeks assurance in the primacy of the "economic infrastructure." By focusing on these relationships, *Subaltern Studies* opens up once more the thorny question of "consciousness" and how Marxists might study it. For Marxist historians of India, the task today is not to repeat the

---

[264] Chakrabarty, "Death of history," and *idem*, "Subaltern studies and the critique of history," *Arena* 96 (1991) pp. 105–20.

[265] Said, "Orientalism reconsidered," p. 101.

[266] For the inaugural manifesto of *Subaltern studies*, see Ranajit Guha, "On some aspects of the historiography of colonial India," in *idem* (ed.), *Subaltern studies I: Writings on South Asian history and society* (Delhi: Oxford University Press, 1982) pp. 1–8. This is of course a politico-intellectual project. As Said puts it, *Subaltern studies* attempts to seize control of the Indian past "from its scribes and curators in the present, since much of the past continues into the present." See Edward Said, "Foreword," in Ranajit Guha and Gayatri Chakravorty Spivak (eds.), *Selected subaltern studies* (New York: Oxford University Press, 1988) pp. i–x; the quotation is from p. vii. This collection includes Guha's original statement of the project.

received orthodoxies of Marxism, but to restore to Marx's thought the tensions original to them. For it is only by accentuating these tensions that we may be able to extend the Marxian problematic to cover the peculiar problems thrown up by our experience of capitalism.[267]

There has been considerable controversy over whether the group has been able to rework historical materialism without repudiating it, and this has recently spilled over into a wider (and angrier) debate about ethnocentrism and the hand-wringing that supposedly accompanies bourgeois liberalism's "bad conscience" into the postcolonial world.[268] The exchange has been so thoroughly badtempered that Spivak's reading of *Subaltern Studies* is all the more instructive. I should say at once that she is not so much concerned to "extend the Marxian problematic," as Chakrabarty would prefer, but to reconfigure it, and it would be quite wrong to think of her interventions as being solely directed toward any putative "reconstruction" of historical materialism; she draws upon Marx's writings to enlarge a wider horizon of concern. In particular, and as I want to show, she articulates Marxism with poststructuralism and with feminism, and uses each to disrupt the privileges the other two typically arrogate to themselves. What makes her work both complicated and constructive is the way in which she uses this triangle as an "epistemic membrane" to think through the postcolonial condition and its dialectical imbrication in Western ethnocentrism. The work of the *Subaltern Studies* group provides a particularly useful focus for her because its members work in institutions on both sides of the membrane, not only in India but also in Australia, Britain, and the United States. In engaging with their writings in this way, Spivak is able to conduct a critique that

---

[267] Dipesh Chakrabarty, "Invitation to a dialogue," in Ranajit Guha (ed.), *Subaltern studies IV: Writings on South Asian history and society* (Delhi: Oxford University Press, 1985) pp. 364–76; the quotation is from p. 376.

[268] I am referring to the exchange between Gyan Prakash, writing from Princeton, and Rosalind O'Hanlon and David Washbrook, writing from two British universities. I have noted their locations because these were made central to the exchange. In the original essay Prakash sought to displace capitalism as the single metanarrative of comparative history in order to avoid "homogenizing the histories that remain heterogeneous within it." His critics claimed that he had merely substituted one metanarrative for another, and that his postfoundational history produced colonial subjects who were simply mimics of the concerns of North American intellectuals: "The true underclasses of the world are only permitted to present themselves as victims of the particularistic kinds of gender, racial and national oppression which they share with preponderantly middle-class American scholars and critics." There is something very odd about that "sharing," but in any case Prakash objected to the Anglocentrism of its formulation: "Britain becomes the unacknowledged privileged space from which universalism fixes its stern gaze across the Atlantic (the empire strikes back?)." See Gyan Prakash, "Writing post-orientalist histories of the Third World: Perspectives from Indian historiography," *Contemporary studies in society and history* 32 (1990) pp. 383–408; Rosalind O'Hanlon and David Washbrook, "After Orientalism: Culture, criticism and politics in the Third World," *loc. cit.*, 34 (1992) pp. 141–67; Gyan Prakash, "Can the subaltern ride? A reply to O'Hanlon and Washbrook," *loc. cit.*, pp. 168–84.

continuously interrupts its own theoretical geometry and that, as I propose
to show, revolves around the vexed question of liberal humanism.[269]

As this implies, Spivak's argument is dense and complicated. I can only
suggest its basic outlines here, but I feel distinctly uneasy about rendering
Spivak's work in this way: and I realize that "render" has the double sense
of delivering and tearing. Not only is there an important difference between
the oral presentation of her arguments and their written, often reworked
form, but my re-presentation will erase their subtleties and complexities. I
should particularly like to dislodge the idea that Spivak's work can be
summarized in a simple and straightforward manner. It cannot. As MacCabe
suggests in his introduction to a collection of her essays, the difficulty of
reading Spivak resides in the very issues she is trying to explicate, problems
that "do not yet have the clarity of the already understood."[270] And Spivak's
way of struggling with them is inseparable from her way of writing about
them and her profound sense of her own "privileged vulnerability."

Spivak evidently has great sympathy with the collective project of *Subaltern
Studies*, but one of her central claims turns on the argument that the group
disavows Western humanism in one register only to reintroduce it in another.
She locates the critical force of the project in post-Nietzschean European
thought, including the work of Foucault, Barthes, and Lévi-Strauss, all of
whom "question humanism by exposing its hero – the sovereign subject as
author, the subject of authority, legitimacy and power." Following from this,
she continues, the group reaches a crucial but aporetic conclusion:

There is an affinity between the imperialist subject and the subject of humanism.
Yet the crisis of anti-humanism – like all crises – does not move our collective
"fully." The rupture shows itself to be also a repetition. They fall back upon
notions of consciousness-as-agent, totality, and upon a culturalism, that are discontinuous with the critique of humanism.[271]

I realize that this might seem opaque, so here is O'Hanlon's gloss:

At the very moment of this assault upon western historicism, the classic figure of
western humanism – the self-originating, self-determining individual, who is at once

---

[269] She is not alone. Chakrabarty now talks of "provincializing Europe" as a necessary corollary
of subaltern studies. Philosophically, he suggests, such a project should ground itself in a
transcendence of liberalism, and he identifies the same theoretical geometry as Spivak: "a ground
that ties Marx shares with certain moments in poststructuralist thought and feminist philosophy."
See Dipesh Chakrabarty, "Postcoloniality and the artifice of history: Who speaks for 'indian' pasts?"
Representations 37 (1992) pp. 1–26; the quotation is from p. 20.

[270] Colin MacCabe, "Foreword" to Spivak, *In other worlds*, pp. ix–xix; the quotation is from p. x.
Cf. Trinh T. Minh-ha's discussion of "clarity [as] a means of subjection" in her *Woman, native,
other*, pp. 16–17.

[271] Gayatri Chakravorty Spivak, "Subaltern studies: Deconstructing historiography," in Guha
(ed.), *Subaltern studies IV*, pp. 330–63; reprinted in her *In other worlds*, pp. 197–221; the quotation
is from p. 202.

a subject in his [sic] possession of a sovereign consciousness whose defining quality is reason, and an agent in his power of freedom – is readmitted through the backdoor in the figure of the subaltern himself, as he is restored to history in the reconstruction of the Subaltern project.[272]

This may well be the "cognitive failure" that Spivak says it is, but it is nonetheless located several removes from Said's conjunction of antihumanism and humanism (above, p. 177), because humanism is reintroduced here not to empower the author *but to recuperate the subaltern*. In one sense, this echoes (in displaced form) E. P. Thompson's recovery of the poor stockinger from the condenscension of posterity – the famous claim that "the English working class made itself as much as it was made" – and his construction and elevation of the artisan to the status of the universal working man. But in another sense it is distinctly different, partly because the subaltern is born out of a *critique* of humanism – Thompson's sensibilities are almost wholly on the other side – and partly because its birth is registered through the *differences* within this "composite culture" (which Thompson only notes in the margins of his account).[273]

Even so, the incoherence remains, and Spivak suggests that the retrieval of subaltern consciousness, the focus of the historiographical project, might be more usefully read as "the charting of what, in poststructuralist language, would be called the subaltern subject-effect." She explains:

A subject-effect can be briefly plotted as follows: that which seems to operate as a subject may be part of an immense discontinuous network . . . of strands that may be termed politics, ideology, economics, history, sexuality, language and so on. (Each of these strands, if they are isolated, can also be seen as woven of many strands.) Different knottings and configurations of these strands, determined by

---

[272] Rosalind O'Hanlon, "Recovering the subject: *Subaltern studies* and histories of resistance in colonial South Asia," *Modern Asian Studies* 2 (1988) pp. 189–224; the quotation is from p. 191.

[273] Elsewhere I have described this aspect of Thompson's project thus:

One of the most recurrent elements in Thompson's account of the making of the English working class is the artisan displaying an extraordinary resilience, at once cultural and political, in the face of the mounting exploitation and oppression of the early Industrial Revolution. We encounter him in the dressing shops in Huddersfield, in the weavers' cottages strung out along the roads to Leeds, at midnight meetings on the moors around Sheffield, and during riots and attacks all over the Midlands and the north of England: so often, in fact, that he assumes the status of the universal working man, a ritualized expression of traditionally ascribed working-class sensibilities and solidarities, with the result that the way in which individual men and women, living and laboring in different towns and villages, came to feel and to articulate this community of interest is never allowed to become a problem. In short, we are invited, however tacitly, to conflate an image of striking literary potency with a figure of comparable historical efficacy.

See Derek Gregory, *Regional transformation and industrial revolution* (London: Macmillan, 1984) p. 175. I have discussed Thompson's humanism already; but I should note that Spivak has also contested what she sees as his attempt "to establish the disciplinary privilege of history over philosophy": Spivak, *In other worlds*, p. 284 n. 14.

heterogeneous determinations which are themselves dependent upon myriad circumstances, produce the effect of an operating subject.[274]

In other words, the subject is constituted at the intersection of multiple
and competing discourses. Such a model is not confined to subaltern
studies; it is also one of the central propositions of Laclau and Mouffe's
more general formulation of post-Marxism which, interestingly, also takes
its point of departure from Gramsci.[275]

Subaltern subjectivities

The filiations with Laclau and Mouffe's post-Marxism are striking – and
all the more so since these two authors have been taken to task for their
treatment of colonialism and postcolonialism[276] – but here I want to confine
my comments to three observations which bear directly on Spivak's (I
assume independent) proposal: these concern gender, spatiality and resistance. As I work with these three points, I hope it will be clear that I am
also moving around the three sides of Spivak's epistemic membrane: moving
between feminism, poststructuralism, and historical materialism.

First, Spivak emphasizes that the subaltern is *gendered* and insists that this
impacts on both historical eventuation and historical reconstruction. Although the *Subaltern Studies* group is not silent about subaltern women,
Spivak draws attention to the ways in which their agency has frequently
been subsumed under notations of class (so that the specificity of gender
slides out of the frame) and also, more generally, to the ways in which
"woman as concept-metaphor" – femininity as a discursive field – has

---

[274] Spivak, "Subaltern studies," p. 341. O'Hanlon, "Recovering the subject," p. 197 similarly
urges the rejection of "humanism's obsessive invocation of origins as its ultimate legitimation and
guarantee": that is,

> the myth, which gives us the idea of the self-constituting subject, that a consciousness or being
> which has an origin outside itself is no being at all. From such a rejection we can proceed to
> the idea that though histories and identities are necessarily constructed and produced from many
> fragments, fragments which do not contain the signs of any essential belonging inscribed in them,
> this does not cause the history of the subaltern to dissolve once more into invisibility.

[275] Laclau and Mouffe, *Hegemony and socialist strategy*. Barrett suggests that Gramsci represents for
them "the furthest point that can be reached within Marxism and the intrinsic limitations of [its]
theoretical problematic." Their project is thus to radicalize Gramsci's thematic and to extend it
beyond the perimeter of (Western) Marxism. See Michèle Barrett, *The politics of truth: From Marx
to Foucault* (Cambridge: Polity Press, 1991) p. 63.

[276] Landry and Maclean, "Rereading Laclau and Mouffe," pp. 53–59, object that Laclau and
Mouffe doubly marginalize capitalism's "periphery" by using the political space of colonization as
an abstract illustration of overdetermination (in which differences are reduced to equivalences)
without considering the concrete struggles against colonialism and neocolonialism: "History is in
danger of being dematerialized rather than deconstructed." These objections make Spivak's
scrupulous attention to the *Subaltern studies* project all the more important.

functioned in many of the group's essays. Too often, Spivak concludes, "male subaltern and historian" have been "united in the common assumption that the procreative sex is a species apart, scarcely if at all to be considered a part of civil society."[277] This is not peculiar to subaltern studies either, and other nominally critical theories are open to the same objection: Spivak is well aware of what she calls "the masculine frame within which Marxism marks its birth."[278] But her point is that this is not an empirical failing – a silence in the archives or an inattentiveness to empirical materials – but an epistemological one:

The question is not of female participation in insurgency, or the ground rules of the sexual division of labor, for both of which there is "evidence." It is, rather, that, both as object of colonialist historiography and as subject of insurgency, the ideological construction of gender keeps the male dominant. If, in the context of colonial production, the subaltern has no history and cannot speak, the subaltern as female is even more deeply in shadow.[279]

These scattered remarks also sow the seeds of a critique of Occidental feminism:

Between patriarchy and imperialism, subject-constitution and object-formation, the figure of the women disappears, not into a pristine nothingness, but into a violent shuttling which is the displaced figuration of the "third-world woman" caught between tradition and modernization. These considerations would revise every detail of judgments that seem valid for a history of sexuality in the West....[280]

The addressee here is not only the Foucault of the *History of sexuality*, I think, but also what Spivak describes, in a trenchant critique of Kristeva, as "the inbuilt colonialism of First World feminism toward the Third." In its place, she urges the possibility of learning from (not merely about) women who live "in other worlds," who speak in different voices, and who articulate different postcolonial positions. This is not a "nativist" claim, she argues, but rather resides in an appreciation of "the immense heterogeneity

---

[277] Spivak, "Subaltern studies," p. 217. In her revised version of this essay – reprinted in Guha and Spivak (eds.), *Selected subaltern studies*, pp. 3–32 – Spivak notes that the group "is now including more overt feminist material" (p. 26) and cites as an example the moving investigation of the politics of patriarchy and female sexuality in rural Bengal in Ranajit Guha, "Chandra's death," in *idem* (ed.), *Subaltern studies V: Writings on South Asian history and society* (Delhi: Oxford University Press, 1987) pp. 135–65. Even here, however, she believes that the patriarchal system "gets in behind [Guha's] back" and that the subaltern woman is called into presence "through a patriarchally gendered vision": see Gayatri Chakravorty Spivak, "Feminism in decolonization," *differences: a journal of feminist cultural studies* 3 (1991) pp. 139–70; the quotation is from p. 153.

[278] Spivak, "Can the subaltern speak?" p. 278.

[279] *Ibid.*, p. 287.

[280] *Ibid.*, p. 306.

of the field" which, in its turn, requires "the First World feminist [to] learn to stop feeling privileged *as a woman*."[281] Her purpose, I take it, is to challenge any (nonstrategic) essentialism that assumes a homogeneity of the feminine through which women in the "First World" might feel or claim an immediate and unproblematic access to the experiences, concerns, and oppressions of women in the "Third World." As my quotation marks indicate, however, the two locations are themselves the product of an essentialism, and it is by no means clear that Spivak is able to overcome the separations that they impose in a *non*imperialist way. One reader of her critique of Kristeva claims that Spivak overlooks time- and place-specificities to such a degree that she ends up repeating "the same cultural violation" she was originally supposed to be contesting.[282] I am not sure that this judgment can be extended to some of her later essays, but in the circumstances it might be more helpful to revert to the poststructuralist suggestion that subjects are constituted at the intersection of different discourses as a way of registering the complex and differential topographies within which gender is situated.

Second, in her later essays Spivak shows an acute sensitivity to the *territoriality* (or more generally, the spatiality) of subject-constitution. It follows from the compound and fractured regional geographies of India that the constitution of the subaltern cohort and thus of the subaltern subject will vary over space. "The word 'India' is sometimes a lid on an immense and unacknowledged subaltern heterogeneity," Spivak writes, and when this is released it has the capacity to "make visible the fantasmatic nature of a merely hegemonic nationalism."[283] The *Subaltern Studies* group is aware of this, of course, and many of their essays reconstruct the local territorial structures of colonial India as a necessary moment in the prosecution of their case. Following Foucault – and a number of the subaltern historians have been influenced by his ideas – one might see this as charting what he called "the insurrection of subjugated knowledges."[284] But Spivak is plainly interested in mapping more than the contours of so many local knowledges – and here the term is again Foucault's, not Ryle's or Geertz's – because she wants to identify what could be called (and perhaps should

---

[281] *Idem*, "French feminism in an international frame," in her *In other worlds*, pp. 134–53; the quotations are from pp. 136, 153. For an example of what Spivak has in mind, see her translations and readings of the work of Mahasweta Devi: *idem*, "A literary representation of the subaltern: A woman's text from the Third World," in *ibid.*, pp. 241–68; *idem*, "Women in difference: Mahasweta Devi's 'Delouti the Bountiful,'" in Andrew Parker, Mary Russo, Doris Sommer, and Patricia Yeager (eds.), *Nationalisms and sexualities* (New York: Routledge, 1991) pp. 96–117.

[282] Silvia Tandeciarz, "Reading Gayatri Spivak's 'French feminism in an international frame': A problem for theory," *Genders* 10 (1991) pp. 75–90. See also Benita Parry, "Problems in current theories of colonial discourse," *Oxford Literary Review* 9 (1–2) (1987) pp. 27–58.

[283] Spivak, "Women in difference," pp. 98–99.

[284] Foucault, "Two lectures," pp. 81–82.

be called) a "global localism," the universalizing claims of Eurocentrism. As Dhareshwar reminds us, Eurocentrism is not merely the ethnocentrism of people located in Europe or even the West more generally, but rather a congeries of assumptions and perspectives that now "permeates the cultural apparatus in which we [all] participate."[285] For this reason, among others, the (new) international division of labor is a constant reference point in several of Spivak's later essays, and she is always alert to the unwitting complicity of critique in its perpetuation. I have already noted the way in which her recognition of heterogeneity bears upon her engagements with feminism, and in exactly the same way she is concerned "to open up the texts of Marx beyond his European provenance, beyond a homogeneous internationalism, to the persistent recognition of heterogeneity."[286] But it is, in fact, Foucault's problematic that is a special concern, and when she urges the realization of "space-specific subject-production" – and the phrase *is* hers, not mine[287] – she is drawing attention to the European provenance of poststructuralism. She offers this warning:

The theory of pluralized "subject-effects" gives an illusion of undermining subjective sovereignty while often providing a cover for this subject of knowledge. Although the history of Europe as Subject is narrativized by the law, political economy and ideology of the West, this concealed Subject pretends it has "no geopolitical determinations." The much-publicized critique of the sovereign subject thus inaugurates a Subject.[288]

She explains what she means in a brief but revealing critique of Foucault (and I should make it clear that her purpose is not to dismiss his work; she describes him and Deleuze as among "our best prophets of heterogeneity and the Other"). One of Foucault's most imaginative contributions has been his critique of the repressive hypothesis, and his (contrary) claim that power is productive, constitutively involved in the double process of *assujetissement*: subjection and "subjectification." But Foucault concedes that his attempts to elucidate the productive capacities of power have had recourse to "the metaphor of the point which progressively irradiates its surroundings."[289] Spivak's concern is that, without great care, "that radiating point, animating an affectively heliocentric discourse, fills the empty place of the agent with the historical sun of theory, *the Subject of Europe*" (my

---

[285] Dhareshwar, "Predicament of theory," p. 235.

[286] Spivak, "Subaltern studies," p. 211.

[287] *Idem*, "Scattered speculations on the question of value," in her *In other worlds*, pp. 154–75; the quotation is from p. 171.

[288] *Idem*, "Can the subaltern speak?" pp. 271–72.

[289] Spivak does not give a source for this "admission," but it was made in the course of a conversation with a group of psychoanalysts over Foucault's *History of sexuality*: see his "The confession of the flesh," in *Power/knowledge*, pp. 194–228; the quotation is from p. 199. But the comment is more complicated than Spivak allows. Foucault does admit to using "the metaphor of

emphasis). In her view, such a possibility is ever-present in the architecture of Foucault's project. "Foucault is a brilliant thinker of power-in-spacing," she declares, but he tacitly confines those spaces to particular maps. "The awareness of the topographical inscription of imperialism does not inform his presuppositions," and in consequence his "self-contained version of the West ... [ignores] its production by the imperial project."[290]

Revealed here, with great clarity, is Spivak's ability to position herself in an ambivalent relation to the theoretical method that is being deployed and to use that position, as Young says, "so as to disconcert and disorient the reader from the familiar politico-theoretical structures which it seems to promise." Young suggests that this strategy of supplementarity (which he thinks Spivak shares with Bhabha but not, significantly, with Said) holds out the prospect "of providing a critique in which both theory and detailed historical material can be inflected toward an inversion of the dominant structures of knowledge and power without simply reproducing them."[291] If such is possible, then Spivak's critical appropriation of poststructuralism is not immediately condemned to reproduce endless alterities in the mirror image of the Western subject. As a matter of fact, in a close reading of Foucault's *The order of things* Bhabha has himself argued that the colonial and postcolonial are indeed present in the margins of the closing essay, and he concludes that this implies and imposes the need to reinscribe Foucault's perspective, thus:

His description of the dehistoricized emergence of the human sciences in the nineteenth century would have to be seen in relation to those "objects" of that

---

the point which progressively irradiates its surroundings" – in the first volume of the *History of sexuality* – language which, as his interlocutor suggests, implies "a power beginning from a single center which, little by little, through a process of diffusion, contagion or carcinosis, brings within its compass the minutest, most peripheral details." But Foucault goes on to say that he was talking there about "a very particular case, that of the Church after the Council of Trent" and that, in general, his own inclination is to conduct an "ascending" analysis of power in which such a metaphor would be quite out of place. Matters are still more complicated, however, because Foucault's preferred analytics of power recognizes an anonymous, decentered, capillary power that, nonetheless, appears to invade sector after sector of social life: and that, one might conclude, is indeed an allegory of European colonial power. Certainly, Spivak's point remains: that Foucault's focus is unwaveringly on Europe, on what she identifies several pages later as the Subject *as* Europe (pp. 280–81).

[290] Spivak, "Can the subaltern speak?" pp. 274, 290–91. The same might be said of Philo's sympathetic construction of "Foucault's geography." He wants to call into question those readings of Foucault that threaten to turn him into "the same" – which register him in some more general school – and instead to "recognize (and perhaps marvel at) the 'otherness' of his perspective of geography. . . ." I sympathize, though I am less interested than Philo in claiming Foucault for postmodernism; but there is in any case quite another sense in which Foucault's project is shaped by a particular politico-intellectual geography that Philo only registers in the margins of his essay. See Chris Philo, "Foucault's geography," *Environment and Planning D: Society and Space* 10 (1992) pp. 137–61; the quotation is from p. 140.

[291] Young, *White mythologies*, p. 173.

disciplinary gaze who, at that historical moment, in the supplementary spaces of the colonial and slave world, were tragically becoming the peoples without history.[292]

Foucault could not have written *Orientalism*, it seems to me, not so much because of his antihumanism but more significantly because his focus of interest is on the specific sites at which power is exercised and where it has immediate effects. Said, by contrast, says little about the effects of Orientalism on the people who were its object; his primary concern is with the effects of the discourse in Europe. Put like that, it would have been odd if Foucault had *not* described the constitution of the European subject (suspending for a moment the essentialism of the phrase), because he was inspecting some of the most strategic sites at which the European subject was constituted. It is true that he did not attend to its geographical particularity – Spivak's heterogeneity – and true too that the process of subject-constitution in Europe was yoked to the process of colonialism beyond Europe. But I do not think that these comments disqualify his work from a wider sphere of intelligibility.

Third, it follows directly from the heterogeneity of subject-positions that *resistance* can also take a multiplicity of different forms. I am not sure that "resistance" is quite the word I want, at least, not if it implies what O'Hanlon calls a Swiss cheese theory of hegemony in which a matrix of superordinate power is created and "resistance can [then] only crawl through the holes."[293] In some situations it may be misleading to speak of hegemony at all. In many colonial societies the reach of the state seems to have been so uneven that European powers achieved domination not hegemony. There was a jagged disjuncture between the state and different sectors of "civil society" whose boundary was policed by military, para-military and judicial force.[294] These are necessary qualifications because they underscore the heterogeneity of struggles against colonial impositions that were not always caught in the mirror of colonial power. They were not confined to ostensibly masculinist forms, for example:

It is one of the deepest misconstructions of the autonomous subject-agent that its own masculine practice possess a monopoly, as the term signifies, upon the heroic:

---

[292] Homi Bhabha, "Postcolonial authority and postmodern guilt," in Grossberg, Nelson and Treichler (eds.), *Cultural studies*, pp. 50–66; the quotation is from p. 64.

[293] O'Hanlon, "Recovering the subject," p. 222.

[294] Cf. Ranajit Guha, "Dominance without hegemony and its historiography," in *idem* (ed.), *Subaltern studies VI: Writings on South Asian history and society* (Delhi: Oxford University Press, 1989) pp. 210–309. One should also note that "civil society" is itself a European construction: see Partha Chatterjee, "A response to Taylor's 'Modes of civil society,'" *Public Culture* 3 (1990) pp. 119–32. Lata Mani uses these claims to urge further caution in transferring poststructuralist ideas from European societies. "Unlike in postmodern, late capitalist societies, one must, in colonial contexts, concede the possibility of a radical disjuncture between particular cultural texts and the wider social text": Lata Mani, "Cultural theory, colonial texts: Reading eyewitness accounts of widow burning," in Grossberg, Nelson, and Treichler (eds.) *Cultural studies*, pp. 392–405; the quotation is from p. 394.

that effort and sacrifice are to be found nowhere but in what it holds to be the real sites of political struggle.[295]

The heterogeneity of "sites" also indicates the importance of heteroclite political geographies, but the same rider applies: the politics of space was not always and everywhere conducted within (nor could it be brought within) the spatial imaginary of colonial power.

Sometimes these struggles involved strategies and counterstrategies in something like de Certeau's sense of the term. They were, in effect, contests over what was deemed a "proper" place for a people. An example will show what I mean, borrowed from the superb account of the missionary colonization of southern Africa provided by the Comaroffs. In the early decades of the nineteenth century, when the advance guard of the London Missionary Society (LMS) set up its first encampment among the Tlhapinga, its missionaries were acutely aware of the symbolic importance of seizing the center of the community. But so too, it seems, were the Tlhaping:

The letters of the persevering Reverend Read confirm that the struggle between him and [Chief] Mothibi was framed almost entirely in terms of location. Each sought to place the other in a map of his own making. The chief wished to keep the evangelist on the borders of his realm, close enough to be a source of valued goods but too far away to have direct access to his subjects. Read, on the other hand, wanted to situate his station where it might insinuate itself into the moral landscape of the Tlhaping.[296]

Soon after the missionaries were granted a precarious foothold in the royal capital, however, they began to press Mothibi to move his seat from Lattakoo (Dithakong) to the banks of the Kuruman River. Not only was this the site to which he had previously wanted to confine them; it was also one that promised to open new channels of colonial power by wresting control of precious water resources away from the tribal rainmaker and delivering it into the hands of the missionaries with their network of wells and irrigation ditches. This was an attempt to install what Foucault would call "pastoral power," to replace the spiritual politics of the chieftainship with the Christian politics of the mission. The British lost no time in selecting what they judged to be a suitable site and started work. But when Mothibi and his entourage eventually arrived, the missionaries were set aside.

There was no doubt who had seized the intiative in creating the settlement, or upon whom it was centred. Although close to the Kuruman mission station, it

---

[295] O'Hanlon, "Recovering the subject," p. 223.

[296] Jean and John Comaroff, *Of Revelation and Revolution: Christianity, colonialism and consciousness in South Africa* Volume 1 (Chicago: University of Chicago Press, 1991) p. 201.

was definitively set apart and given its own name, Maruping. The churchmen found themselves once again at the periphery, having to go on Sundays to "Mothibi's tree" in search of an audience.[297]

The politics of space were more complicated than my anecdote can convey, but what does emerge clearly is the contest of two different geographies.

In the politics of space that surrounded the establishment of the LMS among the Southern Tswana, then, each party tried to draw the other into its own scheme of things. Just as Mothibi sought to make evangelists' actions "take place" on a terrain he controlled, Read set out to encompass the chief's court in the rationalized ground plan of European civilization.[298]

But resistance was often (and after dispossession perhaps more often) articulated through what de Certeau would call *tactics*, those insurgent social practices that eluded "proper places," mocked established grids, and defied imperial cartographies of power.[299] In a way, perhaps, tactics mark the preservation of time against the encroachments of disciplinary space. More than a culture of consolation, they retain in the collective memory traces of other ways of being in the world and carry within them promises of other spatialities. I cannot develop these propositions in any detail here, but I want to try to connect the labile, restless nature of de Certeau's tactics to Spivak's discussion of the subject-effect. I can do so, I think, because her outline sketch allows the process of subject-constitution to be conceived as an inherently contradictory one, involving not only different *but also competing* subject-positions. And it may be through the creative dissonance set in motion by these compositions, through what Smith calls "the simultaneous nonconsistency of subject-positions" – a situation that is likely to be particularly pronounced in colonial and postcolonial societies – that the subject is "dis-cerned." Again, the term is Smith's. To "cerne" means both to accept an inheritance and to encircle or enclose, and Smith trades on both meanings. He intends "dis-cernement" to convey both his refusal of the Western philosophical tradition of the Subject and also the release of that subject from its "perfect self-identity, homogeneity and fixity." But he also argues that the subject cannot subsist in such radical hetero-geneity, and in his view it is through the conscious and unconscious

---

[297] *Ibid.*, p. 204.

[298] *Ibid.*, p. 204. The politics of space was gendered in complex ways too: through the proclaimed virility of the chief, through the Tlhaping women who as primary producers were most intimately involved in the everyday politics of water, and through the masculinist iconography of the missionaries who sought to fructify Africa by watering "her vast moral wastes" with "the streams of life" (p. 207).

[299] "They return to us from afar, as though a different space were required in which to make visible and elucidate the tactics marginalized by the Western form of rationality": de Certeau, *Practice of everyday life*, p. 50.

negotiation of these differences and dissonances that a space is opened for resistance. From this perspective, therefore, resistance "may be glimpsed somewhere in the interstices of the subject-positions which are offered in any social formation." This is a dense summary of a difficult argument, but I hope I have said enough to show that this proposal advances neither the self-sufficient subject of Western humanism nor the de-centered subject of poststructuralism, but it is instead articulated around what Flax calls a "core self" that is nonetheless "differentiated, local, historical."[300] This is, I think, consistent with Spivak's concerns. What her own contribution adds to the argument, through its recognition of the differential spaces in which subject-constitution takes place, is a sense of the variable topography – the intensely human geographies – in which dis-cernement occurs.

Identity, difference, geography

Let me try to make this more concrete. In a tantalizing series of essays Michael Watts suggests that time-space compression can be made to speak to the question of difference only if human geography and cultural studies are brought into dialogue with one another. Otherwise how is it, he asks, that so many writers are able to invoke with such imagination "the dissonances and heterogeneities" that compose the larger structures of modernity, and yet fail to consider how, in these global whirlpools, people nevertheless manage to cling on to a (precarious) sense of themselves: "How these identities are cobbled together and contested in order to act?" Without mapping "the labile, sliding identities forged in specific yet localized sites," he demands, how is it possible to approach the experience of "the Korean Buddhist chemical engineer, recently arrived from three years in Argentina, who becomes a Christian greengrocer in Harlem?" Or, to put the question of hybridity more formally and less rhetorically:

How exactly are individuals interpellated by multiple and often contradictory cultural and symbolic practices rooted in historically constituted communities and

---

[300] Smith, *Discerning the subject*, pp. xxx, 118, 150; I derive the notion of a core self not from Smith – though I think his account requires it – but from Jane Flax, *Thinking fragments: Psychoanalysis, feminism and postmodernism in the contemporary West* (Berkeley: University of California Press, 1990) pp. 218–19, 232. I should also note that the connections made here between subjectivity, spatiality, and resistance are distinctly different from the discussion in James Scott, *Weapons of the weak: Everyday forms of peasant resistance* (New Haven: Yale University Press, 1985) and *idem*, *Domination and the arts of resistance: Hidden transcripts* (New Haven: Yale University Press, 1990). A characteristic assumption of Scott's work – and of some of the early essays in *Subaltern Studies* – is that of "a subjectivity or selfhood that pre-exists and is maintained against an objective, material world, and a corresponding conception of power as an objective force that must somehow penetrate this non-material subjectivity": Mitchell, "Everyday metaphors," p. 562.

places? What are the processes by which a sense of self-construction is shared with others? Why and in what ways are such representations made more or less appealing, and how are they contested?[301]

To sketch some possible answers I want to turn to West Africa. My first vignette is drawn mainly from Cameroon, which Mbembe takes to be typical of what he calls "the postcolony," a society recently emerging from direct colonial rule. I am not sure about this – claims of typicality are notoriously difficult to sustain – but if the idioms of power and identity that Mbembe describes do not have a universal application (even in sub-Saharan Africa) they do, I suspect, have a more general one. In his concern to transcend the conventional oppositions between subjection and autonomy, hegemony and counter-hegemony, Mbembe provides a discussion that exemplifies several of the ideas set out in the previous paragraphs. He begins with a simple outline:

The postcolony is made up not of one coherent "public space," nor is it determined by any single organizing principle. It is rather a plurality of "spheres" and arenas, each having its own separate logic yet nonetheless liable to be entangled with other logics when operating in specific contexts: hence the postcolonial "subject" has had to learn to continuously bargain and improvise. Faced with this plurality of legitimizing rubrics ... the postcolonial "subject" mobilizes not just a single "identity" but several fluid identities which, by their very nature, must be constantly "revised" in order to achieve maximum instrumentality and efficacy as and when required.[302]

This state of affairs is hardly peculiar to postcolonial societies, of course, but Mbembe is quick to give it a more particular inflection. One of the most important characteristics of the postcolony, so he argues, is the centrality accorded to the body: in the physical execution of postcolonial power, through its requisitioning and abusing of bodies, and in the imaginary of postcolonial power, through its exaggerated representations of anatomies and bodily functions. This positionality also marks the body itself as a site of resistance, and Mbembe suggests that the postcolonial subject becomes publicly visible at the point where two sets of social practices overlap: on the one side, the everyday rituals that seek to ratify the power of the state (its capacity for violence and its command of consumptive excess); and on the other side, the subject's own powers of

---

[301] Michael Watts, "The shock of modernity: Petroleum, protest and fast capitalism in an industrializing society," in Pred and Watts, *Reworking modernity*, pp. 20–63; the quotations are from pp. 194–96; see also *idem*, "Mapping meaning, denoting difference, imagining identity: dialectical images and postmodern geographies," *Geografiska Annaler* 73B (1991) pp. 7–16.

[302] Achille Mbembe, "The banality of power and the aesthetics of vulgarity in the postcolony," *Public Culture* 4 (1992) pp. 1–30; the quotation is from p. 5.

transgression through play and parody, gesture and mimicry. At these intersections the fantasms of the grotesque and the obscene are invoked *simultaneously* and *on both sides*, so that they at once inscribe *and* deconstruct the regime of domination.[303] I take those twin emphases to be required by the logic of Mbembe's argument. He claims that Bakhtin, on whom he draws for some of these ideas, tied the grotesque too closely to the practices of the dominated. However true this might have been in European societies, Mbembe insists that the grotesque is of formidable importance to the postcolonial regime of domination as a whole: to *both* rulers *and* ruled.[304] The result of this doubling, so he concludes, is a strange "hollowing out" of postcolonial power:

Peculiar also to the postcolony is the fact that the forging of relations between those who command and their subjects operates, fundamentally, through a specific pragmatic: the simulacrum. This explains why dictators can go to sleep at night lulled by roars of adulation and support only to wake up the next morning and find their golden calves smashed and their tablets of law overturned.... [P]eople whose identities have been partly confiscated have been able, precisely because there was this pretense [simulacre], to glue back together the bits and pieces of their fragmented identities.[305]

This is, I think, a provocative analysis. It establishes the physicality and the "anxious virility" of the (phallocratic) regime of domination; and, against the grain yet imbricated in the very textures of power, it glimpses – sights and sites – the tactical: the "poaching of meanings," "the myriad ways in which ordinary people bridle, trick, and actually *toy* with power instead of confronting it directly," the means by which "the people who laugh kidnap power and force it, as if by accident, to contemplate its own vulgarity."[306] Although Mbembe does not refer to de Certeau, it should be clear from these remarks that he speaks from much the same vocabulary. It should be equally clear that his objective is different. Through a series of theoretical moves that have their origins in Europe but most of which remain in the margins of his text, Mbembe seeks to establish the *specificity*

---

[303] *Ibid.*, pp. 5–6, 7–12 and *passim.*

[304] Cf. Peter Stallybrass and Allon White, *The poetics and politics of transgression* (London: Methuen, 1986), which contrasts a distinctively bourgeois and desperately sanitized conception of self in post-Renaissance European culture with an underbelly of corporeal, carnivalesque "popular culture" that was rooted in the grotesque body. These two authors also show how these distinctions were transcoded through the production of space, and a similar process seems to characterize the postcolony: "Thinking the body is thinking social topography, and vice versa" (p. 192).

[305] Mbembe, "Banality of power," p. 15. To appeal to the simulacrum is perhaps to invoke Baudrillard (though Mbembe does not do so directly) but it is, I think, Bataille who is the more decisive influence on his analysis. Cf. Georges Bataille, *Visions of excess: Selected writings* (Minneapolis: University of Minnesota Press, 1985).

[306] Mbembe, "Banality of power," pp. 8, 12, 14, 22.

of a (particular) postcolonial regime of domination – the particular imbrications of power, knowledge, and resistance in the body and the subject – and its distance from European models. What he does not consider, however, and partly in consequence, is the way in which colonial power provided at least some of the preconditions for (this) postcolonial power: He does not register the continued presence of the colonial past, if in displaced, sometimes repressed, and even inverted form, in the Janus-like postcolonial present. Neither does he explore the ways in which this "local" constellation of power-knowledge is articulated with translocal and global processes.

My second vignette addresses these concerns much more directly. I have taken it from Watt's own study of a series of Islamic insurrections, inspired by Alhaji Mohammed Marwa Maitatsine, that took place in several cities in northern Nigeria during the 1980s. Maitatsine was in fact born in Cameroon, but he lived in Nigeria for most of his life. Although he was killed during the first major clash with Nigerian security forces in Kano in December 1980, along with at least 5,000 and perhaps as many as 10,000 other people, his followers, the 'Yan Tatsine, rose again in other northern cities over the next five years. I should say at once that Watts frames his study in terms drawn more from Taussig – and in particular *The Devil and Commodity Fetishism in South America* – than from Spivak. Like Taussig, he insists on reckoning with the globalization of capitalism; like Taussig, he is interested in its local, intrinsically cultural inflections (at one point he invokes a description of the petroleum that fueled the Nigerian boom as "the devil's excrement"); and like Taussig, he uses Benjamin to suggest some of the ways in which the millenarian spirit of the 'Yan Tatsine movement was conveyed in a series of dialectical images.[307] But I want to show that these concerns can also be made to speak to Spivak (as I think he intends).

In the first place, Watts wires these local and regional struggles to the differential geographies of time-space compression within the global economy. In doing so, he maks a number of connections with Spivak's more impressionistic account of the "apparent instantaneity" of "micro-electronic capitalism" and its global circuits of capital accumulation. She notes that the ticker-tape machine, which played an important part in the formation of international stock markets, dates back to 1867: if it could be speeded up to match today's volume of trade, she realizes, it would be a blur. But the first volume of *Capital* also dates back to 1867:

When it is expanded to accommodate the epistemic violence of imperialism as crisis-management, including its current displacements, it can allow us to read the

---

[307] Michael Taussig, *The devil and commodity fetishism in South America* (Chapel Hill: University of North Carolina Press, 1980); see also his *Shamanism, colonialism and the wild man*.

text of political economy at large. When "speeded up" in this way it does not allow the irreducible rift of the international division of labor to blur.[308]

Neither does Watts. He goes to some considerable pain to situate the 'Yan Tatsine disturbances within what he describes as "the rhythms and jagged breaks of a frenzied Nigerian capitalism." This roller-coaster economy of "fast capitalism" was fueled by petrodollars that were pumped through a bloated and corrupt state and then surged through the urban economy to explosive effect. The population of Kano soared from around 40,000 at the end of the Civil War (c. 1970) to over 1.5 million by the end of the decade. This anarchic metropolis became at once "an enormous construction site and a theater of orgiastic consumption," Watts says, awash in commodities, racked by land speculation and caught in a spiraling price inflation.[309]

In the second place, these changes had a direct impact on the field of possible subject-positions and hence on the construction of the subaltern as what Spivak calls a "subject-effect." These politico-economic ruptures left particularly deep scars on Maitatsine's followers, many of whom had been uprooted from the countryside and found work as casual laborers in the interstices of the city's informal economy. Some of this movement was no doubt seasonal, most pronounced during the dry season, but many of these young male migrants were Koranic scholars (*gardawa*), who were tied by tradition into floating, peripatetic networks and supported by alms of food and shelter in the entryways of houses. This was a moral economy of considerable antiquity, which even colonialism had not displaced, but it was now increasingly called into question: the *gardawa* found that their food supply was disrupted by inflation, speculation and hoarding, and that the traditional entryways were barred to them or removed altogether by urban redevelopment and gentrification.[310]

In sum, Watts argues that the *gardawa* formed a critical segment of a subaltern class, pressing against customary obligations and expectations, and beset by the impositions of international circuits of capital. But as he also recognizes, that class was distinctly heterogeneous:

If Maitatsine's followers occupied a particular class position within the structure of Nigerian petrolic accumulation, they brought with them to Kano a conflation of other identities. An ethnic identification as Hausa, a Muslim identity rooted in

---

[308] Spivak, "Scattered speculations," pp. 166, 170–72 and *passim*. For an earlier version see her "Speculations on reading Marx: After reading Derrida," in Attridge, Bennington, and Young (eds.), *Poststructuralism and the question of history*, pp. 30–62.

[309] Watts, "Shock of modernity," pp. 27–28; *idem*, "State, oil and accumulation: From boom to crisis," *Environment and Planning D: Society and Space* 2 (1984) pp. 403–28.

[310] Paul Lubeck, "Islamic protest under semiindustrial capitalism: 'Yan Tatsine explained,'" *Africa* 55 (1985) pp. 369–89.

a particular exposure and interpretation of Islam through the exegesis of the *qu'uran*, a legacy of animist belief in the spirit world, and what I can only describe as a commoner consciousness (in Hausa the term is *talaka*) inherited from the precolonial Muslim Caliphate. In this sense, their experience of proletarianization under a corrupt and flabby form of oil-based capitalism was extremely heterogeneous and fragmented.[311]

In the face of these fragments, Maitatsine's achievement was to articulate a common identity from within Islam, to interpellate a particular Muslim subject amidst the clamor of other, competing Muslim voices. Watts suggests that this interpellation resided in a wish-image, what Benjamin described in another context as an image in the collective consciousness "in which the new and the old are intermingled," but here given a radically new inflection in a series of what Watts takes to be "dialectical images of African modernity."[312] Maitatsine constructed these images out of a critique of the commodity culture of global capitalism – a world whose crazed logic had been translated by the *gardawa* into experiential categories of exclusion, corruption and moral decline – and a populist or "Jacobin interpretation" of Islam that was equally antagonistic toward the power, practices and profligacy of the established Muslim hierarchy. From these global axes, capitalism and Islam, crossed in intricate antagonism, Maitatsine fashioned what Watts describes as a local *bricolage* that was able to contest "the dominant map of Nigeria's cultural economy."[313]

And it did so from sites that challenged the dominant spatialities. I doubt that the activities of the 'Yan Tatsine were confined to the tactical, though many of them had something of the same guerilla-like quality, but it is difficult to avoid seeing their almost liminal physical location as a transcoding of their social positionality. Maitatsine established an enclave within the old walled city of Kano, in an area that Watts describes as a transition zone between the old and new Kano: the 'Yan Awaki quarter in Fuskar-Abbas, between the old city market, the mosques and the prayer grounds on the one side and the new settlement of Fagge, cinemas, stadia and burgeoning industrial estates on the other (figure 16). Around 10,000 strong by 1980, this was a "lumpen, border community," according to Watts, and its physical expansion provoked a series of angry complaints. As urban gardens were appropriated, houses built on neighbors' lots, and merchants expelled from their shops, the 'Yan Tatsine responded with a single answer: "All land plots in this world belong to Allah." The Governor eventually ordered them to vacate the quarter, and Maitatsine responded with a call

---

[311] Watts, "Mapping meaning," p. 13.

[312] Walter Benjamin, *Charles Baudelaire: A lyric poet in the era of high capitalism* (London: Verso, 1973) p. 12; Watts, "Mapping meaning," p. 12.

[313] *Ibid.*, pp. 13–14.

to appropriate all land in the name of Allah and to seize control of the Friday mosque. When the police and security forces moved in it was, significantly, the mercantile and petty bourgeois quarters – those of the "purveyors of modernity" – which felt the brunt of the counter-attack.[314] If 'Yan Tatsine constituted a community of the imagination, as Watts (following Anderson) repeatedly suggests, it was evidently one that was inscribed in and constituted through a complex human geography.

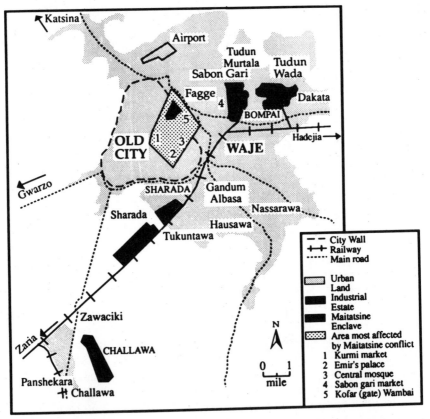

Figure 16   The Maitatsine movement, Kano City, December 1980 [map: Michael Watts, by permission].

I have spent some time on Watts's study partly because it seems to me to speak so directly to the present condition of human geography. His project seeks to break open a particular "knotting" within the local-global dialectic: to show how difference and identity are produced within constellations of power earthed (so to speak) in interconnected spaces and wired

---

[314] Watts, "Shock of modernity," pp. 21–26, 50–54; Lubeck, "'Yan Tatsine."

together by political and economic relations; how difference and identity are contested, negotiated, and shaped through cultural struggles; and how "identity which rests on difference" can produce a common ground for politics.[315] These are not purely abstract concerns, though academics can easily make them seem so: They are the stuff of people's everyday lives. But this does not make their examination a purely empirical matter, changing places and trading experiences without theoretical mediation. As Watts also makes clear, the politics of difference that they embody derives from contrasting and often conflicting geographical imaginations that do indeed "make a difference."

Imaginative geographies and geographical imaginations

At one level, these imaginative geographies are uneven "local knowledges" with their distinctive silences, blank spaces, and distortions. When I first read Watts's work, and the writings of the anthropologists, geographers, and historians who have helped me to formulate my arguments, I soon realized how ill-prepared I was; how desperately little I knew of the world beyond the West. When I was an undergraduate in England in the early 1970s, there were courses available on Africa, South America, and South Asia but, like most of my contemporaries, I did not elect to take them. The fault was in large measure my own; but the other, supposedly more general courses in spatial science seemed to promise so much more. Theirs was not the instrumentalist language of the 1980s, with its characteristic emphasis on the marketplace and the cash register, but it put in place an economy of power-knowledge of the most emphatic (and no doubt masculinist) finality. Its generalizations were not only "rigorous" and "object-ive," I learned, but "powerful." I now realize just how particular were those generalizations and how insidious their powers, but to a working-class kid at Cambridge they seemed highly seductive. This process of interpellation – for that is what it was – worked through another version of the world-as-exhibition, in which scenes were projected in order to marvel at the optics as much as to interpret the images. By this means, the geographer was (again) constituted as a spectator-scientist.

I doubt that mine was a particularly unusual experience, but it is made all the more telling by the continued imperialism of the Western discipline. Its interpellations have a global resonance, orchestrated through a host of international conventions and protocols that maintain its norms and police its practices. This intersects with another, more general circumstance, because our imaginative geographies (inside and outside the academy) are

[315] Watts, "Mapping meaning," p. 14.

global as well as local. They articulate not simply the differences between this place and that, inscribing different images of here and there, but they also shape the ways in which, from our particular perspectives, we conceive of the connections and separations between them. The "global" is not the "universal," in other words, *but is itself a situated construction.* Here the geographies of interpellation can be particularly powerful, not least because they are so often concealed behind the supposedly neutral screens of global science and technology. In a stunning essay on the mediatization of the Gulf War in 1991, for example, Stam reveals the highly particular perspectives brought to bear through the intimate conjunction of intelligence-gathering satellites and planetary television networks, and the highly particular spectatorships that they constructed. "Not only were we able to imagine ourselves anywhere in the world," he observes, "we were also encouraged to spy, through a kind of pornographic surveillance, on a whole region, the nooks and crannies of which were exposed to the panoptic view of the military and the spectator." The use of the collective pronoun is deliberate. The media constructed what Stam calls a "fictive We" – interpellated an imaginary community – whose vantage point was carefully established to both privilege and protect the (American) viewer through the fabrication of ("Allied") innocence and the demonization of the (Iraqi) enemy. By conferring an "instantaneous" ubiquity upon the spectator, the circumference of this Americanocentric vision seemed to be projected from an Archimedian point in geosynchronous orbit above all partisan interests: another universal projection. And since most American viewers "lacked any alternative grid" – one rooted in comprehension of the Middle East, for example, and the multiple legacies of colonialism – Stam concludes that they (we) were able "to believe the worst about our 'enemies' and the best about ourselves."[316]

If this war was, perhaps, the first "postmodern war," its dispositions and displacements did not arrive unannounced. It traded on the mediatization of the Falklands/Malvinas conflict; it clearly learned its lessons from the war in Vietnam; but here is Saint-Exupéry, reflecting on his experience as a pilot during an even earlier war:

All I can see on the vertical is curios from another age, beneath clear, untrembling glass. I lean over crystal frames in a museum; I tower above a great sparkling

---

[316] Robert Stam, "Mobilizing fictions: The Gulf War, the media and the recruitment of the spectator," *Public culture* 4 (1992) pp. 101–26; the quotations are from pp. 102, 113, 124. The reference to pornography is also deliberate: Stam shows how the process of interpellation traded on the ideologies of Orientalism to establish the "fictive We" in a masculine subject-position. "The sodomizing and rape of Kuwait – 'the sexual violation of an innocent, passive, essentially feminine persona' – became the pretext for a 'manly penetration' (some would say gang rape) of Iraq" (p. 121).

pane, the great pane of my cockpit. Below me are men – protozoa on a microscope slide. . . . I am an icy scientist, and for me their war is a laboratory experiment.[317]

The Gulf War generalized that individual experience and, in doing so, turned "experiment" into grotesque "entertainment." As one commentator put it, it is not truth that is the first casualty of war, but the concept of reality. And yet, as he also goes on to argue, even postmodern publics are not passive consumers of spectacle: the meanings of the Gulf War were produced through contestation and elaboration among differently situated subjects. It was possible to glimpse the construction of counter-public spheres and to conceive of other ways of articulating those distant "others."[318]

I rehearse these considerations because they return to the question of authorization. By what right and on whose authority does one claim to speak for those "others"? On whose terms is a space created in which "they" are called upon to speak? How are "they" (and "we") interpellated? These are urgent questions, as important in the classroom as they are in the field. They do not merely speak *to* human geography (though they certainly do that) as have geographies inscribed *within* them. And, as I have indicated, those geographies are constructed at the intersections of the local and the global, so that the sensibilities that these questions necessarily and properly invoke must not condemn us to our own eccentric worlds. To assume that we are entitled to speak only of what we know by virtue of our own experience is not only to reinstate an empiricism: it is to institutionalize parochialism. Most of us have not been very good at listening to others and learning from them, but the present challenge is surely to find ways of comprehending those other worlds – including our relations with them and our responsibilities toward them – without being invasive, colonizing and violent. If we are to free ourselves from universalizing our own parochialisms, we need to learn how to reach beyond particularities, to speak to larger questions without diminishing the significance of the places and the people to which they are accountable. In doing so, in enlarging and examining our geographical imaginations, we might come to realize not only that our lives are "radically entwined with the lives of distant strangers," but also that we bear a continuing and unavoidable responsibility for their needs in times of distress.[319]

---

[317] Antoine de Saint-Exupéry, *Pilote de guerre*, cited in Paul Virilio, *War and cinema: The logistics of perception* (London: Verso, 1989) p. 74.

[318] Robert Hanke, "The first casualty?" *Public* 6 (1992) pp. 135–40; see also David Tomas, "Polytechnical observation," *loc. cit.*, pp. 141–54.

[319] The phrase comes from Michael Ignatieff, *The needs of strangers* (London: Chatto and Windus, 1984); see also Stuart Corbridge, "Marxisms, modernities and moralities: Development praxis and the claims of distant strangers," forthcoming.

PART II

Capital Cities

Introduction

From the number of imaginable cities we must exclude those whose elements are assembled without a connecting thread, an inner rule, a perspective, a discourse. With cities it is as with dreams: everything imaginable can be dreamed, but even the most unexpected dream is a rebus that conceals a desire or, its reverse, a fear. Cities, like dreams, are made of desires and fears, even if the thread of their discourse is secret, their rules are absurd, their perspectives deceitful, and everything conceals something else.

Italo Calvino, *Invisible cities*

We feel justified in advancing the claim – with apologies to Walter Benjamin – that [Los Angeles] has now become the very capital of the twentieth century. ... The social and spatial fabric of contemporary Los Angeles is filled with paradox and contradiction, with an insistent uniqueness as well as with premonitory generality.

Allen Scott and Edward Soja, *Los Angeles: Capital of the late twentieth century*

In these two essays I explore some of the connections between spatiality and representation, between politics and poetics, and between the city and modernism. This is a dense nexus of overlapping and interweaving threads, and for this reason I think it helpful to unravel three of the meanings woven into my title: "Capital cities."

In the first place, modernism was an art of cities, of what Malcolm Bradbury calls "culture-capitals," and its history resonates with the urban geographies of Europe and North America in particular.

In these culture-capitals, sometimes but not always the national political capitals ... a fervent atmosphere of new thought and new arts developed, drawing in not only young native writers and would-be writers, but artists, literary voyagers and exiles from other countries as well. In these cities, with their cafés and cabarets, magazines, publishers and galleries, the new aesthetics were distilled; generations argued, and movements contested; the new causes and forms became matters of struggle and campaign.[1]

From this perspective, Bradbury continues, "the city has *become* culture." To identify the city with culture (or, by no means the same thing, with "civilization") is neither a uniquely modern nor modernist conceit, but when Bradbury offers a rough sketch-map of the shifting cultural geography of Western modernism it becomes clear that the relations forged between the city and high culture in the course of the nineteenth and early twentieth centuries were indeed distinctive. There were many different modernisms, of course, and they reached beyond configurations of class to work with representations of race, sexuality, and gender, but many of them took as their central concern the disruption and dislocation of male bourgeois identities. I insist on the salience of both adjectives: If modernism was a refusal of what Bürger calls "the bourgeois everyday," it was a refusal which – like its provocation – was profoundly gendered. Those experiences, at once disturbing and exhilarating, achieved a particular intensity in the cities of Europe and North America, which were the epicenters of waves of time-space compression.[2] These circles spiraled far beyond the West. Many of the attendant anxieties were projected onto non-Western peoples, who were displayed in the West in novels and ethnographies, on canvases and photographs, in cabinets and exhibitions. Often they were installed within a complex conjunction of exoticism and eroticism, by means of which a stark primitivism was conjured to lay the ghost of a lost subjectivity. These colonial engagements had a peculiar force in nineteenth-century France, where the experience of modernity – and the disorientation of identity and imaginary geography – was perhaps especially radical.[3]

At any rate, the center of Bradbury's map was Paris, Benjamin's "capital of the nineteenth century," and it is there that I open my first essay. There, in the company of David Harvey and Walter Benjamin, I try to capture another sense in which it is possible to speak of "capital cities": one which resonates with Marx as much as with modernism. The connections between the two

[1] Malcolm Bradbury, "The cities of modernism," in Malcolm Bradbury and James McFarlane (eds), *Modernism 1890–1930* (London: Penguin Books, 1976) pp. 96–104; the quotation is from p. 96.

[2] See Scott Lash, "Modernism and bourgeois identity: Paris/Vienna/Berlin," in his *Sociology of postmodernism* (London: Routledge, 1990) pp. 201–36 and David Harvey, "Time-space compression and the rise of modernism as a cultural force," in his *The condition of postmodernity: An enquiry into the origins of cultural change* (Oxford: Blackwell Publishers, 1989) pp. 260–83.

[3] See Chris Bongie, *Exotic memories: Literature, colonialism and the fin de siècle* (Stanford: Stanford University Press, 1992) pp. 14–15; more generally, see Marianna Torgovnick, *Gone primitive: Savage intellects, modern lives* (Chicago: University of Chicago Press, 1990).

are important, of course, and while I think Marshall Berman claims too much when he treats Marx *as* a modernist, the energies of modernism plainly invigorated the subsequent development of Western Marxism.[4] Their imbrications are captured with a particular sensitivity by the art historian, Timothy Clark. Here he is reflecting on Paris in the 1860s, the Paris of Baudelaire, Hausmann, and Manet, but also a Paris remade in the image of capital:

Capitalism was assuredly visible from time to time, in a street of new factories or the theatricals of the Bourse; but it was only in the form of the city that it appeared as what it was, a shaping spirit, a force remaking things with ineluctable logic – the argument of freight statistics or double-entry book-keeping. The city was the *sign* of capital: it was there one saw the commodity take on flesh – take up and eviscerate the varieties of social practice, and give them back with ventriloquial precision.[5]

I do not mean to endorse Clark's reading in its entirety, but his spirited refusal to reduce culture to an epiphenomenon and his determination to confront it in all its materiality and plot its intersections with modalities of class and capital is, I think, exemplary. These concerns can be read in Harvey's Paris too, and they are vividly present in Benjamin's Arcades Project, but they are also illuminated by Allan Pred's extraordinary study of nineteenth-century Stockholm. Pred is acutely aware of the shifts in the ground of capitalist modernity during the last fin de siècle, but he is most concerned to plot the path of those shock waves through the spheres of everyday life and to chart the (re)construction of intensely human geographies in the middle of the maelstrom.

I should perhaps make it plain that the purpose of the first essay is not to place Harvey, Benjamin, and Pred on pedestals – nor to dash them to the ground – but simply to explore some of the ways in which their writings address questions of spatiality and representation. They do so from avowedly critical perspectives which intersect in various ways with historical materialism, but the (historical) geography of critical theory is not confined to Europe. In the 1930s and 1940s some of the most influential members of the Frankfurt School, along with countless other artists, critics, and intellectuals, found refuge in the United States. Benjamin committed suicide before he could escape, but his friend Brecht spent the war in Santa Monica (and hated it) while Adorno and Horkheimer wrote the *Dialectic of enlightenment* in Los Angeles. That city had already been described by the geographer Anton Wagner as an attempt "to create the Paris of the Far

---

[4] Marshall Berman, *All that is solid melts into air: The experience of modernity* (London: Verso, 1982); Eugene Lunn, *Marxism and modernism* (London: Verso, 1985).

[5] T. J. Clark, *The painting of modern life: Paris in the art of Manet and his followers* (Princeton: Princeton University Press, 1984) p. 69.

West." He was as entranced as he was dismayed by its "façade landscapes," "mimics" and "fakes," but those exiles who arrived in his wake were in little doubt that the city in which they now found themselves had left Europe and its version(s) of modernity far behind. Adorno later thought it "scarcely an exaggeration to say that any contemporary consciousness that has not appropriated the American experience, even if in opposition, has something reactionary about it."[6] Whether those appropriations could be justified is a moot point. Mike Davis is skeptical:

In Los Angeles ... the exiles thought they were encountering America in its purest, most prefigurative moment. Largely ignorant of, or indifferent to, the peculiar historical dialectic that had shaped Southern California, they allowed their image of first sight to become its own myth: Los Angeles as the crystal ball of capitalism's future. And, confronted with this future, they experienced all the more painfully the death agony of Enlightenment Europe.[7]

Much the same might be said of some late twentieth-century obsessions with Los Angeles, particularly those that repeat and enlarge Wagner's grand language:

By hyping Los Angeles as the paradigm of the future (even in a dystopian vein), they tend to collapse history into teleology and glamorize the very reality they would deconstruct. Soja and Jameson, particularly, in the very eloquence of their different "postmodern" mappings of Los Angeles, become celebrants of the myth.[8]

Perhaps they do; my second essay rereads Edward Soja's provocative "mapping" of *Postmodern geographies*. But what interests me as much (there) is the return glance: the West Coast fascination with Western Europe. When Scott and Soja claim Los Angeles as "the capital of the late twentieth century," the echo of Benjamin is deliberate and the theoretical style that Soja in particular adopts is distinctly French: Aglietta, Baudrillard, Foucault, Lefebvre. Indeed, in an essay that is not included in his book Soja conducts a sort of *tour raisonné* around an exhibition staged in his own department at UCLA to commemorate the bicentennial of the French Revolution. This installation sought to link late nineteenth-century Paris and late twentieth-century Los Angeles; its centerpiece was a model of the "Bastaventure," a building that appeared as the Bastille from one perspective and the Bonaventure Hotel from another.[9] I am not altogether sure what (post)modern

---

[6] I derive these vignettes from Mike Davis, *City of Quartz: Excavating the future in Los Angeles* (London: Verso, 1990) pp. 48–49.

[7] *Ibid.*, p. 48.

[8] *Ibid.*, p. 86.

[9] Allen Scott and Edward Soja, "Los Angeles: Capital of the late twentieth century," *Environment and Planning D: Society and Space* 4 (1986) pp. 249–54; Edward Soja, "Heterotopologies: A remembrance of other spaces in the Citadel-LA," *Strategies* 3 (1990) pp. 6–39.

point was being made, but I suspect that such a project also says something about the way contemporary theory has itself become a form of symbolic capital whose colonizing powers remake there and then in the image of the here and now. I have commented on these possibilities already. But these metropolitan preoccupations with theory – the third set of meanings contained within my title – do not imply that theory is nothing *more* than symbolic capital, a profitable investment on the trading floor of the international academy. It has to be possible – and it is an important part of my project, in these essays and elsewhere – to rise above the cynical disparagement of theoretical work to interrogate its *other*, creative, imaginative and productively political values.

3

City/Commodity/Culture: Spatiality and the Politics of Representation*

Lucidity came to me when I at last succumbed to the vertigo of the modern. This last word, no sooner formulated, melts in the mouth. The same thing happens with the whole vocabulary of life. . . . However, the path I was following was such that I could no longer avoid consulting the map of its territory.

Louis Aragon, *Paris paysan*

I was afraid my friends would laugh at me, or my family would think I was going crazy, but I simply couldn't stay away from that imaginary map, so rich and ripe with geographic incident.

Pierce Lewis, *Beyond description*

Maps of modernity

Aragon's dilemma is not an oddity of early twentieth-century surrealism; it is, rather, a symptomatic condition of modernity. When Baudelaire reflected on "The painter of modern life" and called for an art capable of registering the passing moment without destroying its transient passage, it was already possible to sense the pull of what Connor calls that "irrevocable tension between the way human beings felt they lived and the forms used to render that sensation."[1] The tension was aggravated by the pulsating drive of

*This essay was originally written for the Vegasymposium in Stockholm, April 1991, when the Anders Retzius medal was presented to Allan Pred by the Svenska Sällskapet för Antropologi och Geografi.
[1] Charles Baudelaire, "The painter of modern life"; the original essay was written in 1859–60 and first published in 1863, but it is reprinted in his *The painter of modern life and other essays* (London: Phaidon, 1964) pp. 12–15; Steven Connor, *Postmodernist culture: An introduction to theories of the contemporary* (Oxford: Blackwell Publishers, 1989) p. 4.

technical and scientific change through the nineteenth and into the twentieth centuries, especially in Europe and North America, but it was also
heightened – and, I think, generalized in significantly new ways – by the
turmoil of the First World War and the explosive force of the Russian
Revolution.[2] With the First World War, wrote Benjamin on the eve of the
Second,

a process began to become apparent which has not halted since then. Was it not
noticeable at the end of the war that men returned from the battlefield grown
silent – not richer, but poorer in communicable experience?... For never has
experience been contradicted more thoroughly than strategic experience by tactical
warfare, economic experience by inflation, bodily experience by mechanical warfare,
moral experience by those in power. A generation that had gone to school on a
horse-drawn streetcar now stood under the open sky in a countryside in which
nothing remained unchanged but the clouds, and beneath these clouds, in a field
of force of destructive torrents and explosions, was the tiny, fragile human body.[3]

More recently, Eksteins has emphasized not only the importance of the
war, but of Germany's particular role within it. And when he argues that
it was Germany (as much as Russia) that "represented the idea of revolution," he clearly has in mind something other than the narrowly political
meaning of the word: He invokes instead a distinctively modern sense of
a whole world giving way, a permanent impermanence. "If newness had
been a strong German concern prior to 1914 and during the war," he
suggests, "it became a universal preoccupation in the west after the war."
But it was rooted in a profound mood of tragedy, of a loss that could not
be redeemed. After the war, he writes, "the devastation was so wide and
the task of reconstruction so staggering that notions of how this was to
be accomplished dissolved often into daydream."[4] If this was quintessentially Benjamin's world, however, I want to show that he set about mapping
it in a distinctly original way.

I say "mapping" deliberately, because the task of resolving contradictions
and dilemmas like these – or, at any rate, of representing them, to oneself
and to others – not only called into question ordinary conceptions of time:
It also shattered conventional conceptions of space. By the opening decades
of the twentieth century, the disassociation between "structure" and "lived

[2] For the coordinates of modernism, see Perry Anderson, "Modernity and revolution," *New Left Review* 144 (1984) pp. 96–113 and the elegant extension in Alex Callinicos, "Modernism and capitalism," in his *Against postmodernism: A Marxist critique* (Cambridge: Polity Press, 1989).

[3] Walter Benjamin, "The storyteller: Reflections on the works of Nicolai Leskov," in his *Illuminations: Essays and reflections* (ed. Hannah Arendt) (New York: Schocken Books, 1968) pp. 83–109; the quotation is from p. 84. The essay was first published in 1936.

[4] Modris Eksteins, *Rites of Spring: The Great War and the birth of the modern age* (Toronto: Lester and Orpen Dennys, 1989) pp. 169, 257–58.

experience," as Jameson calls it, had been transcoded into a radically new relation between space and place: by then, "the truth of [daily] experience no longer coincide[d] with the place in which it [took] place."[5] Early modernism was so deeply implicated in this multiple and compound crisis of representation that I do not think it possible to say (as some commentators have claimed) that fin-de-siècle art or social thought – the distinctions between them were often highly ambiguous – somehow privileged time over space.[6] The characteristically modernist gesture was to *disrupt* narrative sequence, to *explode* temporal structure, and to accentuate *simultaneity* as a way of declaring, and indeed demonstrating, that "things [did] not so much fall apart as fall together."[7] What lay behind many of these modernist experiments was not so much an insistence on the nonrepresentability of art, therefore, or a defiant commitment to *l'art pour l'art*, as the belief that the "vertigo of the modern," in Aragon's apt phrase, could be figured through an exploration of its spatiality, through a reading of what he called "the map of its territory."

In fact, I suspect that it is high modernism – those shifts in cultural and intellectual registers that were installed after World War II – that most clearly betrays the subordination of space in social thought. It was during those decades that a wedge was driven between C. P. Snow's two cultures, the "arts" and the "sciences"; and in the Anglophone social sciences time was valorized (most obviously in the plenary trajectory of models of "modernization") and space reduced to an isotropic plane (most obviously in the geometric lattices of spatial science).[8] One implication of this proposal – which I advance with considerable caution since I am aware of how uneven cultural and intellectual histories are: how misleading it is to slice some "essential section" through them – is that the present preoccupation with the postmodern and postmodernism signals a break with *high* modernism while simultaneously maintaining a deep affinity with that earlier, essentially fin-de-siècle modernism. Let me repeat: I am hesitant about this, and I do not mean to suggest that postmodernism is simply the return of any earlier modernism. There are important differences

[5] Stephen Kern, "The Cubist war," in his *The culture of time and space, 1880–1918* (Cambridge: Harvard University Press, 1983) pp. 287–312; Fredric Jameson, *Postmodernism or, the cultural logic of late capitalism* (Durham: Duke University Press, 1991) p. 411.

[6] Hence my disagreement with Edward Soja, *Postmodern geographies: The reassertion of space in critical social theory* (London: Verso, 1989); see Chapter 4, this volume.

[7] Eugene Lunn, *Marxism and modernism* (London: Verso, 1985) p. 35.

[8] I say "Anglophone" for two reasons: The German distinction between the *Geisteswissenschaften* and the *Naturwissenchaften* cannot be mapped directly onto the English-language distinctions between the arts or humanities, the social sciences, and the sciences; and the various Francophone structuralisms were arguably less interested in time and more interested in forms of "spatial thinking" than the Anglophone social sciences.

between them. Still, I want to use this tense relation to frame the discussion that follows. For it seems to me that the postmodern critique of social theory – of its metanarratives and their various imperialisms, of its assertive generalities that conceal its intrinsic particularities – has once again brought the politics of social theory and the poetics of social inquiry into the same discursive space. If we are to "unlearn our privileges," to "unmaster our narratives" as it were, then we will have to examine our textual strategies – and in particular the duplicities of narrative and image – because it is through these modes of representation that many of our most common-place privileges are put in place.

This is the pivot around which the rest of the essay moves. I want to make my discussion as concrete as possible by considering some of the recent writings of David Harvey and Allan Pred. Their reconstructions of two nineteenth-century European cities (Paris and Stockholm) torn apart by the convulsions of modernity have much to tell us about these matters, but for them to speak to one another I think it is necessary to (re)introduce another voice to mediate between them: that of Walter Benjamin. Through his extraordinary interventions in what I regard as the historical geography of modernity, we can begin to glimpse some of the deeper resonances between politico-intellectual concerns at the beginning and the end of the twentieth century. It is also possible to tease out some of the more intricate relations between spatiality and the politics of representation. For not only does Benjamin provide a bridge between Harvey's Marxism and Pred's materialism, but his graphical sensibility illuminates the contrasts between their geographical imaginations. My readings of Harvey, Benjamin, and Pred are thus intended to bring into focus the politics and the poetics of three historical geographies, each of which seeks, in somewhat different ways, to provide a "map" of a particular (European) arena of modernity. Since the metaphor of the map has become increasingly common – and increasingly contentious – in discussions of this sort, I should say that I use the term with two caveats. First, I realize that the supposed objectivity of "maps" is an effective fiction: that their texts and images are as vulnerable to deconstruction as any others. Second, the "maps" that I discuss here are something more than purely metaphorical devices: their accounts of the inscription of social life in space means that they also have a substantial materiality.

The literary diver: David Harvey and Second Empire Paris

I find myself most deeply impressed by those works . . . that function as both literature and social science, as history and contemporary commentary.
        David Harvey, *Consciousness and the urban experience*

Harvey's investigations of *Consciousness and the urban experience* are focused
on a dazzling reconstruction of Second Empire Paris (figure 17).[9] His
interest in that city is twofold. He wants to use its turbulent historical
geography to illuminate the relations between capitalism, modernity, and
urbanization – or, to put this the other way round, to bring the general
propositions of historical materialism to bear upon the particular case of
Benjamin's "capital of the nineteenth century" – and also to clarify the
politics of space or what he calls, more specifically, "the urbanization of
revolution." Largely for this reason, Harvey's major essay in that collection
moves determinedly toward (though it does not quite reach) the Paris
Commune of 1871. "We have much to learn from the study of such
struggles," he concludes. It will be an important part of my argument that
the shape and direction of this "learning" – and of his politico-intellectual

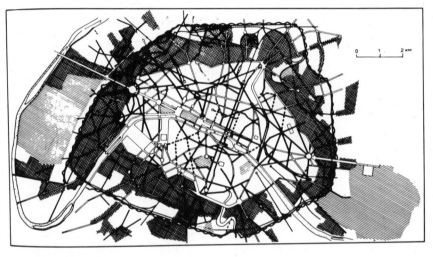

▬▬▬new streets
▓▓▓new districts

Figure 17   The Hausmannization of Paris [source: Leonardo Benevolo, The history
of the city*].*

project as a whole – implicates Harvey in an essentially *progressive* conception
of history in which the past is able to illuminate the present precisely
because each is conceived as a moment in the unfolding of a single
master-narrative.[10]

[9]   David Harvey, "Paris, 1850–1870," in his *Consciousness and the urban experience* (Oxford: Blackwell
Publishers, 1985) pp. 62–220. I have borrowed "The literary diver" from the epigraph that heads
Harvey's essay. "Paris is indeed an ocean," says Balzac, so deep that "there will always be something
extraordinary, missed by the literary diver."
[10]   Harvey, *Consciousness*, p. 220.

Harvey's portrayal of the events leading up to the Commune is, in part, framed by the work of Henri Lefebvre.[11] This Marxist outlaw, if I may so call him, has provided the inspiration for at least two other reconstructions that can bring Harvey's image into a sharper and more distinctive focus. For Castells, the Commune was not the harbinger of any working-class revolution; it was, rather, a protest against the instruments of speculation, "the manipulator of the rules of exchange, not the one who appropriated the means of production," and against the institutions of the ancien régime, "the accountants of the old morality." Harvey differs from Castells not so much in the targets he identifies – on the contrary, he makes much of the ravages of speculation and the involvement of the state in underwriting the transformation of urban space – but in the theoretical armature he deploys. Castells plainly departs much further from historical materialism than Harvey would like. "For the Commune of Paris," Castells declares, "surplus value was an historical abstraction."[12] Harvey, in contrast, keeps close to the labor theory of value and so penetrates much further into the world of work and the process of production. In doing so, his trajectory is paralleled (though I think on a different level) by Ross's contextual reading of Rimbaud's poetry. She sees the Commune as a series of anti-hierarchical gestures and improvisations, as "a revolt against deep forms of social regimentation," which (as Harvey insists) included the imposition of capitalist conceptions of "work"; but she gives much greater prominence to the ways in which these tactics were elaborated within different *cultural* spaces.[13] This is emphatically not to say that Harvey is indifferent to culture and consciousness, but it is to say (as I now want to show) that his presentation of these themes is made through an almost subterranean textual strategy that contrasts markedly with the way in which he presents the basal logic of the capitalist mode of production.

Marxism and metanarrative

The metanarrative that makes Harvey's primary presentation possible is determined by what is in many ways an immensely productive rereading

---

[11] Not only Lefebvre's own investigations of the Commune but also his more general theoretical work: see in particular his *The production of space* (Oxford: Blackwell Publishers, 1991); this was originally published in French in 1974.

[12] Manuel Castells, "Cities and revolution: The Commune of Paris, 1871," in his *The city and the grassroots: A cross-cultural theory of urban social movements* (Berkeley: University of California Press, 1983) pp. 15–26. Elsewhere Harvey reads this as a sign of "Castells's apparent defection from the Marxist fold": David Harvey, *The urbanization of capital* (Oxford: Blackwell Publishers, 1985) p. 125.

[13] Kristin Ross, *The emergence of social space: Rimbaud and the Paris Commune* (Minneapolis: University of Minnesota Press, 1988).

of Marx. Harvey repeatedly draws attention to his previous account of *The limits to capital* and the containing contours of late nineteenth-century Paris are accordingly mapped by the instruments of historico-geographical materialism that are calibrated by the concept of the *commodity* inscribed at its center.[14] He once described to me how he thought of peeling away the layers of the commodity like the layers of an onion, and the image has more than analytical force, since both processes are likely to bring tears to the eyes. Here Harvey's leading claim, I take it, is that the rationalization of urban space depended on the mobilization of finance capital – on a new prominence for money, credit, and speculation – which installed both *spaces as commodities* and *commodities in spaces.*

On one side of this coin, there was a much tighter integration of finance capital and landed property, to such a degree that "Parisian property was more and more appreciated as a pure financial asset, as a form of fictitious capital whose exchange value, integrated into the general circulation of capital, entirely dominates use value."[15] Not surprisingly, as its pulverized spaces were commodified, so Paris became a divided city:

That Paris was more spatially segregated in 1870 than in 1850 was only to be expected, given the manner in which flows of capital were unleashed to the tasks of restructuring the built environment and its spatial configuration. The new condition of land use competition organized through land and property speculation forced all manner of adaptations upon users. Much of the worker population was dispersed to the periphery . . . or doubled up in overcrowded, high-rent locations closer to the center. Industry likewise faced the choice of changing its labor process or suburbanizing.[16]

On the other side, and literally so, the center of the city was increasingly given over to the conspicuous commodification of social life:

Hausman[n] tried . . . to sell a new and more modern conception of community in which the power of money was celebrated as spectacle and display on the *grands boulevards,* in the *grands magasins,* in the cafés and at the races, and above all in those spectacular "celebrations of the commodity fetish," the *expositions universelles* . . . [Those] World Exhibitions, as Benjamin put it, were "places of pilgrimage to the fetish Commodity," occasions on which "the phantasmagoria of capitalist culture attained its most radiant unfurling."[17]

---

[14] David Harvey, *The limits to capital* (Oxford: Blackwell Publishers, 1982).

[15] Harvey, *Consciousness,* p. 82.

[16] Harvey, *Consciousness,* p. 95. It is this stark geography that the Communards challenged: "The Commune of 1871 will be in large part the retaking of central Paris, the true Paris . . . by the exiles of the *quartiers extérieurs,* of Paris by its true Parisians, *the reconquest of the City by the City*": Jacques Rougerie, cited in T. J. Clark, *The painting of modern life: Paris in the art of Manet and his followers* (Princeton: Princeton University Press, 1984) p. 276.

[17] Harvey, *Consciousness,* pp. 103, 200.

Much of Harvey's essay, as he teases out and traces through one thread after another, is clearly structured by the circuits of capital that are wired together in this way to illuminate the totality of capitalist modernization in Second Empire Paris. And make no mistake: Harvey is determined to present his account *as* a totalization. "To dissect the totality into isolated fragments," he warns, is "to lose contact with the complex interrelations that intertwine to produce the simple narrative of historico-geographical change that must surely be our goal."[18]

Must it? Harvey's primary motifs give his project a certain robustness but, I think it fair to say, a certain rigidity too. Yet many of the events and themes that he describes quite clearly refuse to be "mastered" by his leading narrative. Most recalcitrant are the images of women that occur again and again as the text unfolds: the city of Paris, which appeared to Balzac as a "mysterious, capricious and often venal" woman and to Zola as a "fallen and brutalized" one; the "powerfully formed, bare-breasted" Liberty of Delacroix and the "terrifying" prostitute posing as Liberty by the Tuileries in Flaubert; the "voluptuous sensuality" of the Orient, supposedly "the locus of irrational and erotic femininity."[19] These are all highly specific images of female sexuality that seem to have a double function. For contemporary men, as Harvey recognizes, they provided ways of debasing the struggles of women and so domesticating resistance to the established patriarchal order. For Harvey, himself, however, they become a license for bringing gender and sexuality within the frame of meaning provided by the master-narrative of capitalist production and reproduction. With money as the "universal whore," Harvey can declare that "the city itself has become prostituted to the circulation of money and capital."[20] The coding is unexceptional – one can find it in Marx, in Baudelaire and Benjamin[21] – and it is not without value. Yet the iconographic tradition that these images embody cannot be assimilated as directly into the logic of capital and class as Harvey seems to think. One example must suffice. When Harvey watches the way in which Frédéric Moreau, the bourgeois hero of Flaubert's *L'éducation sentimentale*, "glides as easily from space to space and relationship to relationship as money and commodities change

---

[18] *Ibid.*, p. 68.

[19] *Ibid.*, pp. 180, 191–94, 201.

[20] *Ibid.*, p. 177.

[21] In Benjamin's case, however, representations of femininity and sexuality have a complex genealogy and Rauch argues that their deconstruction functions as an allegory of (aesthetic) history that both unmasks the *appearance* of femininity as a projection of the male subject and marks the site of a radical inscription of the *corporeality* of the female body. See Angelika Rauch, "The *Trauerspiel* of the prostituted body, or woman as allegory of modernity," *Cultural Critique* 10 (1988) pp. 77–88. Other feminist critics have read Benjamin in other ways: see, for example, Christine Buci-Glucksmann, "Féminité et modernité: Walter Benjamin et l'utopie du féminin," in Heinz Wismann (ed.), *Walter Benjamin et Paris* (Paris: Éd. du Cerf, 1986) pp. 403–20.

hands," what he fails to notice is that Frédéric was not only capitalizing on the freedom conferred by his class position: for, as Massey remarks, "he did have another little advantage too."[22]

Modernity and the midway: interiors and exteriors

Toward the closing sections of Harvey's essay, however, and overlapping with these images, glimpses of another historical geography come to the surface: one that is looser, more open in texture, and perhaps even more politically charged. Many of these moments are occasioned by engagements with Baudelaire – "that apostle of modernity" – and Benjamin. In my view, these passages reveal Harvey at his very best: and a best that is, thematically and stylistically, far removed from the somewhat wooden prose that supports his earlier narrative. With Benjamin looking over his shoulder, often guiding his pen, Harvey offers a series of arresting and often moving readings of Baudelaire's poems. Slowly, he begins to inaugurate a different rhetorical and representational strategy: one that drives many of his abstract claims into the concrete of everyday experience. Modernity, so he seems to say, is like a swingboat on the midway, poised in an agonizing stasis between past and future.

There is . . . a contradiction in Baudelaire's sense of modernity after the bittersweet experience of "creative destruction" on the barricades of 1848. Tradition has to be overthrown, with violence if necessary, in order to grapple with the present and create the future. But the loss of tradition wrenches away the sheet anchors of our understanding and leaves us drifting, powerless. The aim of the artist, he wrote in 1860, is "to extract from fashion the poetry that resides in its historical envelope," to understand modernity as "the transient, the fleeting, the contingent" as against the other half of art, "the eternal and immovable." The fear . . . is "of not going fast enough, of letting the spectre escape before the synthesis has been extracted and taken possession of." But all that rush leaves behind a great deal of human wreckage. "The thousand uprooted lives" cannot be ignored.[23]

Reflecting on the complex allegories in Baudelaire's *Paris spleen*, Harvey subtly yet powerfully alters his understanding of Hausmann, who becomes something more than capitalism's puppet-master:

---

[22] Harvey, *Consciousness*, p. 204; Doreen Massey, "Flexible sexism," *Environment and Planning D: Society and Space* 9 (1991) pp. 31–57; the quotation is from pp. 47–49. She is in fact referring to Harvey's discussion of Frédéric in his *The condition of postmodernity: An inquiry into the origins of cultural change* (Oxford: Blackwell Publishers, 1989) pp. 263–64. It is, I think, necessary to emphasize that "another" in Massey's sentence. The relations between gender and class were complicated and cannot be folded into a single point: Massey's argument was about the importance of gender, *not* the unimportance of class.

[23] Harvey, *Consciousness*, p. 175.

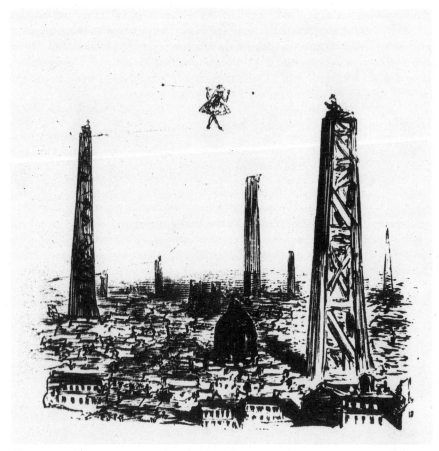

*Figure 18   The triangulation of Paris. Cham's cartoon shows Madame Saqui, the cel-
ebrated aerialist, maintaining communications with the surveyors during the triangulation
project [source: David Pinkney,* Napoleon III and the rebuilding of Paris*].*

The towers from which the triangulation of Paris proceeded symbolized a new
spatial perspective on the city as a whole [figure 18], as did his attachment to the
geometry of the straight line and the accuracy of leveling to engineer the flows of
water and sewage. The science he put to work was exact, brilliant, and demanding;
"the dream" of Voltaire and Diderot had learned to calculate. But there was ample
room for sentiment – from elaborate street furnishings (benches, gas lights, kiosks)
and monuments and fountains (like that in the Place Saint-Michel) to the widespread
planting of trees along the boulevards and the construction of gothic ghettos in the
parks, everything reimported romance into the details of a grand design that spelled
out the twin ideals of Enlightenment rationality and imperial authority. The modernity that Hausman[n] created was powerfully rooted in tradition.[24]

[24] *Ibid.*, p. 178.

And gradually Harvey's awareness of the overdetermination of the politics of Paris during the Second Empire prompts him to reinscribe his earlier and somewhat stylized account of the geometry of the divided city within the fleeting encounters and multiple spheres that make up the geographies of everyday life:

Paris experienced a dramatic shift from the introverted, private, and personalized urbanism of the July Monarchy to an extroverted, public, and collectivized style of urbanism under the Second Empire.... Public investments were organized around private gain, and public spaces appropriated for private use; exteriors became interiors for the bourgeoisie, while panoramas, dioramas and photography brought the exterior into the interior. The boulevards, lit by gas lights, dazzling shop window displays, and cafés open to the street (an innovation of the Second Empire), became corridors of homage to the power of money and commodities, play spaces for the bourgeoisie. When Baudelaire's lover suggests the proprietor might send the ragged man and his children packing, it is the sense of proprietorship over public space that is really significant, rather than the all-too-familiar encounter with poverty....

It was for this reason that the reoccupation of central Paris by the popular classes took on such symbolic importance. For it occurred in a context where the poor and the working class were being chased, in imagination as well as in fact, from the strategic spaces and even off the boulevards now viewed as bourgeois interiors. The more space was opened up physically, the more it had to be partitioned and closed off through social practice. Zola, writing in retrospect, presents as closed those same Parisian spaces that Flaubert had seen as open.[25]

If this provides a gloss on my previous discussion of what Harvey describes elsewhere as Frédéric's "casual penetration" of Parisian space, it still holds to capital and class as the primary axis through which power is inscribed in the city; but it would not, I think, be difficult to open it up to quite other modalities of power. The complex distinctions between "private" and "public," "interior" and "exterior" that run through this passage were also enforced through a series of patriarchal codes whose spatialities were not a reflection of – were sometimes even orthogonal to – those produced through the grid of class relations. While the *flâneur* – the stroller – is a central figure in Benjamin's essays on Baudelaire and Paris, Janet Wolff argues that this was an intrinsically masculine subject-position. The possibility of a *flâneuse* was closed off by the gender divisions of the nineteenth century. "The *flâneur* is the modern hero," she writes, and his freedom to move about the public spaces of the city was simply denied to most women.[26] That said, and as Wolff acknowledges in some

[25] *Ibid.*, p. 204.
[26] Janet Wolff, "The invisible *flâneuse*: Women and the literature of modernity," in her *Feminine sentences: Essays on women and culture* (Berkeley: University of California Press, 1990) pp. 34–50; the

of her qualifying remarks, the distinctions between "private" and "public" did not correspond in any simple fashion to the "feminine" and the "masculine." Although the so-called "private sphere" was coded as feminine, as the seat of domesticity and the woman's workplace, it was plainly a domain of phallocratic power; and although middle-class men did as much as they could to confine and restrict the movement of women through public spaces, (some) women were nonetheless emerging into those spaces (at specific sites). The parentheses indicate the complexities of the situation: the varying topographies of class and gender. Matters are made still more complicated by Elizabeth Wilson's arresting rereading of the *flâneur*:

> The *flâneur* represented not the triumph of masculine power, but its attentuation. A wanderer, he embodies the Oedipal under threat . . . a shifting projection of angst rather than a solid embodiment of male bourgeois power.[27]

Whatever we make of this, it is evidently possible to transpose the partitions between "private" and "public," "interior" and "exterior" into different and irreducible registers (as contemporaries did themselves) and read their coincidences, oppositions, and transgressions as what Benjamin would have called a tense *constellation* of different human geographies.[28] Indeed, Benjamin's twin exposés of the Arcades Project, which were both published as "Paris, capital of the nineteenth century," prefigure the distinctions that Harvey makes in his dialectical rendering of the interiors and exteriors of Second Empire Paris.[29]

I hope it will be obvious that the two textual strategies I am trying to separate here do not correspond to an economic "base" and a politico-cultural superstructure. Harvey's primary narrative is structured by the basal

---

quotation is from pp. 38–39 and p. 47; see also Griselda Pollock, "Modernity and the spaces of femininity," in her *Vision and difference: Femininity, feminism and histories of art* (London: Routledge, 1988) pp. 50–90.

[27] Elizabeth Wilson, "The invisible *flâneur*," *New Left Review* 191 (1992) pp. 90–110; the quotation is from p. 109. Harvey describes himself as a "restless analyst" who likes to "wander the streets, play the *flâneur*" (Harvey, *Consciousness*, p. xv): and if Wilson's reading of the *flâneur* is accepted, then it provides an arresting gloss on the critique of Harvey set out in Rosalyn Deutsche, "Boys town," *Environment and Planning D: Society and Space* 9 (1991) pp. 5–30.

[28] According to Adorno, Benjamin thought of the constellation as "a juxtaposed rather than integrated cluster of changing elements that resist reduction to a common denominator, essential core, or generative first principle": Martin Jay, *Adorno* (Cambridge: Harvard University Press, 1984) p. 15. I will have much more to say about this later.

[29] Benjamin's 1935 exposé has been translated from German into English as "Paris, capital of the nineteenth century," in his *Reflections*, pp. 146–62; the 1939 exposé was originally written in French and may now be found as "Paris, capitale du XIXe siècle," in his *Paris, capitale du XIXe siècle: le livre des passages* (Paris: Éd. du Cerf, 1989) pp. 47–59. The dialectical structure of the exposés is clarified in Anne, Margaret, and Patrice Higonnet, "Façades: Walter Benjamin's Paris," *Critical Inquiry* 10 (1984) pp. 391–419, but their commentary is insufficiently attentive to the differences between the two versions.

logic of the capitalist mode of production, but this secondary text cannot
be compressed into the superstructure. It stubbornly refuses to respect
such a classical dualism. To be sure, the commodity remains at the center
of this "other" historical geography, but its multiple meanings constantly
threaten to escape the narrow-gauge wires of Harvey's circuit diagrams.
"Then, as now," he says at one point,

> the problem was to penetrate the veil of fetishism, to identify the complex of
> social relations concealed by the market exchange of things. . . . It takes experience
> and imagination to get behind the fetishism, and imagination is as much a product
> of interior needs as it is a reflection of external realities.[30]

So it is. But I sometimes worry that a problematic like this can too readily
marginalize material culture, pushing its "surface forms" to one side in its
haste to disclose the social relations so deeply inscribed in the process of
commodification. One of Benjamin's most prescient insights was to establish an interest in cultural artifacts as a legitimate (though hardly self-
sufficient) moment of a critical history. Still more important, if the social
relations that are symbolized – objectified, naturalized, and fetishized – in
this way are not fully transparent, then neither can they be made visible
from a single perspective: experience and imagination are not sutured
around a single (masculine) subject-position. Harvey seems to concede as
much when he agrees that "each particular perspective tells its own
particular truth." And yet, he continues, these perspectives

> scarcely touch each other and they come together on the intellectual barricades
> with about the same frequency as urban uprisings like the Paris Commune. The
> intellectual fragmentations of academia appear as tragic reflections of the confu-
> sions of an urbanized consciousness; they reflect surface appearances, do little to
> elucidate inner meanings and connections, and do much to sustain the confusions
> by replicating them in learned terms.[31]

I am not so sure (or perhaps I am just confused); but I do think that any
attempt to fold "the complex of social relations" concealed by commodity
fetishism into the envelope of a "simple narrative" ought to be resisted.
One way of doing so is by deliberately opening the text to multiple ways
of knowing, refusing to round the edges and plane the differences into a
single integrated setting. In this sense, Harvey's reference to the barricades
can be turned against him. Ross has suggested that their subversive
character derived not only from the way in which they demarcated insurgent

---

[30] Harvey, *Consciousness*, p. 200. It is hardly necessary to point out that the imagery of "penet-
rating the veil of fetishism" is freighted with many of the same assumptions to which I have
already objected.

[31] *Ibid.*, p. 263.

spaces of political action but also from the way in which they were constructed: "Monumental ideals of formal perfection, duration or immortality, quality of material and integrity of design [were] replaced by a special kind of *bricolage* – the wrenching of everyday objects from their habitual context to be used in a radically different way."[32] As I now want to suggest, it was Benjamin's lasting achievement to have realized this "tactical mission of the commonplace," as Ross calls it, in his own luminous investigations of nineteenth-century Paris.

Passages: Walter Benjamin and the Arcades Project

At the center of this world of things stands the most dreamed-of of their objects, the city of Paris itself.
> Walter Benjamin, *Surrealism: The last snapshot of the European*
> *intelligentsia*

I approach Benjamin with considerable caution. In her thoughtful essay on his contemporary significance for cultural studies, McRobbie describes him as an experimentalist at nonfiction. But for all its resonance the term is an awkward one, not least because, as she says later, Benjamin occupied that space where

criticism and creative writing merge into each other and dissolve as separate categories, where fiction and nonfiction also overlap, and where the death of the author coincides with the birth of a different kind of writing and writer. The *Passagenwerk* should be read then like a modernist novel whose major influences were the growth of cinema and photography, a series of speculations and reveries on urban culture, a surrealist documentary, a creative commentary.[33]

McRobbie is right to warn against those appropriations of Benjamin's work that render it in overly analytical terms and risk imposing a misleading conceptual clarity upon it. She claims that in Buck-Morss's imaginative reconstruction of the Arcades Project in particular,

his work is not seen as it might be, that is as a kind of performance art. . . . Without this [Buck-Morss] is forced into seeking some kind of intellectual resolution in the

---

[32] Ross, *Emergence*, p. 36.

[33] Angela McRobbie, "The *Passagenwerk* and the place of Walter Benjamin in cultural studies: Benjamin, cultural studies, Marxist theories of art," *Cultural Studies* 6 (1992) pp. 147–69; the quotation is from pp. 157–58. The *Passagenwerk* refers to the notes and drafts that formed Benjamin's Arcades Project; I elaborate on its provenance below.

concepts and ideas found in the *Passagenwerk* which the work itself is incapable of providing.[34]

And yet Buck-Morss does not confine herself to the *Passagenwerk* but draws on a remarkable knowledge of Benjamin's other writings. I am not the person (and this is not the place) to comment on these differences in detail; all I will say is that no matter how much Benjamin needs to be approached as a creative artist – which he plainly was and which I think Buck-Morss plainly does – it would be misleading to distance him from the equally creative development of a theoretical discourse about culture and cultural history.

Let me try to show how these differences have arisen through a summary of Benjamin's own biography: Autobiography was one of his favored modes of exploration and discovery.[35] Benjamin was born in 1892, the son of prosperous German-Jewish upper middle-class parents. He spent his childhood in Berlin, a period and place recalled in a typically artful way in his *Berlin chronicle*.[36] His parents supported him through university, but in 1925, shortly after his committee refused to accept his *Habilitationschrift* – a postdoctoral requirement for a university teaching post – they lost most of their money and Benjamin was obliged to make ends meet through his writing. In 1926 he moved to Paris, where he embarked upon a series of notes that would eventually thread their way into a vast project woven around the Paris arcades or *Passages*. This undertaking was to occupy him, discontinuously and in different forms, until his death in 1940: It became known as the *Passagenwerk* or *Passagenarbeit*.[37] It was during this first extended period in Paris that Benjamin fell under the spell of surrealism.

[34] *Ibid.*, pp. 156–57; Susan Buck-Morss, *The dialectics of seeing: Walter Benjamin and the Arcades Project* (Cambridge: MIT Press, 1989).

[35] For fuller biographies, see Julian Roberts, *Walter Benjamin* (London: Macmillan, 1982) and Bernd Witte, *Walter Benjamin: An intellectual biography* (Detroit: Wayne State University Press, 1991).

[36] "A Berlin Chronicle," in Walter Benjamin, *Reflections: Essays, aphorisms, autobiographical writings* (ed. Peter Demetz) (New York: Schocken Books, 1989) pp. 3–60. *Berliner Chronik* was edited posthumously by Gershom Scholem, but a fuller (though fragmentary) account is contained in Benjamin's *Berliner Kindheit um Neunzehnhundert, Gesammelte Schriften* IV, 1 (Frankfurt am Main: Suhrkamp Verlag, 1972): see Burkhardt Lindner, "The *Passagen-Werk*, the *Berliner Kindheit* and the archaeology of the 'recent past,'" *New German Critique* 39 (1986) pp. 25–46.

[37] Rolf Tiedemann, the editor of Benjamin's collected works, consistently refers to the Arcades Project – Volume V of the *Gesammelte Schriften* – as *Das Passagen-Werk* (Frankfurt am Main: Suhrkamp Verlag, 1982), but several commentators have objected to this description for its implicit closure. It is, says Sieburth, "an editorial invention that runs the risk of reifying an exploratory process of writing into an inert textual artifact. Benjamin, by contrast, describes his [project] not as a work but as an ongoing event, a peripatetic meditation or *flânerie* in which everything chanced upon en route becomes a potential direction his thoughts might take." See Richard Sieburth, "Benjamin the scrivener," in Gary Smith (ed.), *Benjamin: Philosophy, aesthetics, history* (Chicago: University of Chicago Press, 1989) pp. 13–37; the quotation is from pp. 26–27.

Although he soon disavowed its immanent idealism – he eventually participated in the *Collège de sociologie*, the circle of dissident surrealists founded just before the war by Bataille, Caillois, and Leiris – he always credited Aragon's surrealist novel *Le paysan de Paris* as the most direct inspiration for the Arcades Project. By 1928, however, Benjamin was back in Berlin, contributing to literary journals and newspapers and writing and presenting radio programs. There he witnessed the gradual collapse of the Weimar Republic until, in 1933, the Nazi seizure of power forced him to flee to France. He returned to Paris as an exile – "condemned to a way of life closely resembling that of the emigré extras in Rick's Café in *Casablanca*"[38] – but still managed to live off his writing. From 1935 his modest income was supplemented by a stipend from the Institute of Social Research. Although he had become keenly interested in the writings of the Frankfurt School, as the Institute became known, the relation between his work and that of Adorno or Horkheimer (its principal architects) can perhaps be best described as one of creative tension; certainly, he was no disciple. As far as Adorno was concerned, it was probably the other way round. His debates with Benjamin were an attempt "to establish himself on an equal footing with this man whose disciple he had become in 1929."[39] In any event, if, as Arendt remarks, Benjamin was the most peculiar Marxist ever to be associated with the Frankfurt School – itself hardly a bastion of orthodoxy – there can surely be no doubt that his later writings were saturated in the tonalities of historical materialism.[40]

As this summary perhaps indicates, and as the tussle between McRobbie and Buck-Morss confirms, locating Benjamin within an intellectual landscape is far from straightforward. He has to be placed in relation to, if never entirely within, a Jewish intellectual tradition (Kabbalism) which conceived of the cosmos as an endless network of correspondences and symbolic connections; and yet as late as 1933 he could still remind his close friend Gershom Scholem of "the abyss" of his lack of knowledge of the Kabbala. He was fascinated by surrealism and intensely sympathetic to its attempts to disrupt the conventional expectations of a bourgeois consciousness; and yet he despaired of the surrealists ever taking leave of "the world of dreams" and jolting their audience awake. He was deeply interested in

[38] Peter Demetz, "Introduction," in Benjamin, *Reflections*, p. xiii. McRobbie, "*Passagenwerk*," p. 155 suggests that "precisely because none of the 'advantages' of being white, middle class and male saved Benjamin from the forces of history and from fascism, he might be seen as a symbol of modernity in all its complexity."

[39] Susan Buck-Morss, *The origin of negative dialectics: Theodor Adorno, Walter Benjamin and the Frankfurt Institute* (New York: Free Press, 1977) p. 165.

[40] Hannah Arendt, "Introduction," in Benjamin, *Illuminations*, pp. 1–55; the quotation is from p. 11. For perceptive discussions of Benjamin's relation to historical materialism, see Susan Buck-Morss, *Negative dialectics*; idem, "Walter Benjamin: Revolutionary writer," *New Left Review* 128 (1981) pp. 50–75, 129 (1981) pp. 77–95; Lunn, *Marxism and modernism*, Part III.

developing historical materialism, perhaps even to the point where it would assume a symbolist or surrealist form; and yet he remained uncomfortable at what he took to be its continued proximity to a bourgeois conception of reason and "progress."

These three coordinates can be made to intersect in different ways and, not surprisingly, the contemporary recuperation of Benjamin has allowed itself considerable license. To Eagleton, for example, Benjamin appears as "an archaeologist *avant la lettre*" – where the *lettre* is plainly Foucault's – whose work prefigures many of the current motifs of deconstruction and poststructuralism.[41] Eagleton is not alone in reading Benjamin's work as an archaeology (of sorts), but for my present purposes the affinities that he claims with contemporary critical practice are of more moment.[42] They are further enriched by what Eagleton also calls the "dialectical impudence" of this "Marxist rabbi" – that is to say, Benjamin's attempt to conjure a revolutionary aesthetics *from the commodity form itself.*[43] I think it possible to use this interpretation (in outline, at least) to show that although Benjamin came to place the deconstruction of commodity fetishism at the center of his work, he proceeded in a radically different direction from Harvey. One word of caution is necessary: The Arcades Project is a notoriously difficult textual terrain and what follows is only the roughest map of one possible route through it.

Image and narrative: the enchantment of modernity

Unlike Harvey, Benjamin's conception of history was a profoundly tragic one. He did not see the project of modernity as a process of progressive "disenchantment," in which Reason had gradually released humankind from

[41] Terry Eagleton, *Walter Benjamin or, towards a revolutionary criticism* (London: Verso, 1981) pp. 56, 131. Other commentators have read Benjamin's work as an archaeology of modernity too. To Frisby, for example, Benjamin was "the archaeologist of modernity" who excavated "three spatial, labyrinthine layers of reality" – the arcade, the city, and the mythological underworld that lay beneath the city (represented by the catacombs) – in order to "cut through yet another labyrinth ..., that of human consciousness." David Frisby, "Walter Benjamin: The prehistory of modernity," in his *Fragments of modernity* (Cambridge, England: Polity Press, 1985) pp. 187–265; the quotation is from pp. 210–11; see also Marc Sagnol, "La méthode archéologique de Walter Benjamin," *Les temps modernes* 40 (444) (1983) pp. 143–65.

[42] For a sympathetic discussion of the relations between deconstruction and Benjamin's (early) writings, see Christopher Norris, "Image and parable: Readings of Walter Benjamin," in his *The deconstructive turn: Essays in the rhetoric of philosophy* (London: Methuen, 1983) pp. 107–27. Derrida draws attention to the affinities between his own work and some of Benjamin's later ideas in Jacques Derrida, *The truth in painting* (Chicago: University of Chicago Press, 1987) pp. 175–81; this was first published in French in 1978.

[43] Terry Eagleton, "The Marxist rabbi: Walter Benjamin," in his *The ideology of the aesthetic* (Oxford: Blackwell Publishers, 1990) pp. 316–40; the quotation is from p. 326.

the snares of myth and superstition.[44] If, as Harvey says, Marx was a child of the Enlightenment, then Benjamin's own childhood memories suggested that the modern world was a place of reenchantment. This was not, I think, a nostalgic vision of an age of lost innocence, as it was for so many German intellectuals who mourned the encroachments of a modern and intensely rational *Zivilisation* upon a traditional and essentially spiritual *Kultur*, or even the neo-Romantic prospect of a revolutionary utopia that animated Bloch and Lukács.[45] It was, rather, a highly distinctive way of thinking about commodity fetishism, which decisively shaped the form in which Benjamin wanted to present his account. I will focus on two features in particular: Benjamin's fascination with the image and his objections to linear narrative.

In his early drafts, Benjamin proposed to read the Paris Arcades (figure 19) as a dream world – "a dialectical faery scene" – furnished by objects that were simultaneously desired and commodified. Awakening would constitute the moment when "the spell or illusion of reconciling a desire for fulfillment with a structure of exploitation and alienation [could] be broken"[46] In his more developed sketches, which he worked on following his return to Paris, Benjamin both deepened and widened the project by representing the cultural landscape of nineteenth-century Paris (and, by extension, Europe) as a *phantasmagoria*. The phantasmagoria was a magic lantern that became popular in the early decades of the nineteenth century. Painted slides were illuminated in such a way that a succession of ghosts ("phantasms") was paraded before a startled audience. But the phantasmagoria was no ordinary lantern, because it used back-projection to ensure that the audience remained largely unaware of the source of the image: Its flickering creations thus appeared to be endowed "with a spectral reality

---

[44] Here I follow Michael Jennings, *Dialectical images: Walter Benjamin's theory of literary criticism* (Ithaca: Cornell University Press, 1987). It is only fair to add that other commentators portray Benjamin in a somewhat more optimistic light: notably Jürgen Habermas, "Walter Benjamin: Consciousness raising or rescuing [redemptive] critique," in his *Philosophical-Political profiles* (Cambridge: MIT Press, 1983) pp. 131–63 and Richard Wolin, *Walter Benjamin: An aesthetic of redemption* (New York: Columbia University Press, 1982).

[45] Cf. Michael Löwy, "Revolution against 'Progress': Walter Benjamin's romantic anarchism," *New Left Review* 152 (1985) pp. 42–59.

[46] Helga Geyer-Ryan, "Counterfactual artifacts: Walter Benjamin's philosophy of history," in Edward Timms and Peter Collier (eds.), *Visions and blueprints: Avant-garde culture and radical politics in early twentieth-century Europe* (Manchester, England: Manchester University Press, 1988) pp. 66–79; the quotation is from p. 67. The emphasis on "re-awakening" indicates that Benjamin had already moved sufficiently far from surrealism to use Proust rather than Aragon as his model; the moment of awakening was analogous to the shock of recognition produced by Proust's *mémoire involontaire*. See Sieburth, "Benjamin the scrivener," p. 18; Ackbar Abbas, "Walter Benjamin's collector: The fate of modern experience," in Andreas Huyssen and David Bathrick (eds.), *Modernity and the text: Revisions of German modernism* (New York: Columbia University Press, 1990) pp. 216–39; see especially pp. 226–29.

Figure 19   Passage de l'Opéra, Galerie du Baromètre, Paris [photograph: Marville, ca. 1870].

of their own."[47] Benjamin uses the phantasmagoria as an allegory of modern culture, which explains both his insistence on seeing commodity culture as a projection – not a reflection – of the economy, as its mediated (even mediatized) representation, and also his interest in the visual, optical, "spectacular" inscriptions of modernity. Indeed, Benjamin was one of the earliest commentators to understand the centrality, the constitutive force, of the image within modernity. What he proposed to do, in effect, was to harness the latent energy of the modern image, to turn it back on itself and thereby use that image as "a critique of reason."[48]

This project coincided with Benjamin's critique of narrativity. He believed that one of the most powerful ways in which the modern world had become enchanted was through the operation of Reason itself. In its name, meta-narratives had been imposed upon human history to purge it of its specificities and present historical eventuation as a continuous, organic, and homogeneous progression toward the present. It was not merely the "progressive" conception of history to which Benjamin took exception – that this is reinforced by the use of a conventional narrative form is self-evident[49] – but also its "homogenization." In a superbly sustained reading of Benjamin, and one to which I shall want to return, Eagleton suggests that he saw the circulation of the commodity as paradigmatic of this process:

The commodity, which flaunts itself as a unique, heteroclite slice of matter, is in truth part of the very mechanism by which history becomes homogenized. As the signifier of mere abstract equivalence, the empty space through which one portion of labor-power exchanges with another, the commodity nonetheless disguises its virulent anti-materialism in a carnival of consumption. In the circulation of commodities, each presents to the other a mirror which reflects no more than its own mirroring; all that is new in this process is the very flash and dexterity with which mirrors are interchanged.... The exchange of commodities is at once smoothly continuous and an infinity of interruption: since each gesture of exchange is an exact repetition of the previous one, there can be no connection between them. It is for this reason that the time of the commodity is at once empty and homogeneous: its homogeneity is, precisely, the infinite self-identity of a pure recurrence which, since it has no power to modify, has no more body than a

[47] Margaret Cohen, "Walter Benjamin's Phantasmagoria," *New German Critique* 48 (1989) pp. 87–107; see also Terry Castle, "Phantasmagoria: Spectral technology and the metaphorics of modern reverie," *Critical Inquiry* 15 (1988) pp. 26–61 and Jonathan Crary, *Techniques of the observer: On vision and modernity in the nineteenth century* (Cambridge: MIT Press, 1990).

[48] Ackbar Abbas, "On fascination: Walter Benjamin's images," *New German Critique* 48 (1989) pp. 43–62; the quotation is from p. 52. See also Benjamin's celebrated essay on "The work of art in the age of mechanical reproduction," in his *Illuminations*, pp. 217–51.

[49] "Narrative was useless for his purpose because the notion of progress is built into its structure of continuity, the form whereby the later not only grows out of the earlier utterance but subsumes, surpasses or completes it in complexity and inclusivity": Brigitte Frase, "Raising debris," *Hungry Mind Review* 16 (1990-91) pp. 20–21. For a more general discussion, see Hayden White, *The content of the form: Narrative discourse and historical representation* (Baltimore: Johns Hopkins University Press, 1987).

mirror-image. What binds history into plenitude is the exact symmetry of its repeated absences. It is because its non-happenings always happen in exactly the same way that it forms such an organic whole. Since the significance of the commodity is always elsewhere, in the social relations of production whose traces it has obliterated, it is freed ... into polyvalence, smoothed to a surface that can receive the trace of any other commodity whatsoever. But since these other commodities exist only as traces of yet others, this polyvalence is perhaps better described as a structure of ambiguity – an ambiguity that for Benjamin is "the figurative appearance of the dialectic, the law of the dialectic at a standstill."[50]

Benjamin sought to interrupt this process by calling into question its endless suppression of difference beneath repetition. What was distinctive about his attempt to do so was that it went beyond disclosure of the logic of capital *to assault the modalities of representation*. A concern with what Wolin calls the "image-character" of truth became a vital moment in Benjamin's work, therefore, "for in this way he sought to confer equal rank to the spatial aspect of truth and thereby do justice to the moment of representation that is obscured once truth is viewed solely as a logical phenomenon."[51] In other words, Benjamin effectively "spatialized" time, supplanting the narrative encoding of history through a textual practice that disrupted the historiographic chain in which moments were clipped together like magnets.

In practice, this required him to reclaim the debris of history from the matrix of systematicity in which historiography had embedded it: to blast the fragments from their all-too-familiar, taken-for-granted and, as Benjamin would insist, their *mythical* context and place them in a new, radically heterogeneous setting in which their integrities would not be fused into one.[52] This practice of montage was derived from the surrealists, of course, who used it to dislocate the boundaries between art and everyday life. Benjamin displayed much the same interest in the commonplace in his *Einbahnstrasse* – which Bloch saw as an attempt to give philosophy a surrealist form[53] – and in the notes and drafts for the Arcades Project itself. His intention was "to carry the montage principle over into history" and use it to present nineteenth-century Paris – that glittering world which all the world came to see – as "a rubble heap of found objects" – an image that he hoped would dispel the established vision of the imperial past.[54] In many ways montage was another version of that antihierarchical gesture that Ross found in the strategies of the Commune, but it is important to see that in Benjamin's hands its transgressive power was also

---

[50] Eagleton, *Walter Benjamin*, pp. 28–29.

[51] Wolin, *Walter Benjamin*, p. 100.

[52] Geyer-Ryan, "Counterfactual artifacts," pp. 66–68; Jennings, *Dialectical images*, p. 51.

[53] Walter Benjamin, *Einbahnstrasse* (Berlin: Ernst Rowohlt, 1928) is available in English translation as "One Way Street" in his *One way street and other writings* (London: New Left Books, 1979) pp. 45–106.

[54] Frisby, *Fragments*, p. 215; Frase, "Raising debris," p. 20.

used to challenge the idea of an Olympian metanarrative through what Sieburth perceptively identifies as an attempt "to inscribe citation and commentary, the text and its interpretation on the same plane of the page."

By situating its "primary" and "secondary" reflections on the same textual . . . level, Benjamin's manuscript places into radical question the very possibility of metalanguage, that is, of a discourse that might somehow stand above, outside or beyond that of which it speaks. . . . [His handwritten pages] democratize the traditional hierarchies separating author from reader, original from copy, citation from commentary – like the passage itself, the text that thus emerges has no outsides. In this it resembles the polyphonic play of language which Bakhtin terms heteroglossia – different voices, different discourses refracting each other in dialogue.[55]

These concerns struck at the heart of the traditional aesthetic paradigm, whose strictures refused to allow what Eagleton calls "the specificity of the detail" any genuine resistance to "the organizing power of the totality." Benjamin's way of working – of seeing? – was directed, in contrast, toward the construction of a *constellation* that would embody "a stringent economy of the object which nevertheless refuses the allure of identity, allowing its constituents to light each other up in all their contradictoriness."[56]

These are not easy ideas to grapple with, but I hope it is now possible to glimpse the field in which Benjamin's fascination with the image and his critique of narrativity could be brought together. As Buck-Morss puts it in her own version of the book that Benjamin never actually wrote – which she presents as "a picture book of philosophy" putting into practice Benjamin's "dialectics of seeing" – he was convinced that

what was needed was a visual, not a linear logic. The concepts were to be imagistically constructed, according to the cognitive principles of montage. Nineteenth-century objects were to be made visible as the origin of the present, at the same time that every assumption of progress was to be scrupulously rejected.[57]

The conceptual form that Benjamin eventually gave to this textual strategy was the construction of what he called the "dialectical image."

The dialectical image

I should say at once that the noun is as difficult as the adjective. Many commentators do indeed draw attention to Benjamin's preoccupation with

---

[55] Sieburth, "Benjamin the scrivener," p. 33. This rebounds nicely against Richard Dennis's objections that Harvey's essay on Paris is an exercise in "trading and speculating on the labor of others": Richard Dennis, "Faith in the city?" *Journal of Historical Geography* 13 (1987) pp. 310–16; the quotation is from p. 311.

[56] Eagleton, "Marxist rabbi," p. 330.

[57] Buck-Morss, *Dialectics of seeing*, pp. ix, 218.

the *visual* image: with that sense of the optical and spectacular that is preserved so emphatically in the title of Buck-Morss's account of the Arcades Project – "the dialectics of seeing" – and that is captured, somewhat more loosely, in Jay's meditation on what he calls the "scopic regimes of modernity."[58] But others are more reluctant to conceive of the image in strictly visual terms. In his discussion of the dialectical image, for example, Jennings emphasizes not so much the optical implications of the concept as Benjamin's rehabilitation of allegory as a "revelatory instrument" that by virtue of its "brokennness," its resistance to mimesis, was "the only form of language capable of resisting the allure of the commodity."[59] Eagleton agrees that this break between signifier and signified at once mimed the circulation of the commodity and released "a fresh polyvalence of meaning, as the allegorist grubs among the ruins of once integral meanings to permutate them in startling new ways." But, like Buck-Morss, he also suggests that in Benjamin's later work the sense in which the allegorical signifier could inscrible "its own network of 'magical' affinities across the space of an inscrutable history" was given a new inflection in the dialectical image itself. And if Benjamin was deliberately reviving allegorical emblematics in the Arcades Project, one might argue that its subversive inscriptions were reinforced in avowedly iconographic terms through a reading of photographs, exhibitions, architectural forms, and the like.[60] The filiations between the visual image and the linguistic image are likely to be extremely complicated, therefore, and I think one has to recognize that there may well be something essential to Benjamin's project that cannot be grasped through the visual image alone.[61]

Buck-Morss argues that Benjamin produced the dialectical image by wrenching the fragments of the past out of their usual context and placing them on the conceptual grid shown in figure 20. She identifies the two axes – waking/dreaming and petrified/transitory – in conventional Hegelian terms – consciousness and reality – but one could also suggest that the lower arc describes the trajectory of surrealism, the resolution of dream and reality into what Breton called "surreality," while the upper arc describes the trajectory of historical materialism, the physical inscriptions of a political consciousness.[62]

---

[58] *Ibid.*, Martin Jay, "Scopic regimes of modernity," in Hal Foster (ed.), *Vision and visuality* (Seattle: Bay Press, 1988) pp. 3–23; an extended version appears in Scott Lash and Jonathan Friedman (eds.), *Modernity and identity* (Oxford: Blackwell Publishers, 1992) pp. 178–95, in which Jay seeks to connect these scopic regimes to distinctive urban spatialities.

[59] Jennings, *Dialectical images*, pp. 110, 170–71; see also Lloyd Spencer, "Allegory in the world of the commodity: The importance of *Central Park*," *New German Critique* 34 (1985) pp. 59–77.

[60] Eagleton, "Marxist rabbi," pp. 326–27; Buck-Morss, *Dialectics of seeing*, pp. 71, 170–76.

[61] See Shierry Weber Nicholsen's review of Buck-Morss, *Dialectics of seeing*, in *New German Critique* 51 (1990) pp. 179–88.

[62] The proximity of Hegel to historical materialism or surrealism is far from simple. Jay's description of Hegel as a thinker "whose importance for the Western Marxist tradition was second

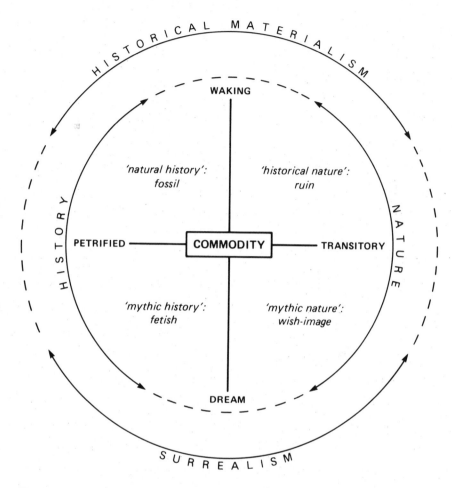

Figure 20   Benjamin's dialectical image [after Buck-Morss].

only to that of Marx himself" is unexceptional: Indeed, in what is often construed as one of the founding texts of Western Marxism the young Lukács announced his desire to achieve "a Hegelianism more Hegelian than Hegel." Nevertheless, as Jay's remarkable intellectual odyssey demonstrates, the reception of Hegel's ideas has been highly uneven. Many writers in the turbulent world of the 1920s and 1930s thought it impossible "to see things whole," as Jay puts it, and the architects of the Frankfurt School (notably Horkheimer and Adorno) were particularly instrumental in sounding the retreat from Hegelian Marxism. Jay notes that the Surrealists "were among the first in France to recover the importance of Hegel for revolutionary thought": but even there he recognizes substantial differences between them that revolved, precisely, around their contrasting views of the prospects for a rationalist resolution of contradictions in some final synthesis. See Martin Jay, *Marxism and totality: The adventures of a concept from Lukács to Habermas* (Cambridge, England: Polity Press, 1984); the quotations are from pp. 53, 219, 285.

If this is an acceptable approximation, then the two arcs can be folded over one another in a relation of extreme tension, as Benjamin sought to give surrealism a materialist form and to develop historical materialism in a surrealist form. Another, equally tense relation can be produced by reading Benjamin through Adorno's eyes and seeing the left-hand arc as "history" and the right-hand arc as "nature." To the two friends, these were mutually determining concepts constitutively involved in the formation of historical materialism.

History and nature were not abstract "invariant concepts" but "arrange themselves around the concrete historical facticity," forming a constellation which released in the phenomena the moment of transitoriness which might break their mythical spell over the present.[63]

Other interpretations of this grid are no doubt possible, but what matters is its dialectical structure – the way in which Benjamin sought to exploit the interpenetrations of successive conceptual couples. At any rate, Buck-Morss suggests that this screen of polarities may be regarded as the "invisible, inner structure" of the Arcades Project.[64] Benjamin was not, I think, much interested in the fragments in themselves – he was certainly no empiricist – and neither did he concern himself with elaborating the grid in any systematic way. If I understand Buck-Morss properly, Benjamin's hope was rather that when the fragments were placed on this conceptual grid a moment of "profane illumination" would be generated. The dialectical image would appear, however fleetingly, at the null point of the grid, as what today we might think of as a sort of philosophico-political hologram. By this means, so Benjamin believed, the commodity (which had come to occupy a central place within the Arcades Project) would be seen as an historical object doubly constituted in the act of representation: *doubly* constituted because it was both displayed for sale (there and then) and disclosed in its duplicities (here and now).[65] The historicity of the object

[63] Buck-Morss, *Negative dialectics*, p. 57; for a fuller discussion, see pp. 52–62.

[64] Buck-Morss, *Dialectics of seeing*, pp. 210–12. I cannot imagine that "structure" is being used here in any very technical sense. Buck-Morss's rendering of the Arcades Project undoubtedly risks imposing an imaginary coherence upon Benjamin's corpus, but I am at a loss to understand how Jameson can describe the constellation as "virtually Althusserian *avant la lettre*; it also still retains something like a nostalgia for centeredness and for unified (if not necessarily organic) form": Fredric Jameson, *Late Marxism: Adorno, or the persistence of the dialectic* (London: Verso, 1990) p. 244.

[65] "Each field of the co-ordinates can...be said to describe one aspect of the physiognomic appearance of the commodity, showing its contradictory 'faces' ": *ibid.*, p. 211. Buck-Morss devotes four of her chapters to discussions of the four fields, but I should emphasize that this grid is a *plenary* conception of the Arcades Project that was put into practice through *multiple* dialectical images: arcades and exhibitions, wax figures and mechanical dolls, the *flâneur* and the prostitute, and so on. For a more focused discussion, see *idem*, "The *flâneur*, the sandwichman and the whore: The politics of loitering," *New German Critique* 39 (1986) pp. 99–140.

thus referred not to the passage of time but to what McRobbie calls "the pacing of capitalist production manifest in the language of consumer culture": not the old but the outmoded.[66]

Benjamin achieved his emphasis on the *constitution* of the historical object *as* a dialectical image through the sheer force of the adjective.[67] The dialectical nature of the image, the profanity of its illumination, required a break with the conventional historiographic pietism that the present had to be seen in the light of a sepulchral past whose beams would usher historical eventuation into the future. Instead, Benjamin sought to bring about an explosion that would bring down the Dream House of History by forcing a discarded, forgotten, even repressed past into an unfamiliar, unreconciled constellation with the present. Like the surrealists, his intention was to shock: to stage a "confrontation in which time is arrested to a compact monad, spatialized to a shimmering field of force, so that the political present may redeem an endangered moment of the past by wrenching it into illuminating correspondence with itself."[68]

Unlike the surrealists, however, Benjamin's constellation was not an arbitrary construction: the commodity was not placed at the center of the grid by chance. Indeed, Buck-Morss contends that Benjamin's sense of objective necessity distances his way of working from allegory and deconstruction too. She claims that the meaning of the allegorical image remains an expression of subjective intention and so "is ultimately arbitrary" – in contrast to the dialectical image – and that the carnival of deconstruction reduces revolutionary criticism to an endless parade of new interpretations: "Fashion masquerades as politics."[69] If Benjamin's project can be seen as a deconstruction of commodity fetishism, therefore, this has to be a deconstruction of a very different kind: one which somehow invests the text with new yet *necessary* meaning. But Benjamin's attempt to square this particular circle was not uniquely determined by the conceptual thematics of historical materialism (important though these were). It depended, crucially, on the present being understood as a moment of revolutionary possibility, on the notion of a revolutionary "now time" or *Jetztzeit*, which acted as what Buck-Morss calls "a lodestar for the assembly of historical fragments." Without its "power of alignment," she continues, "the possibilities for reconstructing the past [would have been] infinite and arbitrary."[70]

---

[66] McRobbie, "*Passagenwerk*," p. 166.

[67] "The historical object is... something constituted in the act of writing history or criticism. It is in fact identical with... the dialectical image": Jennings, *Dialectical images*, p. 205.

[68] Eagleton, "Marxist rabbi," p. 327.

[69] Buck-Morss, *Dialectics of seeing*, pp. 241, 339.

[70] *Ibid.*, pp. 338–39; see also Jürgen Habermas, *The philosophical discourse of modernity* (Cambridge, England: Polity Press, 1987) pp. 11–16. It is for this reason that Ivornel elects to write of "Paris, Capital of the Popular Front": as a way of "explicitly [introducing] Benjamin's own present into

And yet the constellation was not a means of closure; "necessity" was not transmuted into teleology. Benjamin's purpose was to prise open the texture of historical eventuation and create a space for revolutionary political action. The moment of *Jetztzeit* was supposed to contain within itself, as an immanent possibility, an interruption in the empty, homogeneous time of conventional history. Benjamin's belief was as complex as it was controversial, deriving as it did from both surrealism and Jewish mysticism, but a rough approximation of what he had in mind can be gained from his final thesis on the philosophy of history:

> The Jews were prohibited from investigating the future. The Torah and the prayers instruct them in remembrance, however. This stripped the future of its magic . . . [but] does not imply . . . that for the Jews the future turned into homogeneous, empty time. For every second of time was the strait gate through which the Messiah might enter.[71]

I have no wish to minimize the difficulties of Benjamin's strategy, which inhere in its mythical-Messianic genealogy and its radicalization of modern conceptions of temporality and historical consciousness. Nevertheless, the (weak) Messianic power that Benjamin invoked was, I think, a distinctly worldly one.[72] In his eyes, within his "materialist optics," the moment of profane illumination – the spark of the dialectical image – had the power to intervene in the directionality imposed upon human history by tradi-

---

the work of disengaging the past." Ivornel is aware that such a title signposts only one constellation. "It should not be forgotten that this moment and this place (Paris 1934–40) form the center of a vortex . . . from which concentric circles expand to include the contemporary history of the whole of Europe." See Phillipe Ivornel, "Paris, Capital of the Popular Front or the posthumous life of the nineteenth century," *New German Critique* 39 (1986) pp. 61–84; the quotation is from pp. 69–70. Buck-Morss attempts a similar genealogical mapping in her *Dialectics of seeing*, pp. 8–43 and when she identifies Naples, Moscow, Paris and Berlin as "points of [Benjamin's] intellectual compass" (p. 40) she presumably prefigures her discussion of the "lodestar" which orientates the construction of the constellation.

[71] Walter Benjamin, "These on the philosophy of history," in his *Illuminations*, pp. 253–64; the quotation is from p. 264. This essay was intended to serve as the preface to the Arcades Project; it was completed in the spring of 1940.

[72] Cf. Wolin, *Walter Benjamin*, p. 264: "Benjamin's relevance for historical materialism is to be found in [his] late attempt to secularize the notion of redemptive criticism." Even so, Wolin's emphasis on the secular and my description of this Messianic power as "worldly" does not mean that Benjamin's project is without important implications for political theology: see Thomas McCarthy, "Critical theory and political theology: The postulates of communicative action," in his *Ideas and illusions: On reconstruction and deconstruction on contemporary critical theory* (Cambridge: MIT Press, 1991) pp. 200–15. See also Stéphane Mosès, "The theological-political model of history in the thought of Walter Benjamin," *History and Memory: Studies in Representation of the Past* 1 (1989) pp. 5–33; Rolf Tiedemann, "Historical materialism or political messianism? An interpretation of the Theses 'On the Concept of History,' " in Smith (ed.), *Benjamin*, pp. 175–209; Irving Wohlfarth, "Re-fusing theology: Some first responses to Walter Benjamin's Arcades Project," *New German Critique* 39 (1986) pp. 3–24; Wolin, *Walter Benjamin*, pp. 107–37.

tional, bourgeois narratives of progress. To put it still more concretely, conceiving of the "history of the present" in this way (and by this means) was a way of empowering the emancipatory production of human geographies. It would surely be difficult to think of a more brilliant illustration of the politics of representation.

The Vega cap: Allan Pred and fin-de-siècle Stockholm

A Vega *cap finally tops his head.*

Allan Pred, *Lost words and lost worlds*

Pred's account of Stockholm centers on the period between 1880 and 1900:

A twenty year span during which modernity made its many guised entrance, during which consumption came to be marked by a whirl of quickly passing fashions and fads, during which commodity fetishism took hold, during which the circulation of money accelerated, during which the iron cage of bureaucratic regulation and surveillance dropped down over previously ungoverned details of daily existence, during which everyday life on the streets was characterized by restless and anonymous movement, by fleeting, fragmented impressions.[73]

I want to show how Pred brings these restless geographies into focus by making a series of comparisons between his project and the reconstructions of Second Empire Paris offered by Harvey and Benjamin. I begin by situating his work in relation to "mainstream" historical materialism. There are many points of contact between them, if sometimes at several removes, but Pred is clearly unconvinced by the closures of classical Marxism – in particular its silence over space – and so, like Harvey, he endorses the development of a materialist analysis of the production of space.[74] But it is equally obvious that Pred also wants to go beyond historico-geographical materialism – largely, I think, because he regards its spatial analytics as too unyielding to accommodate the contingencies of everyday life as they unfold in place.[75] This sense of uncertainty has more than a theoretical valency,

---

[73] Allan Pred, *Lost words and lost worlds: Modernity and the language of everyday life in late nineteenth-century Stockholm* (Cambridge: Cambridge University Press, 1990) p. xiii.

[74] This is true in both a general and a specific sense. Most obviously, Pred constantly accentuates the materiality of social life – its "material continuity" and "physicality" – but there is another thread running through his writings in which he more obliquely positions his work in relation to historical materialism. This is usually accomplished with reference to particular traditions within Western Marxism, broadly conceived, and most concretely through those whose work has influenced social history in seminal ways: E. P. Thompson and Raymond Williams. See, for example, Allan Pred, "Social reproduction and the time-geography of everyday life," *Geografiska Annaler* 63B (1981) pp. 5–22.

[75] See Allan Pred, "Place as historically contingent process: Structuration theory and the time-geography of becoming places," *Annals of the Association of American Geographers* 74 (1984)

however, because Pred increasingly argues that it has textual implications too. He is particularly disconcerted by the imposition of a single meta-narrative on what he takes to be the multiple constellations of the past:

There cannot be one grand history, one grand human geography, whose telling only awaits an appropriate metanarrative. Through their participation in a multitude of practices and associated power relations, through their participation in a multitude of structuring processes, people make a plurality of histories and construct a plurality of human geographies.[76]

My main concern will thus be the ways in which Pred elects to represent the concrete particularities of fin-de-siècle Stockholm, where the parallels with Harvey and Benjamin seem to me particularly instructive.

Materialism and modernity

Like Harvey, Pred charts the rising tide of commodification: "the carousel of consumption and commodity fetishism – the merry-giddy-go-round and around and around of money in circulation ... [that] hailed the entrance of modernity in the capital cities of Europe." And he too sees the city pounded by waves of speculative investment and plunged into a maelstrom of creative destruction. "From the early 1880s onward, when there was a high-tempo speculative investment of overaccumulated capital in the city's construction sector, Stockholm was more than figuratively a place where all that was solid melts into air." The allusion is to Marx's description of modernity as a world in which "all fixed, fast-frozen relations, with their train of ancient and venerable prejudices and opinions are swept away, all new-formed ones become antiquated before they can ossify," a world in which "all that is solid melts into air ... "[77] Where Harvey draws on Marx for a more formal rendering of these transformations, however, Pred offers only impressionistic outlines. He does not attempt to reconstruct the circuits of capital through which these pulsating changes were set in motion and brought so crashingly together. There is little sign of the theoretical notes that score Harvey's recital of Second Empire Paris, so to speak, only their echoes in the murmurs of everyday life.

---

pp. 279–97; *idem, Place, practice and structure: social and spatial transformation in southern Sweden 1750–1850* (Cambridge, England: Polity Press, 1986).

[76] Allan Pred, "Making histories and constructing human geographies," in his *Making histories and constructing human geographies* (Boulder, Colo.: Westview Press, 1990) pp. 3–40; the quotation is from p. 14.

[77] Pred, *Lost Words*, pp. 51, 122; Marshall Berman, *All that is solid melts into air: The experience of modernity* (London: Verso, 1982) p. 21. Marx's original account will be found in the *Communist Manifesto*.

Like many others working at the interface of social theory and human geography, Pred invokes Marx's reminder in the *Eighteenth Brumaire of Louis Bonaparte* that "men [and women] make their own history; but they do not make it just as they please; they do not make it under circumstances chosen for themselves, but under circumstances directly encountered, given and transmitted from the past."[78] This is a deceptively simple formulation, however, and in working out its implications Pred borrows from traditions of social theory which, while they certainly have something in common with historical materialism, are nonetheless removed from most mainstream versions of Marxism. Among the most important are Bakhtin's discourse theory, Bourdieu's symbolic anthropology, Foucault's "history of the present" and Giddens's structuration theory. This is a heterogeneous list, of course, but it is more than a collection of materials that just happen to be at hand. Pred strongly implies that there is an elective affinity between these various discourses; or, to put things somewhat differently, that their "multiple voices" articulate a set of common concerns that reach beyond Marx's seminal discussions of praxis to become central to the contemporary reconstruction of critical social theory.[79] He thinks it possible to use the ideas assembled from these various sources in concert with ideas derived from human geography to inscribe propositions about social practice in time and space. His purpose is to illuminate the construction of historical geographies in ways that transcend – and trangress – the logic of capital etching its signature on successive landscapes of accumulation.

The urban morphologies that loom so large in the landscape of Harvey's Paris provide only a skeletal backdrop for Pred's Stockholm. Although he too connects "creative destruction" to the spatial polarization of class relations, Pred is more interested in the ways in which this modern *grille* was imposed on the skeins of interaction that were spun, snapped and spliced together in the course of everyday life. These paths intersected in myriad ways that were not regulated by the residential lattice of class relations. The bourgeoisie and the working class may have lived in different parts of the city but their day-to-day activities brought them into constant, unavoidable and unpredictable contact with one another. If most members

[78] See, for example, Allan Pred, "After words on then and there, here and now, and afterwards," in his *Making histories*, pp. 228–34. The *Eighteenth Brumaire* has a much more direct and concrete relevance to Harvey's account of Second Empire Paris, of course, but he is also alert to its metatheoretical implications. He commends the *Eighteenth Brumaire* as a model of the process of reflection that is indispensable for anyone seeking to negotiate "the path between the historical and geographical grounding of experience and the rigors of theory construction": that is to say, "the evaluation of experience, a summing up that can point in new directions, pose new problems and suggest fresh areas for historical and theoretical enquiry": Harvey, *Consciousness*, p. xvi.

[79] Pred once described this as an "emerging consensus" in social theory whose roots supposedly lay in the marchlands between realism and Marxism: Pred, "Social reproduction," p. 5. I am not sure that he would (or could) still hold to such a view.

of Stockholm's bourgeoisie felt secure enough inside their houses, their vulnerability once they were outside the front-door was almost palpable.

Nowhere did many of the bourgeoisie sense more heightened apprehension or greater threat than amidst the welter of activities and bustling movement of the city's streets. There one might be cast adrift in a ... sea of pedestrian promiscuity, where the banker and the bum, the wholesaler and the whore, the retailer and the ragpicker, the respectable and the disrespectful, the high and the low, the clean and the dirty, flowed and jostled, side-by-side, over the same spaces.[80]

Their characteristic response, as in Second Empire Paris, was to try to turn these savage "exteriors" – the streets – into domesticated "interiors," and it is hardly surprising that Pred should interpret the bourgeois politics of space in Stockholm as another version of Hausmannization.[81] But Pred is less interested in the logistics of this strategy – in its class filiations and geometries – than in its "ideologistics," and in tracing its inscription (and subversion) in the language of everyday life he goes some considerable distance beyond Harvey's "other" historical geography.

His inquiry depends on a close connection between "lost words" and "lost worlds." To ordinary language philosophers, the two are fused in the fluidity of language-games, but Pred sutures them around constellations of power and practice embedded in place. "To uncover lost words," he declares, "is to lift the lid from a treasure chest of past social realities, to reveal fragments shimmering with the reflections of lost worlds of everyday life." Moved by the same instincts that animated Benjamin's collector, Pred unpacks distinctive glossaries of production and consumption, of spatial orientation, of social distinction, and shows how these otherwise unpromising attics contain the discarded furniture of people's everyday lives.[82] His recovery of a "popular geography" existing alongside, underneath, and often in place of the official toponymy of the city is perhaps closest to the spirit of what Sieburth describes as Benjamin's "onomastic inventory."[83] Benjamin was profoundly aware of the linguistic universe conserved in the city and

---

[80] Pred, *Lost Words*, p. 129.

[81] *Ibid.*, pp. 133–34.

[82] *Ibid.*, pp. 7–8. The imagery is mine; in fact Benjamin talks about unpacking cases of books – "what memories crowd in upon you!" he declares – but the *Passagenarbeit* suggests a more general rummaging around the accumulated debris of the past. The figure of the collector is nonetheless apposite, however, because Benjamin argues that "for a true collector the whole background of an [object] adds up to a magic encyclopaedia whose quintessence is the fate of his object." In other words, what matters is not the object alone – Benjamin was dismissive of contemporary cultural history precisely because it fetishised the object – but rather the constellation of past and present fragments in which it is placed. See Walter Benjamin, "Unpacking my library," in his *Illuminations*, pp. 59–67 and Konvulut H of *Das Passagenwerk*; for a discussion, see Abbas, "Walter Benjamin's Collector," especially pp. 231–32.

[83] Sieburth, "Benjamin the scrivener," p. 15. Onomastics is the study of the origin and forms of names of persons or places. "As he charts the cultural topography of nineteenth-century Paris,"

he devoted one of the files of the *Passagenarbeit* (Konvolut P) to a collection-contemplation of the street names of Paris:

All that remains given of the increasingly swift dissipation of perceptual worlds is "nothing other than their names: *Passagen*. . . . The forces of perversion work deep within these names, which is why we maintain a world in the names of old streets."[84]

In much the same way, Pred treats the volatility of naming as an allegory of commodification. The "ephemeral commodities and consumption practices" that punctuated the whirling world of modernity "were also one with a wealth of lost words and meanings . . . that either disappeared after decades or centuries of usage, or briefly appeared some time between 1880 and 1900 only to fade away during that period or shortly after the turn of the century."[85] Benjamin believed that the inventory of these names not only revealed the elements of a "mythical geography," a topographical domain charged with mythical significance, but also simultaneously preserved and prefigured what he called "the habitus of a lived life."[86] It is in this latter sense that Pred seeks to bring to light the undervocabulary, the subversive, rough-and-tumble street names and the scatalogical and unashamedly sexist transpositions that at once undermined and challenged the hegemonic inscriptions of the bourgeoisie on the symbolic landscape of modernizing Stockholm and also reinscribed and reinforced a series of masculinist assumptions about gender and sexuality.[87]

Spaces of representation and everyday life

Although Pred refuses to submit the multiple historical geographies of Stockholm to the discipline of a singular narrative, this does not commit his inquiries to the prison of the parochial. He wants his empirical findings to speak to a series of larger questions, and his purpose is to use Stockholm to bring the wider world into the frame without ever losing sight/site of the city itself. I put it like that because the visual thematic is such an unusually prominent motif in a text that is so attentive to spoken language.

Sieburth remarks, "Benjamin finds himself entering into a landscape of proper nouns." From its early drafts through to its later elaborations, he argues, the Arcades Project maintains "as its basic structural and heuristic device the form of a *list*, a paratactic mapping of cultural traffic, a gazeteer of urban signs."

[84] Winfried Menninghaus, "Walter Benjamin's theory of myth," in Gary Smith (ed.), *On Walter Benjamin: Critical essays and recollections* (Cambridge: MIT Press, 1988) pp. 292–325; the quotation is from p. 311.

[85] Pred, *Lost Words*, p. 53.

[86] Menninghaus, "Theory of myth," pp. 310–11; Sieburth, "Benjamin the scrivener," p. 15.

[87] Pred, *Lost Words*, pp. 136–42.

Indeed, Pred sees the problem of representation in terms that clearly mirror Benjamin's own visual obsessions.

[M]odernity was characterized by an increasingly fragmented social life and rapid changes in impressions and relationships, especially in the streets. Particularly in the case of Stockholm, the experience of modernity was by no means limited to the restless, phantasmagoric movement of people on the city's principal thoroughfares. Social contacts around dwelling and workplace also frequently took on a fleeting, flickering quality because connections to both were so often short-lived. Those living or laboring beside a person, those people who one saw and interacted with on a daily basis, could disappear as quickly and suddenly as they appeared, could leave an unfilled void or be replaced by others who might also evaporate with the same rapidity.[88]

How, then, to convey this "phantasmagoric" world? Let me consider two strategies, which Pred pursues where the echoes of Benjamin are, I think, unmistakable.

First, and most generally, Pred deploys what, following Geertz, he calls "blurred genres": a deliberate erasure of the finely etched line between the academic and the artistic. He insists that he is "free to play the art-ificer, to play upon words."

The repetitive phrase-sentence as driving-pulsating sentence, as pound-pound-pummel sentence, straightforwardly strives to violate reception barriers, to hammer home a message; while the repetitive-phrase sentence as facet-rotating sentence, as prism-turning sentence, by putting a cubistic face on the subject or object around which it circles, strives to seduce, to be (mind's) eye-opening, to whisper sweet (and sour) newthings in the (mind's) ear. The use of poetic forms is not an end in itself, but

an attempt to exploit the physicality of the text,

to exploit the landscape of the page....[89]

This is not an innocent strategy. Pred's purpose is to call attention to the multiple, labile meanings embedded in seemingly the most straightforward and stable of propositions. If the conventional closures of sentences and texts can be opened up in this way then, by virtue of the embeddedness of language in the material flow of power and practice, the taken-for-granted categories of social life become equally vulnerable to displacement and deconstruction. "The poetics of my textual strategy," Pred warns, "are the politics of my textual strategy."[90]

---

[88] *Ibid.*, p. 186.

[89] *Ibid.*, pp. xv-xvi; Clifford Geertz, "Blurred genres: The refiguration of social thought," in his *Local knowledge: Further essays in interpretive anthropology* (New York: Basic Books, 1983) pp. 19–35.

[90] Pred, *Lost Words*, p. xv.

Other writers have made the same connection. Benjamin understood it very well, of course, though he was not very interested in typographical acrobatics and word-play. Harvey is no less sensitive to the demands of dialectical modes of representation. I doubt that he would assent to the formal articulation of historical materialism and deconstruction proposed by some authors – still less to the claim that Marx and Derrida "are on the same side"[91] – but he has often acknowledged that one of the most intriguing (and far from incidental) aspects of Marx's use of language is its polysemy. "Marx's words are like bats: one can see in them both birds and mice."[92] For this reason, Harvey also prefers a nonlinear mode of representation, although this refers more to the logic of the argument than to what Pred calls "the landscape of the page," and he develops a relational style of theorizing that exploits the tensions within and between concepts. "Different starting-points yield different perspectives," he admits, "and what appears as a secure conceptual apparatus from one vantage point turns out to be partial and one-sided from another."[93]

All of this may be granted, but it provides no ready-made solutions: and I have to say that Pred's "poetic forms" do not always enlarge my horizon of meaning. There is no question that his experiments are serious, but somehow his traverse across the typographical high-wire often seems to end in a tangle of thoughts and a pile of words. This almost always happens whenever Pred is trying to convey a series of theoretical ideas that are interconnected in complex ways, when he feels compelled to spatchcock several sentences into one and have their clauses skewer one another from multiple directions.[94] I understand and sympathize with the intention; but I simply do not think the strategy works very well. Yet the images Pred conjures up through his more concrete – and, significantly, more graphical – modes of representation seem to me considerably more accomplished, and it is to these that I now turn.

---

[91] Michael Ryan, "The limits of *Capital*" in his *Marxism and deconstruction: A critical articulation* (Baltimore: Johns Hopkins University Press, 1982) pp. 82–102; the quotation is from p. 102.

[92] The original phrase is Pareto's, but the most sustained discussion of its implications (and one to which Harvey returns again and again) is Bertell Ollman, *Alienation: Marx's conception of man in capitalist society* (Cambridge: Cambridge University Press, 1971) pp. 3–71.

[93] Harvey, *Consciousness*, p. xvi. More particularly, Harvey seems drawn to the way in which Marx's concepts are continually breaking apart and coming together in new combinations. He tries to follow the same strategy in his *Limits*, (see pp. xv–xvi) and in *Condition of postmodernity* he suggests that Benjamin worked in much the same way: "Benjamin . . . worked the idea of collage/montage to perfection, in order to capture the many-layered and fragmented relations between economy, politics and culture . . ." (p. 51).

[94] See also Allan Pred, "The locally spoken word and local struggles," *Environment and Planning D: Society and Space* 7 (1989) pp. 211–33. The obvious parallel is with Olsson's dazzling linguistic experiments, but he somehow glides "behind the back" of language without marking the surface of the text.

Second, then, Pred convenes his separate accounts of the vocabularies used in different spheres of social life within a plenary reconstruction of a single docker's everyday path through the city. This is a telling way to drive language into the concrete, of course, and to reinforce the "ordinariness" of its use; but it is also a way of suggesting a more general "interweaving" of these heteroglossia within the fabric of everyday life. Benjamin is often credited with much the same "micrological" sensivity, the ability to see a large picture in the smallest of details.[95] As I have already indicated, such an emphasis on a "way of seeing" is particularly appropriate, but here I want to draw attention to the graphical imagination at work in Benjamin's reflections on his childhood in Berlin:

I have long, indeed for years, played with the idea of setting out the sphere of life – bios – graphically on a map. . . . I have evolved a system of signs, and on the gray background of such maps they would make a colorful show if I clearly marked in the houses of my friends and girl friends, the assembly halls of various collectives, from the "debating chambers" of the Youth Movement to the gathering places of the Communist youth, the hotel and brothel rooms that I knew for one night, the decisive benches in the Tiergarten, the ways to different schools and the graves that I saw filled, the sites of prestigious cafés whose long-forgotten names daily crossed our lips, the tennis courts where empty apartment blocks stand today, and the halls emblazoned with gold and stucco that the terrors of dancing classes made almost the equal of gymnasiums.[96]

Later in the same essay, Benjamin recalls an afternoon in Paris to which he owed these insights "that came in a flash, with the force of an illumination." It was then (and there), he recalls, sitting in the Café des Deux Magots in St Germain-des-Prés, that he understood why the modern city inscribes itself in memory through "the images of people less than those of the sites of our encounters with others or ourselves":

Suddenly, and with compelling force, I was struck by the idea of drawing a diagram of my life, and knew at the same moment exactly how it was to be done. With a very simple question I interrogated my past life, and the answers were inscribed as if of their own accord, on a sheet of paper that I had with me.[97]

At the time it was "a series of family trees," but now he saw it as a "labyrinth." He said he was not much interested in "what is installed at

---

[95] The term was originally Bloch's, but in praising Benjamin's "sensitivity for individual detail" he was addressing more than an ability to recover overlooked and discarded fragments of the past. He claimed that "Benjamin possessed an unequalled micrological-*philological* sensivity" (my emphasis), so that his point was that Benjamin approached the world – and hence the city – as a *text* whose elements have been scattered and suppressed by the closures of conventional history. See Frisby, *Fragments*, pp. 213–14.

[96] Benjamin, "Berlin Chronicle," p. 5.

[97] *Ibid.*, pp. 30–31.

its enigmatic center" – "ego or fate" – but rather with its many entrances and corridors. For this enabled him to wonder about "the intertwinements of our path through life" and whether there are "paths that lead us again and again to people who have one and the same function for us: passageways that always, in the most diverse periods of life, guide us to the friend, the betrayer, the pupil or the master."[98] That afternoon in Paris, he concludes, "against the background of the city, the people who had surrounded me closed together to form a figure."[99] As one might expect, the graphical impulse was not alien to Benjamin's reconstruction of Second Empire Paris either, but I make so much of these passages because they seem to evoke something of the same sensibility that is displayed in the diagrams of Hägerstrand's time-geography: a sensitivity to the micro-topographies of everyday life.[100] Pred insists that it is essential to use Hägerstrand's graphical notation to represent the path of Sörmlands-Nisse, his imaginary docker (figure 21), because such a mode of representation is able to capture simultaneities and conjunctures that "can easily escape linear language."[101] Again and again, he argues that diagrams of this sort have the unique capacity to make the structuration of social life seen/scene in the double sense of both making the processes *visible* and embedding them in *place*.

But how is such a time-geography to be read? In one sense, I suggest, the line-diagram depicts exactly that *danse macabre* that once so alarmed Hägerstrand's critics (and friends) – an alienated world of bodies in autonomic motion – and yet it is one that also captivated both Baudelaire and Benjamin. "Baudelaire describes neither the Parisians nor their city," wrote Benjamin.

Forgoing such descriptions enables him to invoke the ones in the form of the other.... As Baudelaire looks at the plates in the anatomical works for sale on the

[98] *Ibid.*, pp. 31–32.

[99] *Ibid.*, p. 32.

[100] I have provided a critical discussion of time-geography in my "Suspended animation: The stasis of diffusion theory," in Derek Gregory and John Urry (eds.), *Social relations and spatial structures* (London: Macmillan, 1985) pp. 296–336. Interestingly, Buck-Morss, *Dialectics of seeing*, p. 342, exploits similar ideas in her brief reconstruction of Benjamin's daily path through Paris to the *Bibliothèque Nationale*.

[101] Pred, "Place as historically contingent process," p. 292. Harvey agrees that the Hägerstrand model provides "a useful [description] of how the daily life of individuals unfolds in space and time" but in itself, he objects, it tells us nothing about the production of the locational structures within which these time-space paths are spun nor about the relations of power and domination that are instantiated through them: see Harvey, *Condition of postmodernity*, pp. 211–12. Pred's investigations of Stockholm do not address the first of these objections – nor are they intended to – but they clearly meet the second by making the relations between power and practice focal to any developed version of time-geography: see Allan Pred, "Power, everyday practice and the discipline of human geography," in Allan Pred (ed.), *Space and time in geography: Essays dedicated to Torsten Hägerstrand* (Lund, Sweden: C. W. K. Gleerup; Lund Studies in Geography, Series B, No. 48, 1981) pp. 30–55.

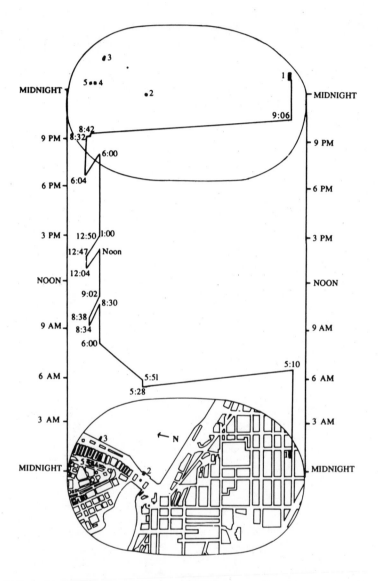

Notes: 1 = residence at *Bondegatan* 17; 2 = café at *Fiskarehamnen*; 3 = ship docked at *Skeppsbron*; 4 = café on *Österlånggatan*; 5 = public bar on *Österlånggatan*.

Figure 21   The daily path of Sörmlands-Nisse [diagram: Allan Pred, by permission].

dusty banks of the Seine, the mass of the departed takes the place of the singular skeletons on these pages. In the figures of the *danse macabre*, he sees a compact mass on the move.[102]

In much the same way, I think, the "singular skeletons" of Hägerstrand's time-geography are able to evoke "the mass of the departed."[103] The bare bones of the line-diagram reveal what one might even call a "post-mortem" geography: a world in which, through the endless repetition of daily paths (time-space routines), the ever-same is reproduced as the ever-new. But this is not an objection to representing the world in this way, as I once thought, because – to recall Eagleton's discussion of Benjamin (above, pp. 233–34) – *it allows the diagram to be read as an allegory of commodity circulation itself.* The figure of the docker then becomes doubly effective because so much of his life-world was anchored at "pivotal points where goods circulated between the city and nonlocal worlds." This is a highly specific figuration, of course: like Flaubert's Frédéric, his masculinity conferred a particular (though evidently class-bounded) mobility upon Sörmlands-Nisse. The generalized nexus of patriarchal relations was inscribed in his particular time-space path through the mediation of what Hägerstrand would call "capability" and "authority" constraints.[104]

But the diagram does not stand alone. Pred intends "the imagery of this entire daily-path depiction [to be seen as] a ... montage."[105] Again, the echoes of Benjamin are clear. In figure 22 I have taken Pred at his word and brought together some of the photographs and word-pictures – the images – which frame and compose the text. It would be unduly fanciful to suppose that the result is a dialectical image, but it nonetheless discloses a field of extreme tensions. In quite another sense, therefore, once the implications of time-geography are internalized, then I think it is possible to achieve something like a moment of profane illumination: to grasp the multiple ways in which the docker made this world meaningful *to himself*, sometimes on terms that replicated the dominant order (xenophobia, patriarchy), to be sure, but sometimes on terms that were hostile to and even, on occasion, subversive of it. In my view, these passages contain some of the finest writing in the book.[106]

There is no extravagant experimentation, no fiddling around with composition, and yet the result is an astonishingly vivid evocation of fin-de-siècle

---

[102] Walter Benjamin, "On some motifs in Baudelaire," in his *Illuminations*, pp. 155–200; the quotation is from p. 168. It was Anne Buttimer who told Hägerstrand that the world he depicted reminded her of a *danse macabre*: see Gregory, "Suspended animation," p. 335.

[103] In this connection I think it significant that Hägerstrand should have originally thought of his project as a "population archaeology" and that he did not conceive of it in purely individualistic terms: see Gregory, "Suspended animation," pp. 305–6.

[104] Pred, *Lost Words*, p. 200.

[105] *Ibid.*, p. 291.

[106] *Ibid.*, pp. 229–35.

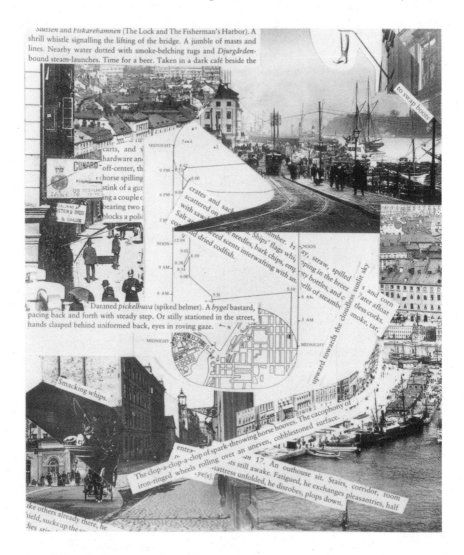

Figure 22   Montage: the time-geography of Sörmlands-Nisse.

Stockholm. Pred achieves this effect through a multilayered series of overlapping images that assault the senses. The stench-drenched visit to the outhouse; the shopkeeper's signs swinging above the street; the smell of sour beer and stale cigar smoke issuing from the open café doors; the ship's flags whipping in the breeze; the thrumming of iron-ringed wheels over cobblestones; the eyes of the policemen, watching and waiting; the bodies dozing on the sun-warmed sacks of grain: All of these word-pictures hang flesh on the bare bones of the line-diagram and bring it explosively to life in an intensity of experience that is quite alien to most conventional human geographies. In this sense, I suggest, Pred's redemption of "the habitus of a lived life" depends as much on the meaning Benjamin gave to that phrase as it does on those subsequently given to it by Bourdieu and Hägerstrand.

And yet this plenary representation, for all its power, does not quite achieve the force of a constellation. To be sure, Pred does not work with a screen of polarities, but his conceptual architecture is not at issue here. More to the point is his *way* of working. Imagine what would have happened had he constructed not one time-geography but several time-geographies: not just Sörmlands-Nisse's daily path to and from the docks, but also the path of one of the merchants he saw at breakfast through the open windows of the Skeppsbro Cellar, the path of the waitress who had to endure his lunch-time assaults, and the path of one of the policemen patrolling the Österlånggatan. Presumably the word-pictures would have been different, both in composition and in detail: But what of the photographs which Pred uses to make the docker's daily path scene/seen? At present, clearly, one is tacitly invited to view these images through Sörmlands-Nisse's eyes, not simply to accompany him through the streets and into the bars but to enter into the fiction of seeing what he saw.[107] But how would one then make these same images scene/seen from *other* points of view? The merchant, the waitress and the policeman would have walked along at least some of the same streets, but did they see – could they possibly have seen – the same things? Would they have seen their surroundings in the ways in which they have been composed by the photographer?

In raising questions like these I am seeking to make two points. On the one hand, photography (even "documentary" photography) is not an innocent art. Pred is endlessly inventive in making sense of the textual archive, revealing the "lost worlds" hidden within its seemingly silent glossaries; and yet, for all his visual sensibility, he does not subject the visual archive to the same process of interrogation. It is necessary to do so, as Tagg suggests, because the nineteenth-century development of photography took place across what he calls "a radiating cultural geography":

A plurality of particular photographies were deployed in a whole range of differently directed practices. Their effects and legitimations were not given in advance, but derived from different discursive economies. They were institutionalized under markedly different terms, discussed and evaluated by different agents, in very different frameworks of reference and accorded quite different statuses.[108]

---

[107] So, for example, plate 10 is not only "*Götagatan*, looking north" but also "the view from Sörmlands-Nisse's eyes shortly after turning off from *Bondegaten*."

[108] John Tagg, "The discontinuous city: Picturing and the discursive field," *Strategies* 3 (1990) pp. 138–58; the quotations are from pp. 147–48, 150. Two indispensable points of entry into

On the other hand, by convening a series of distinctive vocabularies within the lifeworld of one docker, suturing them around the figure of Sörmlands-Nisse, and in particular by passing off the point of view represented in these photographs as a "plenary" perspective, accessible to and even shared by other participants in the making of Stockholm's multiple historical geographies, Pred risks losing the very heterogeneity that is contained within his own remarkable vision of the structuration of social life in time and space. These are not minor quibbles, because they suggest that Pred's poetics *could* be put at the service of a much more radical politics, one which is sensitive to difference without fragmenting politics in endless mutilations.

Archives and archaeologies

It is surprising how long the problem of space took to emerge as a historico-political problem.

Michel Foucault, *Power/knowledge*

I have used Benjamin to mediate between Harvey and Pred, but another figure has been wandering through these passages too: Michel Foucault. In several places the language I have used has been Foucauldian and in several others parallels could be drawn between Benjamin (in particular) and Foucault. I do not mean to imply that one can be assimilated to the other, of course, but there are two affinities between their projects that seem to me particularly telling. The first is their sense of historicity or, as I have called it here, of interventions in historical geography. Foucault once suggested that, as he used the term, "archaeology" could almost mean "description of the archive." By archive, he continued, he meant

the set ... of discourses actually pronounced; and this set of discourses is envisaged not only as a set of events which would have taken place once and for all and which would remain in abeyance, in the limbo or purgatory of history, but also as a set that continues to function, to be transformed through history, and to provide the possibility of appearing in other discourses.[109]

It would not be difficult to interleave this passage with Benjamin's theses on the philosophy of history and, indeed, with the labyrinthine construction

---

contemporary discussions are *idem, The burden of representation: Essays on photographies and histories* (London: Macmillan, 1988) and Carol Squiers (ed.), *The critical image: Essays on contemporary photography* (Seattle: Bay Press, 1990).

[109] Michel Foucault, "The archaeology of knowledge" in his *Foucault live: Interviews 1966–1984* (New York: Semiotext(e), 1989) pp. 45–56; the interview was originally published in 1969.

of the Arcades Project itself. On such a reading, the discourses of the past are reactivated in the present – *our* present – not, as Thompson once put it, in order to "shake Swift by the hand" (or Walpole by the throat) but precisely to call such continuities into question: to disrupt the unilinear, progressive trajectory of historical eventuation and to substitute a much sharper, less self-assured "history of the present."[110] Harvey's sensibilities are clearly much closer to Thompson's than are Pred's. We have much to learn from the struggles of the Commune, so Harvey would argue, not only because they can illuminate the tactics and strategies of an otherwise unremarked politics of space but also because they help to enlarge our own moral space: "We are saying that these values, and not those other values, are the ones which make this history meaningful to us, and that these are the values which we intend to enlarge and sustain in our own present."[111] I do not want this to be misunderstood: Benjamin and Foucault have a highly developed ethical sense too. But neither of them allows hindsight to become a privilege, and perhaps it is *that* privilege that we should also "unlearn as our loss"?

That visual thematic haunts Foucault as much as it does Benjamin. One commentator describes his way of working as an "art of seeing," and another represents Foucault's archaeology – and I would add his genealogy – as a way of opening up what he calls "visibilities":

Visibilities are not forms of objects, nor even the forms that would show up under light, but rather forms of luminosity which are created by the light itself and allow a thing or object to exist only as a flash, sparkle or shimmer.[112]

The prose is elliptical, but Benjamin's heteroclite constellation seems to mine the same historico-geographical vein: in both cases "archaeology" is not so much an excavation, bringing buried or hidden objects to the surface, as a way of showing the particular – anonymous, dispersed – practices and the particular – differentiated, hierarchized – spaces through which particular societies make particular things visible. Deleuze argues that Foucault's archaeology also involves the recovery of a tense, chameleon-like struggle between the visible and the "articulable" (fields of statements): and again, one can read a similar tension in the Arcades Project. But it is not so easy to translate these protocols into Harvey's Paris or Pred's Stockholm. Where Harvey draws on Benjamin, I think it *is* possible to glimpse elements of different – jostling, colliding – human geographies, but these

---

[110] Thompson, p. 234.

[111] *Ibid.*

[112] Gilles Deleuze, *Foucault* (Minneapolis: University of Minnesota Press, 1988) p. 52; see also John Rajchmann, "Foucault's art of seeing," in his *Philosophical events: Essays of the '80s* (New York: Columbia University Press, 1991) pp. 68–102.

are almost always submerged by his insistence on treating power as centered, marshalled, possessed. And one rarely hears in these pages the multiple voices that animated everyday life. Pred's archaeology of Stockholm is more attentive to the fissures and fractures of modern urbanism; he is acutely aware of the relations between the visible and the articulable, and his imbrication of "lost worlds" (heterotopias?) and "lost words" (hetero-glossia) begins to suggest a much more complex human geography than most of us have dreamed of. All the same, its uneven edges and its awkward topographies are finally resolved in what I take to be a thoroughly modern (not modernist) *dénouement*. The time-geography of the solitary docker embodies a world whose different topographies and glossaries are at last made whole and centered around the inscriptions of power and practice in and on the individual. And yet, as I have tried to suggest, it is possible to use his own mode of representation to open this account to the uneven, unequal unfolding of multiple human geographies. That more remains to be done is not Pred's fault: rather it is Pred's challenge. It is also, I suggest, as much the challenge of modernism as it is of any postmodernism.

4

Chinatown, Part Three? Uncovering
Postmodern Geographies*

Bunker Hill is old town, lost town, shabby town, crook town.
                    Raymond Chandler, *The high window*

*The mystery must be unravelled, like the thread that led Theseus away from the
Minotaur, out of the labyrinth. But this unravelling must lead the counterfeit
detective into the contemporary labyrinth: the city that must be the setting for all
such journeys of discovery.*
            Elizabeth Wilson, *Hallucinations: Life in the postmodern city*

Pastiche

Running his fingers lightly over the keyboard, Soja sets out to solve two
closely connected mysteries. The first concerns the subordination of space
in social theory. Like any private-eye operating out of southern California,
he plunges into the past to discover what really happened. Working in the
shadows of the officially sanctioned disciplines, picking up information
wherever he can, Soja is determined to show that things were not quite
what they seemed. If he can pull it off – if he can show that Lefebvre
and Foucault were in reality the legitimate offspring of critical social theory
– then he will have the key to the second case in his hands. Sometimes,
he realizes, spatiality is about concealment and evasion. Like most people,
he's been had by illusions of opacity, of transparency. Lots of times. But

*This chapter is a commentary on Edward Soja, *Postmodern geographies: The reassertion of space in
critical social theory* (London: Verso, 1989). I have taken my title from Mike Davis, "*Chinatown,
Part Two?* The 'internationalization' of downtown Los Angeles," *New Left Review* 164 (1987)
pp. 65–86.

now he knows that sometimes – just sometimes – you can turn it round. You can use spatiality to bring social life into the open. Now he knows enough to break the second case. Pausing only to check his pockets for cigarettes, Soja hits the streets of Los Angeles. He uncovers a trail of deception and distortion that takes him to the pinnacles of power. Brushing aside the veils of illusion, he steps into the room and walks toward the window. He's waited a long time for this. And as the figure in the smart suit and dark glasses turns around, Soja isn't surprised. He knew all along what Marlowe had been told in *The long good-bye*:

There's a peculiar thing about money. . . . In large quantities it tends to have a life of its own. . . . The power of money becomes very difficult to control.

I begin in this way not only because postmodernism often relies on parodic forms – on that "repetition with critical distance that allows ironic signaling of difference at the very heart of similarity"[1] – but also because it conveys, as economically as I can manage, what I take to be the essence of Soja's *Postmodern geographies*. I will return to my counterfeit detective at the end of the essay; everything in between is elaboration. But to begin in this way is *not* to mock the seriousness of Soja's project, and I want to make it plain that I regard *Postmodern geographies* as a brilliant book. The work of a master-craftsman, its intellectual sparkle is the product of a rare and generous critical intelligence. Soja has taken a collection of his previously published essays and reworked every one into a new setting. The result is a carefully polished text, with each word weighed and set in place to bring out the deeper tonalities of the others. Although in what follows I focus on only two of its facets, these are the ones that seem to me to be integral to its design. The first is the intellectual history that makes the project possible – the terrain on which Soja mobilizes his claims for the construction of a distinctly *postmodern* geography – and the second is his creative deconstruction of Los Angeles as "the quintessential post-modern place." It is through these two interventions, more than any of the others, I think, that Soja seeks to reestablish the importance of space within critical social theory.

A history of the present

"A history of the present" is one of the ways in which Foucault described his project, but I use it here for two reasons. In the first place, Soja invokes Foucault on several occasions to argue for an "archaeology" of human

---

[1] Linda Hutcheon, *A poetics of postmodernism: History, theory, fiction* (London: Routledge, 1988) p. 26.

geography, but his own inclinations are, I think, much closer to what Foucault called a *genealogy*. Although his claims are on occasion global – and sometimes unduly so – Soja is not only interested in the continental ruptures between different systems of knowledge; he is also concerned to expose the intricate local connections between power and knowledge. I hope that "a history of the present" allows me to register the second of these terms without altogether erasing the first. In the second place, the phrase draws attention to the critical purpose of histories of this kind. Soja's account is by no means a conventional intellectual history, and – unlike some histories of geography – it does not use the past to legitimize the present. If Soja seeks to call into question the self-legitimating model of "progress" that underwrites the modern discipline, however, he also wants to close off the possibility of any "return": the past provides no solace. His history of the present, like Foucault's, is thus intended to be radically unsettling.

His point of entry is the last fin de siècle (1880–1920), which was an astonishing period in Western social and intellectual history. So many spheres of social life were set spinning by dramatically new experiences of time and space – by time-space compressions of unprecedented intensity – and the templates of social thought were simultaneously set in motion by the shock waves of what Kern calls a new *culture* of time and space.[2] Robert Musil brought these events into a single frame in his remarkable image of "a kind of super-American city where everyone rushes about, or stands still, with a stop-watch in his hand."

Overhead-trains, overground-trains, underground-trains, pneumatic express-mails carrying consignments of human beings, chains of motor-vehicles all racing along horizontally, express lifts vertically pumping crowds from one traffic-level to another.... At the junction one leaps from one means of transport to another, is instantly sucked in and snatched away by the rhythm of it, which makes a syncope, a pause, a little gap of twenty seconds between two roaring outbursts of speed, and in these intervals in the general rhythm one hastily exchanges a few words with others. Questions and answers click into each other like cogs of a machine.[3]

The novel from which this is taken is set in Vienna in 1913, and Timms suggests that Musil's intention was to emphasize the contrast between this

---

[2] Stephen Kern, *The culture of time and space 1880–1918* (Cambridge: Harvard University Press, 1983); I have borrowed "time-space compression" from David Harvey, *The condition of postmodernity* (Oxford: Blackwell Publishers, 1989). Like Kern, Soja's history is confined to the West: but he surely knows, through his earlier work in Africa, that this imperial culture of time and space reached far beyond the West and that its incursions and compressions were considerably more disruptive (and truly shocking) in those "other" places.

[3] Robert Musil, *The man without qualities* (volume I, 1930; reprinted London: Pan, 1979) p. 30.

frenzied, chronometric model of metropolis and the "retarded tempo" of the enervated, claustrophobic capital of the Hapsburg Empire.[4] Yet, as Timms acknowledges, that extraordinary, imperial city was the site of an almost paradigmatic urban modernism. On one side was the romanticism of Camillo Sitte, but on on the other side was the sleek rationalism of Otto Wagner. In the following passage Schorske provides a cameo of a debate from which Musil's caricature of the "super-American city" could easily have been drawn:

Wagner, out of a bourgeois affirmation of modern technology, embraced as essence what Sitte most abhorred in the Ringstrasse: the primary dynamic of the street. The conservative Sitte, fearful of the workings of time, placed his hope for the city in contained space, in the human, socializing confines of the square. Wagner, even more than the Ringstrasse progressivists before him, subjected the city to the ordinance of time. Hence street was king, the artery of men in motion; for him the square could serve best as goal of the street, giving direction and organization to its users. Style, landscaping, all the elements through which Sitte sought variety and picturesqueness in the fight against modern anomie, Wagner employed essentially to reinforce the power of the street and its temporal trajectory.[5]

Wagner was also (appropriately) responsible for planning the city railway system and closely involved in the design of its stations, viaducts, tunnels, and bridges. In any event, it was this chronometric image that was endlessly repeated from one medium to another, page to print, canvas to celluloid, spooled and recycled.

The stopwatch turned out to be the leitmotiv of modern social theory too. Within both historical materialism and conventional social science, so Soja claims, these new ways of experiencing time and space were measured by the hands of the historical clock alone. Countless commentators have wired modernity to a changed consciousness of time, of course, and many of them have given a central place to Baudelaire's essay on "The painter of modern life." In that essay, published in 1863, Baudelaire defined modernity as "the transient, the fleeting, the contingent." It was, so he said, "one half of art, the other being the eternal and the immutable." It was that same sense of restless animation, of the repudiation of tradition, of the celebration of the *new*, that propelled a series of avant-garde movements through the nineteenth and into the early twentieth centuries. Indeed, Habermas reminds us that the very word itself – *avant-garde* – connotes

---

[4] Edward Timms, "Musil's Vienna and Kafka's Prague: The quest for a spiritual city," in Edward Timms and David Kelley (eds.), *Unreal city: Urban experience in modern European literature and art* (Manchester, England: Manchester University Press, 1985) pp. 247–63.

[5] Carl Schorske, *Fin-de-siècle Vienna: Politics and culture* (New York: Random House, 1981) pp. 24–115; the quotation is from pp. 100–101.

invading unknown territory, exposing itself to the dangers of sudden, of shocking encounters, conquering an as yet unoccupied future. The avant-garde must find a direction in a landscape into which no one seems to have yet ventured.[6]

But this is an arresting passage. For unnoticed, unremarked, it has been taken over by an opposing language: a metaphor of territory, occupation, landscape. Foucault captures the difference with forensic precision:

Metaphorizing the transformations of discourse in a vocabulary of time necessarily leads to the utilization of the model of individual consciousness with its intrinsic temporality. Endeavoring on the other hand to decipher discourse through the use of spatial, strategic metaphors enables one to grasp precisely the points at which discourses are transformed in, through and on the basis of relations of power.[7]

The difference is, I suggest, constitutive for Soja: indeed, he claims Foucault as one of the architects of "postmodern geography." He also realizes (as, in a sense, Habermas's remark allows) that the avant-garde was excited by the sheer plasticity of space as well. Thus "many of the avant-garde movements of the fin de siècle – in poetry and painting, in the writing of novels and literary criticism, in architecture and what then represented progressive urban and regional planning – perceptively sensed the instrumentality of space and the disciplining effects of the changing geography of capitalism." But his primary purpose is to show that within social theory and the social sciences, in contrast to the energies of artistic experimentation, the geographical imagination was set on one side. In consequence, the spatial inscriptions of power were virtually erased from the intellectual landscape. "The instrumentality of space was increasingly lost from view in political and practical discourse." In its place (and literally so) social theory was ruled by a triumphant historicism that accorded a special privilege to history and historical contextualization.[8]

In much the same way, but with an inflection that is of particular importance for Soja's critique, Lowe identifies a "despatialization of historiography" as one of the central installations of nineteenth-century bourgeois society:

---

[6] Jürgen Habermas, "Modernity versus postmodernity," *New German Critique* 22 (1981) pp. 3–14.

[7] Michel Foucault, "Questions on Geography," in his *Power/knowledge: Selected interviews and other writings* (Brighton, England: Harvester Press, 1980) pp. 63–77; the quotation is from pp. 69–70. Foucault's comment was a general one and he was not talking about Habermas in particular. For this reason I should perhaps add that I am aware that one of the purposes of Habermas's theory of communicative action is to *transcend* what Foucault calls here "the model of individual consciousness"; but I am also aware that his project remains confined within the horizon of temporality: see in particular Jürgen Habermas, *The philosophical discourse of modernity* (Cambridge, England: Polity Press, 1987).

[8] Soja, *Postmodern geographies*, p. 34.

In bourgeois society, development-in-time despatialized historiography. With the extension of the historical landscape, time now possessed a depth and diversity which it previously lacked.... [B]ourgeois society discovered the concept of historicism.[9]

The class location of historicism – Marx's "annihilation of space by time" in an ideological register – is, I think, an essential subtext of Soja's critique. I say "subtext" because, on the surface, Soja moves to distance himself from Marxism as modernism. *Postmodern geographies* is a sustained challenge to historicism: a remarkable attempt to subvert what Soja takes to be the subordination of space in critical social theory. Concepts of historicity are so deeply implicated in Marxism and modernism, so he claims, that any project to reinstate concepts of spatiality must of necessity be conducted under the sign of a *post* modernism. But the position of Marxism in that sentence is ambiguous: present at the beginning, under erasure at the end, and yet still, as I will try to show, indelibly *there*.

Spatiality is not the only way in which Soja differentiates modernism from postmodernism, but it is the most forceful and probably the most original. Although he identifies several previous writers as "postmodern geographers" *malgré eux* none of them – with the exception of Jameson – identifies postmodernism so closely with spatiality: and, as I will show shortly, there are considerable difficulties in assimilating Jameson to postmodernism *tout court*.[10] But that is not the only reason for my unease at Soja's maneuver, and I want to explore the intellectual history that makes it possible. I should make it clear that I am not setting out with rigid definitions of "modernism" and "postmodernism" as rods to beat Soja's back. Instead, I prefer to allow their meanings – flexible, relational meanings – to emerge in the course of my argument. The looseness of these terms has often been remarked upon and, indeed, Anderson dismisses modernism as "the emptiest of all cultural categories."[11] But his verdict could just as easily be reversed: One of the purposes of his essay was to draw attention to the *multiplicity* of modernisms and in this sense, clearly,

---

[9] Donald Lowe, *History of bourgeois perception* (Chicago: University of Chicago Press, 1982) p. 43. According to one writer, the dual of the "despatialization of history" – of the new depths of time sounded during the nineteenth century – was a new depthless*ness* of landscape. He attributes this, significantly, to the railway, which recomposed the visual field thus: "The views from the windows of Europe have entirely lost their dimension of depth and have become mere particles of one and the same panoramic world that stretches all around and is, at each and every point, merely a painted surface." See Wolfgang Schivelbusch, *The railway journey: The industrialization of time and space in the nineteenth century* (Berkeley: University of California Press, 1986) p. 61.

[10] I presume that it is, in part, Soja's focus on spatiality that prompts him to omit any discussion of Jean-François Lyotard's distinctly a-spatial account of *The postmodern condition: A report on knowledge* (Minneapolis: University of Minnesota Press, 1984).

[11] Perry Anderson, "Modernity and revolution," *New Left Review* 144 (1984) pp. 96–113; the quotation is from p. 112. This is a commentary upon Marshall Berman, *All that is solid melts into*

"modernism" is far too full a category. In any event, the historico-geographical emphasis that Anderson places upon the last fin de siècle in Europe and North America is consonant with Soja's argument and I propose to retain it in what follows. The main difference is that Anderson's span of attention is somewhat longer than Soja's, stretching as it does from 1848 – the famous "springtime of peoples," when the entire heart of the European continent seemed to be aflame – to 1945, when the end of World War II finally cut off the vitality of modernism. I think this periodization is preferable to Soja's, but I will pay particular attention to the decades between 1880 and 1920 in order to conduct my argument on his own terrain. I also retain Soja's insistent focus on space and spatiality, which does, I agree, provide an important perspective on *both* modernism *and* postmodernism. But here too discriminations are necessary. Jameson, who in many other respects closely follows Anderson's account, considers that "a certain spatial turn offers one of the more productive ways of distinguishing postmodernism from modernism proper." But he makes it plain – as I think Soja does not – that the distinction "is one between two forms of interrelationship between time and space, rather than between the two inseparable categories themselves."[12]

On much the same basis I seek to show that Soja's history of the present leaves a series of strategic questions unasked and hence unanswered. These are not lacunae that can be deferred and filled at some future date, because they enter into the very architecture of Soja's argument – precisely because it *is* a history of the present – and they need to be examined as moments of its construction. The scope of the inquiries I have in mind can be indicated by remapping the thematics of modernism through two parallel sets of observations. The first concerns the relations between the aesthetic and the social; the second concerns the relations between the geographical and the historical. These labels cover a lot of intellectual baggage, I realize, and I use them as conveniences (nothing more). In both cases, I intend them to help keep track of Soja's itinerary and to locate his concept of "spatiality" on a wider intellectual landscape.

The aesthetic and the social

Soja is surprisingly reluctant to make anything of the relations between aesthetic theory and social theory. He notes the modern contrast between

---

air: *The experience of modernity* (London: Verso, 1982). Berman's book plays an important part in the architectonics of *Postmodern geographies*, most notably in the provision of the modernism-modernity-modernization triad that forms the basis of Soja's history of the present (Soja, *Postmodern geographies*, pp. 24–31).

[12] Fredric Jameson, *Postmodernism or, the cultural logic of late capitalism* (Durham: Duke University Press, 1991) p. 155. Jameson is another figure in Soja's Pantheon and his influence is even more

their spatial sensibilities more or less in passing but then moves it to the margins of the text. Art and aesthetics are accorded greater prominence – "visibility" would perhaps be a better word – only when he seeks to open a conceptual space for the elucidation of postmodernism. Even then, their treatment remains oblique: we are reminded of Berger's essay on painting and the novel and told to attend to Jameson's reflections on postmodernism as the "cultural logic" of capitalism. It is also puzzling: Both writers certainly drew attention to the importance of what Soja calls "a spatially politicized aesthetic," and Jameson even urged that "a model of political culture appropriate to our own situation will necessarily have to raise spatial issues as its fundamental organizing concern." But Berger was writing about *modern* painting and the *modern* novel and Jameson's essay was a *critique* of postmodernism: He has since made it plain that he objects to being "oddly and comically identified with an object of study."[13] These are particular concerns, of course; my more general argument is that relations between the aesthetic and the social are of such pressing importance to both modernism and postmodernism that a more sustained discussion is required. Indeed, Harvey draws attention to what he regards as "one of the more startling schisms" of intellectual history in precisely these terms. Where social theory has typically privileged time over space, he argues, aesthetic theory has always been deeply concerned with "the spatialization of time." In consequence:

> There is much to be learned from aesthetic theory about how different forms of spatialization inhibit or facilitate processes of social change. Conversely, there is much to be learned from social theory concerning the flux and change with which aesthetic theory has to cope. By playing these two currents of thought off against one another, we can, perhaps, better understand the ways in which political-economic change informs cultural practice.[14]

Let me therefore return to fin-de-siècle Europe and, through two vignettes, suggest some of the ways in which a sense of spatiality was deeply implicated in modernism.

Rimbaud and the emergence of social space　In 1873, when he was still only 18, Rimbaud made his famous declaration about the need to be absolutely modern:

---

pervasive than Berman's. I hope to show that his celebrated reading of postmodernism as the "cultural logic" of late capitalism provides – in a radically simplified form – the basic template for Soja's project and, indeed, Jameson has himself commended Soja's account of "the new spatiality implicit in the postmodern": *ibid.*, p. 418.

[13] John Berger, *The look of things* (New York: Viking Press, 1974); Fredric Jameson, "Postmodernism, or the cultural logic of late capitalism," *New Left Review* 146 (1984) pp. 53–92; reprinted in revised form in his *Postmodernism*, pp. 1–54; see also p. 297.

[14] Harvey, *Condition of postmodernity*, pp. 205–207.

"*Il faut être absolument moderne.*" Putting the accent unequivocally on the present, Rimbaud insisted that this modernist compulsion entailed a search for what Phillipson calls "a language beyond Language," that is to say,

a language of the present, a language that owes nothing to the past, a language without history, without memory, a language for beginning again, a language against repetition.[15]

This new language was, in a sense which I want to explicate further, an intrinsically and fundamentally *spatial* text. Phillipson treats modernism as an attempt to explore and *map* Rimbaud's instruction and claims that Rimbaud "opens up a new *space* in poetry" (my emphasis); but I have in mind something more than a textual strategy.[16] In doing so I draw upon Ross's exhilarating study in which she argues that, for Rimbaud, poetry had to exist as critique – it had to be directed toward emancipation – and, as such, the poet was compelled to alter "the balance away from the textual toward lived practice." Ross therefore makes a parallel turn and formulates the concept of social space as a way "of mediating between the discursive and the event." She traces the outlines of two significant "spatial movements" – the time-space compression of late nineteenth-century French society and, more particularly, the Paris Commune of 1871 ("the first realization of urban space as revolutionary space") – and then seeks to establish a *horizontal* connection, so to speak, a morphic resonance rather than a thematic affinity, between the events of the Commune and Rimbaud's poetry.

The actual, complex links binding Rimbaud to the events in Paris are not to be established by measuring geographic distance. Or, if they are, it is perhaps by considering Rimbaud's poetry, produced at least in part within the rarefied situation of his isolation in Charleville, as one creative response to the same objective situation to which the insurrection in Paris was another. In what way does Rimbaud figure or prefigure a social space adjacent – side by side rather than analogous – to the one activated by the insurgents in the heart of Paris?"[17]

A particularly clear example of Ross's response to this question is to be found in her reading of Rimbaud's *Rêvé pour l'hiver*. She suggests that this early poem invites us

---

[15] Michael Phillipson, *Painting, language and modernity* (London: Routledge & Kegan Paul, 1985) pp. 26–29.

[16] *Ibid.*, p. 27; cf. Fredric Jameson, "Rimbaud and the spatial text," in Tak-Wai Wong and M. A. Abbas (eds.), *Rewriting literary history* (Hong Kong: Hong Kong University Press, 1984) pp. 66–93.

[17] Kristin Ross, *The emergence of social space: Rimbaud and the Paris Commune* (Minneapolis: University of Minnesota Press, 1988) p. 32. Since Soja makes so much of Lefebvre, it is important to notice that Ross draws upon a number of his writings for both her concept of "social space" and her account of the Commune.

to conceive of space not as a static reality but as active, generative, to experience space as created by an interaction, as something that our bodies reactivate, and that through this reactivation, in turn modifies and transforms us. . . . [T]he poem creates a "nonpassive" spatiality – space as a specific form of operations and interactions.[18]

Let me again emphasize that this reading does *not* center on a thematic conjunction between the Commune and Rimbaud's overtly political verse: *Rêvé pour l'hiver* is one of his erotic poems. Ross argues that "in Rimbaud there is little distance between political economy and libidinal economy" – a preoccupation of many twentieth-century French intellectuals – and (equally important) that the significance of the Commune was most evident "in its *displacement* of the political into seemingly peripheral areas of everyday life."[19] This is a particularly compelling thesis, I think, because it clarifies with such originality the critical reception of Rimbaud's specifically "urban" compositions. Blanchard's thoroughly conventional reading of *Illuminations*, for example, is concerned to show that for Rimbaud "[urban] space is never still but always in the process of being generated," that in the city "space, objects, which used to exist solely in relation to an action, a narrative to which they only contributed a fixed content, now appear themselves to constitute the narrative."[20] That much is unexceptional; but Ross enables us to see that these representations form part of a much more general problematic – and one that incorporates a conception of spatiality which is, perhaps, the clearest prefiguration of Soja's own, more formal conception of spatiality.

Simmel and the abstraction of social space In social theory one can detect much the same sensibility: or, at least, one can see it in those writers who elected to struggle *with* the crisis of representation rather than retreat into the false security of positivism. Simmel's sociology provides perhaps the clearest example. Although Simmel was immensely influential during the European fin de siècle – the center of a distinguished intellectual circle, he had a considerable impact on Lukács, Bloch, Mannheim, and many of their contemporaries – it is only recently that his importance has been appreciated by later commentators. Here I am indebted to Frisby's sensitive portrayal of his work. Frisby regards Simmel as "perhaps the first sociologist of modernity in the sense which Baudelaire had originally given

[18] Ross, *Emergence*, p. 35.

[19] *Ibid.*, p. 33. Ross's remark seems to derive from Deleuze and Guattari's attempt to produce a "general economy" in which Marx's political economy and Freud's libidinal economy would be reconciled: Gilles Deleuze and Félix Guattari, *Anti-Oedipus* (Minneapolis: University of Minnesota Press, 1983).

[20] Marc Blanchard, *In search of the city: Engels, Baudelaire, Rimbaud* (Saratoga, Calif.: Stanford French and Italian Studies / Anma Libri, 1985) pp. 128, 148.

[the term]."[21] The appeal to Baudelaire is more than a matter of definition, however, because Frisby means to draw attention to the *aesthetic* dimension of Simmel's writing. His "capacity for capturing the essential nature of modernity was reflected not merely in his substantive analysis," so Frisby claims, "but also in the mode of presentation itself." To Lukács, Simmel was "a Monet of philosophy that has not yet been succeeded by a Cézanne" and, indeed, Frisby describes his way of working as a "sociological impressionism." This carries a number of connotations, some of them extremely problematic, but among the most important is a focus on the *fragment.*[22] Simmel takes great pains to avoid the hypostatization of "society"; he seeks instead to preserve the traces of fleeting threads of social interaction ("fortuitous fragments") which together constitute a "web" *to which there is no "center."*

On every day, at every hour, such threads are spun, are allowed to fall, are taken up again, replaced by others, intertwined with others. Here lie the interactions – only accessible through psychological microscopy – between the atoms of society which bear the whole tenacity and elasticity, the whole colourfulness and unity of this so evident and so puzzling life of society.... We can no longer take to be unimportant consideration of the delicate, invisible threads that are woven between one person and another if we wish to grasp the web of society according to its productive, form-giving forces.[23]

This passage requires a more detailed exegesis than I can provide here, but let me disentangle two themes that bear most directly on Soja's argument.

In the first place, like Baudelaire, Simmel was acutely sensitive to the temporal flux of modern social life. His vision of Berlin – a city indispensable

---

[21] David Frisby, *Fragments of modernity* (Cambridge, England: Polity Press, 1985) pp. 3, 39. According to Liebersohn, Simmel was a "thorough-going modernist" and "no one more deeply imbibed the vitality of German modernism after 1890": Harry Liebersohn, *Fate and utopia in German sociology 1870–1923* (Cambridge: MIT Press, 1988) pp. 126, 131, 137. I enlarge upon these descriptions below.

[22] More detailed discussions will be found in David Frisby, *Sociological impressionism: A reassessment of Georg Simmel's social theory* (London: Heinemann, 1981) pp. 89–101 and *idem, Fragments,* pp. 48–61; see also Wolf Lepenies, *Between literature and science: The rise of sociology* (Cambridge: Cambridge University Press, 1988) pp. 239–244. The impressionist analogy is highly suggestive but Simmel's interest in aesthetics was not bounded by the canvas. In his essay on "sociological aesthetics" he declared (more generally) that "the essence of aesthetic observation and interpretation lies in the fact that the typical is to be found in what is unique, the law-like in what is fortuitous, the essence and significance of things in the superficial and transitory": Georg Simmel, "Soziologische Aesthetik," *Die Zukunft* 17 (1896) pp. 204–16. These sentiments clearly underwrite his focus on the fragment. In the following year Simmel met the poet Stefan George and became an intimate of the George Circle; for some time afterward he described his own work as a philosophical-sociological parallel to George's poetry.

[23] Georg Simmel, *Soziologie* (Berlin: Ducker and Humblot, 1968; first edition 1908) pp. 1026, 1035; cited in Frisby, *Fragments,* p. 55–56.

to the formation of his thought – was reminiscent of Musil's "super-American city"; he once remarked that "if all clocks and watches in Berlin would suddenly go wrong in different ways, even if only by one hour, all economic life and communication of the city would be disrupted for a long time."[24] But this heightened consciousness of time never shaded into an historicism and I frankly do not see how Soja can claim the contrary (though he does). Simmel's analysis was not grounded in any "prehistory of modernity" and he provided no systematic historical analysis of any kind. This is not an idiosyncratic reading. Kracauer and Adorno both made the same observation. Kracauer claimed that none of Simmel's fragments "live in historical time" but each is instead transposed into "eternity, that is, into the sole form of existence in which it can exist as pure essentiality and be contemporary with us at any time." Similarly, Adorno objected to the way in which Simmel's preoccupation with fragments "never leads to their historical concretion but to their reduction to the eternal realm."[25] But one needs to remember Baudelaire: if "the transient, the fleeting, the contingent" make up one half of art, "the eternal, the immutable" make up the other half. As Foucault reads this injunction,

Being modern does not lie in recognizing and accepting this perpetual movement; on the contrary, it lies in adopting a certain attitude with respect to this movement; and this deliberate, difficult attitude consists in recapturing something eternal that is not beyond the present instant, nor behind it, but within it.[26]

If this reading can be sustained – and I take it from one of the architects of Soja's postmodern geography – then Simmel's must be an exemplary project of modernism.

In the second place, Simmel was particularly concerned to plot the contours of social space. Frisby describes him, with some justification, as "the first sociologist to reveal explicitly the social significance of spatial contexts for human interaction."[27] For Simmel, interaction was experienced

---

[24] Georg Simmel, "The metropolis and mental life," in Kurt Wolff (ed.), *The sociology of Georg Simmel* (Glencoe, Ill: Free Press, 1950) pp. 409–24; the quotation is from p. 413. Frisby makes a direct (and much more general) comparison between Simmel and Musil's *Man without qualities* in his *Impressionism*, pp. 157–64.

[25] Frisby, *Fragments*, pp. 41–42, 59, 81; Frisby, *Simmel*, pp. 57–59.

[26] Michel Foucault, "What is Enlightenment?," in Paul Rabinow (ed.), *The Foucault Reader* (New York: Pantheon Books, 1984) pp. 32–50; the quotation is from p. 39. Habermas reads Baudelaire in much the same way: see Jürgen Habermas, *The philosophical discourse of modernity* (Cambridge, England: Polity Press, 1987) pp. 8–10.

[27] Frisby, *Impressionism*, p. 3 and *Fragments*, p. 71. Questions of social space are woven into many of Simmel's writings, but he also wrote several essays specifically on the sociology of space: see David Frisby, "Social space, the city and the metropolis" in his *Simmel and since: Essays on Georg Simmel's social theory* (London: Routledge, 1992) pp. 98–117 and Frank Lechner, "Simmel on social space," *Theory, culture and society* 8 (1991) pp. 195–201.

"as the filling-in of space" – "interaction makes what was previously empty into something for us" – and, reciprocally, space (socially produced space) became both "agent and expression" of social interaction. With this doubling, I suggest, Simmel constructed another version of "nonpassive spatiality."[28] He was concerned to elucidate not only face-to-face inter-action, but also those impersonal, supra-individual forms in which money – and in particular "the power of money to bridge distances" – had a special place. There existed "no more striking symbol of the completely dynamic character of the [modern] world than money," he argued, since it was "the vehicle for a movement in which everything else that is not in motion is completely extinguished." These concerns were constant preoccupations of Simmel's masterwork, *The philosophy of money,* which was first published in 1900. Simmel regarded money as the quintessentially modern form of inter-action (*Wechselwirkung*) and its circuits converged on, looped around, and radiated from the metropolis in webs of ever increasing, constantly changing complexity. In effect, Simmel identified two strategic sites of modernity:

> Whereas the metropolis is, as it were, the point of *concentration* of modernity, the mature money economy (which also has its focal point in the metropolis) is responsible for the *diffusion* of modernity throughout society. Taken together, the two sites signify respectively the *intensification* and *extensification* of modernity.[29]

When Simmel described the city as a "labyrinth," therefore, he meant to convey *both* the existence of "hidden passages" – connections and relations normally concealed from consciousness by the fragmentation of social life – *and* the possibility of discovering strategic "paths" through them.[30] These double meanings marked two moments of social inquiry and they required

[28] David Frisby, *Georg Simmel* (London: Tavistock, 1984) pp. 57–59. I cannot pursue the connections between these ideas and Soja's movement toward a "spatialized ontology" that relies on Buber, Heidegger, and Sartre rather than Simmel (Soja, *Postmodern geographies,* pp. 131–37). But Simmel's charismatic teaching had a great impact on the young Buber during his studies in Berlin and the two became good friends. In any event, there are suggestive parallels between Simmel's concerns and those contained in the following passage from Soja:

> To be human is not only to create distances but to attempt to cross them, to transform primal distance through intentionality, emotion, involvement, attachment. Human spatiality is thus more than the product of our capacity to separate ourselves from the world, from a pristine Nature, to contemplate its distant plenitude and our separateness. In what may be the most basic dialectic in human existence, the primal setting at a distance is meaningless (one of existentialism's most important concepts) without its negation: the creation of meaning through relations with the world. Thus, as Buber argues, human consciousness arises from the interplay – dare I add unity and opposition? – of distancing and relation (Soja, *Postmodern geographies,* p. 133).

[29] Frisby, *Simmel and since,* p. 69; see also Georg Simmel, *The philosophy of money* (London: Routledge, 1978; 2nd ed., 1990).

[30] Frisby, *Impressionism,* pp. 95–96.

close attention to the spatiality of social life. Simmel once compared social forms to geometric forms and one commentator subsequently described his work as "a geometry of the social world." "Just as geometry locates and measures relations in space," he suggests, "so sociology outlines the contours of the social universe which are usually hidden from us."[31] But this is to use "geometry" in a purely analogical sense, whereas Simmel's concept of social space can be given a much more concrete inflection. And it is in precisely *this* sense, I suggest, that "the contours of the social universe" can be mapped and made to reveal the *Wechselwirkungen* that spiral across the turbulent landscape of modernity with a restlessness that confounds more conventional modes of representation. In geography, Harvey confronts the same problem in a brilliant essay in which he both draws upon *The philosophy of money* and seeks to enlarge a number of Simmel's central concerns:

Money is simultaneously everything and nothing, everywhere but nowhere in particular, a means that poses as an end, the profoundest and most complete of all centralizing forces in a society where it facilitates the greatest dispersion, a representation that appears quite divorced from whatever it is supposed to represent.[32]

But it is the analytics of representation that are particularly important to Simmel, and Liebersohn characterizes his double cartography very well, I think, when he notes:

Wechselwirkungen were accelerating at a dizzying rate in Imperial Berlin as it underwent its rapid transformation in the late nineteenth and early twentieth century from provincial capital to world metropolis. Simmel tried to calm his contemporaries' nerves by explaining the logic underlying the seeming chaos they experienced.... [He] delighted in showing how simple truths gave way to ironies under modern conditions. All that was solid melted into air; the ceaseless flux of modernity took on a new solidity.[33]

These two vignettes indicate that modernism was by no means as inimical to the question of spatiality as Soja implies. I do not mean to suggest that either Rimbaud or Simmel anticipated Soja in any substantive sense, but there are formal and thematic affinities between them that are of considerable interest to any critical reading of his text. I want to use these now to clarify the composition of *Postmodern geographies* more directly.

---

[31] Raymond Aron, *German sociology* (New York: Arno Press, 1979) p. 5.

[32] David Harvey, "Money, time, space and the city," in his *Consciousness and the urban experience* (Oxford: Blackwell Publishers, 1985) pp. 1–35; the quotation is from p. 3.

[33] Liebersohn, *Fate and utopia*, pp. 128, 131. The echo of Berman, *All that is solid melts into air*, is deliberate.

Modernity and spatiality I begin with an obvious paradox. Soja's project is avowedly postmodern in its theoretical sensibilities; and yet its construction seems to be predicated on an Enlightenment model of semiotics and a modernist model of the text. At the very beginning of the book Soja tells us that:

The discipline imprinted in a sequentially unfolding narrative predisposes the reader to think historically, making it difficult to see the text as a map, a geography of simultaneous relations and meanings that are tied together by a spatial rather than a temporal logic.... What one sees when one looks at geographies is stubbornly simultaneous, but language dictates a sequential succession.[34]

Ironically, Lessing advanced an identical argument in what became the Enlightenment's most comprehensive statement on the limits of painting and poetry: *Laocoon*. This extraordinary treatise was first published in 1776 and although Lessing's influence waned during the romantic period his assertive distinctions between the spatial and temporal arts have proved to be remarkably durable. Lessing represented space as the sphere of the painter and time as the sphere of the poet. Painting was supposed to express subjects which exist side by side (bodies) whereas poetry expressed subjects that succeed each other (actions), thus:

| Space | Bodies | Painting |
| Time | Actions | Poetry |

Although neither bodies nor actions could stand alone, so that each art could suggest the sphere of the other, Lessing objected to the assimilation of one art form by another. Hence his remark that "the *co-existence* of the physical object comes into collision with the *consecutiveness* of speech." The added emphases are exactly equivalent to Soja's distinctions.[35] Soja is far from being the first to bring them to bear upon what used to be called "the problem of geographical description" and a generation of previous writers endorsed the same, essentially Kantian distinctions.[36] But he is one

---

[34] Soja, *Postmodern geographies*, pp. 1–2.

[35] See David Wellbery, *Lessing's* Laocoon: *Semiotics and aesthetics in the age of reason* (Cambridge: Cambridge University Press, 1984).

[36] Perhaps they endorsed them *because* they were Kantian distinctions: Kant's separations between geography and history, *nebeneindander* and *nacheinander*, had a formative influence on the institutionalization of the modern discipline. Even Darby – an historical geographer – appeared to accept them in his essay on the problem of geographical description. There he drew attention to the "difficulty of conveying a visual impression in a sequence of words" which, so he believed, was "one of the disadvantages of the writer as compared with the painter." And he invoked Lessing to insist that "we can look at a picture as a whole and it is as a whole that it leaves an impression upon us; we can, however, read only line by line." See H. C. Darby, "The problem of geographical description," *Transactions of the Institute of British Geographers* 30 (1962) pp. 1–14.

of the first to use them to try to *enlarge* rather than confine the geographical imagination. Where previous writers usually sentenced themselves to a conventional textual strategy, Soja is intent on escape. "At the core of each essay," he declares, "is an attempt to break out from the temporal prison-house of language [and] to make room for the insights of a spatial hermeneutic."[37]

But where does he escape to? Here is Bradbury and McFarlane:

> Modernist works frequently tend to be ordered . . . not on the sequence of historical time or the evolving sequence of character, from history or story, as in realism or naturalism; they tend to work spatially or through layers of consciousness, working towards a logic of metaphor or form. The symbol or image itself . . . whether it be the translucent symbol with its epiphany beyond the veil, or the hard objective center of energy, which is distilled from multiplicity, and impersonally and linguistically integrates it – helps to impose that synchronicity which is one of the staples of Modernist style.[38]

Theirs is not an eccentric perspective. It is precisely what Berger had in mind when he spoke about modes of narration "constantly having to take into account the simultaneity and extension of events and possibilities." More generally, Lunn remarks that "in much modernist art, narrative or temporal structure is weakened, or even disappears, in favor of an aesthetic ordering based on synchronicity, the logic of metaphor, or what is sometimes referred to as 'spatial form.' "[39] Within modernism synchronicity and simultaneity signal not a stasis, therefore, but a determined attempt to explore (and exploit) the *spatial form* of narrative.[40] Attempts of this kind formed part of a much more general movement, a characteristic impulse of fin-de-siècle modernism to subvert Lessing's distinctions across the

---

[37] Soja, *Postmodern geographies*, p. 1.

[38] Malcolm Bradbury and James McFarlane, "The name and nature of modernism," in Malcolm Bradbury and James McFarlane (eds.), *Modernism 1890–1930* (London: Penguin Books, 1976) pp. 19–55; the quotation is from p. 50.

[39] Berger, *Look of things*, p. 40; Eugene Lunn, *Marxism and modernism: Lukács, Brecht, Benjamin, Adorno* (London: Verso, 1985) p. 39.

[40] The term is usually attributed to Joseph Frank: see Jeffrey Smitten and Ann Daghistany (eds.), *Spatial form in narrative* (Ithaca: Cornell University Press, 1981). At its simplest – and theoretically least forceful – Franks's concept of spatial form connotes merely the textual strategies through which writers subvert the chronological sequence of conventional narrative. But it is necessary to attend to the hermeneutics of reception too: It is of course the case that we *read* line by line, as Soja remarks, but we do not *understand* what we read line by line. According to Mink, narratives derive their power from their success "at producing and strengthening the act of understanding in which actions and events, although represented as occurring in the order of time, can be surveyed as it were in a single glance as bound together in an order of significance." A narrative is thus intrinsically synoptic ("configurational") and derives its intelligibility from the way in which it gradually brings into view a larger whole. See Louis Mink, "History and fiction as modes of comprehension," *New Literary History* 1 (1969–70) pp. 541–58.

whole field of the arts. One may cite, as particularly explicit examples, Klee's orchestration of time and space within his paintings and Apollinaire's determination to subvert the conventional *linéaire-discursive* reading of his poems. Indeed, Apollinaire considered that the modern poet's greatest challenge was "to express the sense of life being lived all over the globe at one and the same time" since that was "the ideal way in which he could convey the swiftness of communication, the intensity of modern life."[41] And, as I have already implied, the novel furnishes still more illustrations. To return to Musil's Vienna once again:

> For the novelists of the early twentieth century, the city has become too amorphous for structured, sequential narrative. In Dickens each chapter corresponds to a clearly defined social milieu and we are offered a coherent map of social interaction. But the dynamic rhythms of the modern metropolis are felt to be too elusive to be fixed in this way, its social relations too fractured and impersonal, its states of consciousness too fluid. The contours of the city streets dissolve in Musil's Vienna as they do in Rilke's Paris, Joyce's Dublin, Woolf's London and Döblin's Berlin.[42]

These remarks open the prospect of a very different reading of Soja's project: one which represents it as a thoroughly *modernist* account of post-modernity. I think such a reading can be sustained, but it does not mean that Soja's Los Angeles mirrors Musil's Vienna. There are, after all, many modernisms. In the passage I have just quoted the superimposition of physical and psychological spaces is a particularly telling description of *The man without qualities*. Musil's doctoral thesis was in fact a study of Ernst Mach, whose work moved between physics and psychology and, indeed, projected them onto the same level. Perhaps even more important, his critique of absolute concepts of time and space had a great influence upon Musil's thinking and, so it seems to me, upon the architectonics of the novel itself.[43] But Soja is doubly distant from these themes. Following Lefebvre, he explicitly rejects the reduction of spatiality to physical objects and forms, apprehended only as a "collection of things" ("the illusion of opaqueness"), and he is equally hostile to the reduction of spatiality to

---

[41] See Margaret Davies, "*Modernité* and its techniques" and Renée Riese Herbert, "Modernism in art and literature," both in Monique Chefdor, Ricardo Quinones, and Albert Wachtel (eds.), *Modernism: Challenges and perspectives* (Urbana, Ill.: University of Illinois Press, 1986) pp. 146–58 and 212–37.

[42] Timms, *Musil's Vienna*, p. 258.

[43] Musil's work was not of course a direct translation of Mach's ideas from "science" into "art" – distinctions that Musil would have found problematic in any case – and he was critical of many of Mach's formulations. But there is no doubt that his study of Mach left a lasting impression upon him nor that it shaped his subsequent treatment of the relations between science and self in the modern world. See David Luft, *Robert Musil and the crisis of European culture 1880–1942* (Berkeley: University of California Press, 1980) pp. 217–68.

psychological constructs, revealed as merely projections of the mind ("the illusion of transparency"). Spatiality "cannot be completely separated from physical and psychological spaces," he concedes, but it must nonetheless be theorized as *socially produced* space.[44] This is still a materialist perspective, of course, and Mach's ideas were endorsed by revisionists within Austrian Marxism and by a number of Bolsheviks who argued that space is "a form of social coordination of the experiences of different people."[45] But Soja clearly remains close to Western Marxism and is so deeply and implacably opposed to positivism (in all its versions) that his views simply cannot be assimilated to those of Mach. To Mach, Habermas tells us, "in its uncon-cealedness, facticity knows no opposition of essence and appearance, of being and illusion, because the facts themselves have been elevated to the status of an essence."[46] Nothing could be further from Soja's project, which is based on a "depth model" and is directed toward using concepts of spatiality in order to disclose – to bring into the open – the sociospatial structures of late capitalism. The construction of a truly postmodern geo-graphy, so he believes, will enable us "to see more clearly the long-hidden instrumentality of human geographies, in particular the encompassing and encaging spatializations of social life that have been associated with the historical development of capitalism."[47]

The geographical and the historical

This is the same maneuver that I attributed to Simmel, in a somewhat different vocabulary, but Soja's models are Lefebvre, Foucault and, more concretely, Jameson. The relations between these three are far from simple, and they are further complicated by the presence of a fourth figure – Sartre – at the center of their discursive triangle. All of these writers are concerned in one way or another with the relations between the aesthetic and the social, but Soja projects the triangle onto a different plane in order to valorize the geographical against the historical. I will therefore move directly to a consideration of this version of the diagram.

The French spatial tradition Soja repeatedly draws attention to the vibrancy of what he calls "the French spatial tradition" which is articulated most powerfully, so he says, by Lefebvre and Foucault.

The very survival of capitalism, Lefebvre argued, was built upon the creation of an increasingly embracing, instrumental and socially mystified spatiality, hidden

[44] Soja, *Postmodern geographies*, pp. 120–31.
[45] The phrase is Bogdanov's: Kern, *Culture of time and space*, pp. 134–35.
[46] Jürgen Habermas, *Knowledge and human interests*, 2nd ed. (London: Heinemann, 1968) p. 85.
[47] Soja, *Postmodern geographies*, p. 24.

from critical view under thick veils of illusion and ideology. What distinguished capitalism's gratuitous [sic] spatial veil from the spatialities of other modes of production was its peculiar production and reproduction of geographically un-even development via simultaneous tendencies toward homogenization, fragmen-tation and hierarchization – an argument that resembled in many ways Foucault's discourse on heterotopias and the instrumental association of space, knowledge and power.[48]

This is a complex passage, though I do not think that Soja draws upon either of these authors in much more than a formal sense. There are, of course, substantial differences between them. Most obviously, Lefebvre is closer to Marxism than Foucault and in consequence binds the production of space more tightly to the structures of capitalism.[49] But there are also important connections and it is these largely formal thematics that Soja draws out to provide the template for his own argument. Both Lefebvre and Foucault accentuate what Soja calls the "instrumentality" of social space – what they would refer to as its *strategic* function – and since this is of such importance to Soja's thesis I need to provide a thumbnail sketch of the arguments involved.

Lefebvre's account is based on the distinction between exchange value and use value that lies at the heart of Marx's analysis of the commodity. Whereas Marx's critique of political economy privileged history over geo-graphy, however, Lefebvre tries to make the "silent spaces" of *Capital* speak. He transforms Marx's original opposition into one between the "abstract space" of capitalism's economic and political systems – externalized, ration-alized, sanitized – and the swirling, kaleidesocopic "lived space" of everyday life. The tension between these spaces transcodes a tension between integration and differentiation that admits of no final solution. In Lefebvre's view, modernity lacks "a unifying direction, a true integration, a style." There is, he insists, no "meta-system" uniting the partial subsystems: "their cohesion is, rather, the object of a strategy."

At the theoretical level this betrays (rather than simply uncovers) a global strategy; it constitutes a new totality, whose elements appear to be both *joined* (joined in space by authority and by quantification) and *disjoined* (disjoined in that same fragmented space and by that same authority, which uses its power in order to unite by separating and to separate by uniting).[50]

---

[48] *Ibid.*, p. 50.

[49] I cannot discuss Foucault's relation to historical materialism here – and apart from a passing admonition of Anderson Soja does not discuss it either – but I do want to insist (like Soja, I think) that it was not a purely antagonistic one. In this connection Lentricchia's careful positioning of Foucault within a horizon of meaning indelibly shaped by Marx seems to me exemplary: Frank Lentricchia, *Ariel and the police* (Madison: University of Wisconsin Press, 1988) pp. 29–102.

[50] Henri Lefebvre, *The survival of capitalism* (London: Alison and Busby, 1976); the quotations are from pp. 27–28, 84–85.

The weight of this passage falls on its first clause: betrays, not uncovers. Lefebvre is claiming here that social space, the conjunction of *l'espace abstrait* and *l'espace vécu*, conceals the strategies that underwrite modern capitalism; but he is also claiming, more subversively, that a theory of the production of space that makes those strategies visible, that renders their contours *en clair*, is capable of underwriting a politics of resistance that must be (and, indeed, can *only* be) a "politics of space." It is because spatiality can be mobilized to subvert (that is, "betray") spatiality in this way that Lefebvre is able to assert that "what is new and paradoxical" in the current situation is that "the dialectic is no longer attached to temporality." "To recognize space," he declares, "to recognize what takes place there and what it is used for, is to resume the dialectic."[51]

The parallels with Foucault are not difficult to see. His terrifying vision of the *société disciplinaire* incorporates a systematic surveillance – a "panopticism" – conducted through a detailed partitioning of space which, in one sense, simply transposes Lefebvre's tension between integration and differentiation into a new register.[52] Certainly, Foucault is adamant that there is no metasystem holding the grid together. He insists that the spatial strategies that underwrite "disciplinary power" – and which he represents as a specifically *modern* form of power – cannot be grasped by positing a central generating mechanism and then conducting a "descending" analysis of power from this supposedly privileged place. On the contrary: these strategies are capillary, diffuse, and they can only be exposed through an "ascending" analysis of power capable of revealing their colonization, displacement, and extension. For Foucault, Deleuze tells us, "Power is characterized by immanence of field without transcendent unification, continuity of line without global centralization, and contiguity of parts without distinct totalization: it is a social space."[53]

Foucault's analysis does not lead as directly as Lefebvre's to a "politics of space," however, and one of the most controversial areas of his work is the limited support it seems to afford for any kind of resistance. But it can be made to intersect with the critical intentions of Soja's project on a more formal level. Foucault constantly represents the construction of social space in terms of what Deleuze and Rajchman call "spaces of visibility" and his writings can be seen, in part, as a "history of the visual unthought."

[51] Lefebvre, *Survival*, p. 17. These remarks distance Lefebvre from Habermas. The distinction between *l'espace abstrait* and *l'espace vécu* clearly mirrors Habermas's distinction between "system" and "lifeworld," but this reflection is immediately shattered by the concept of social space that is absent from the theory of communicative action; see Chapter 6, this volume.

[52] Deleuze provides a more complex discussion of the place of integration and differentiation within Foucault's "topology" – though one is conducted on Deleuze's own terrain, not Lefebvre's – in Gilles Deleuze, *Foucault* (Minneapolis: University of Minnesota Press, 1988).

[53] *Ibid.*, p. 27. Deleuze's most important source is, of course, Michel Foucault, *Discipline and punish: Birth of the prison* (London: Penguin, 1977), but see also his *Power/knowledge*, pp. 78–108.

From this perspective, one of his most subversive counterstrategies is to provide a "description-scene" of "how things were made visible, how things were given to be seen, how things were shown to knowledge or to power." The visual metaphoric is not accidental: Foucault suggests that "we are surrounded by spaces which help form evidences of the ways we see ourselves and one another." In mapping those spaces and *re-presenting* them, therefore, one *de-naturalizes* them. The very act of describing their production – of disclosing their modes of spatialization – not from some privileged vantage point but simply from *another* vantage point dispels what Rajchman calls "the sort of routine, instituted self-assurance" people have about things. And in this sense, I suggest, Foucault's analytics of space also becomes a politics of space.[54]

Soja's own visual metaphoric is equally arresting. He dismisses the reduction of spatiality to physical objects and forms as "a confusing myopia [that] has persistently distorted spatial theorization by creating illusions of opaqueness, short-sighted interpretations of spatiality which focus on immediate surface appearances without being able to see beyond them." Symmetrically, the reduction of spatiality to projections of the human mind becomes

> a hypermetropic illusion of transparency [that] sees right through the concrete spatiality of social life.... Seeing is blurred not because the focal point is too far in front of what should be seen, inducing near-sightedness, but because the focal point lies too far away from what should be seen, the source of a distorting and over-distancing vision, hypermetropic rather than myopic.

These distinctions are derived directly from Lefebvre, who also identifies "the illusion of opacity" (in which social space appears as natural, "substantial," purely physical) and "the illusion of transparency" (in which social space appears as luminous, innocent, freely intelligible).[55]

It should now be clear why Soja claims that Lefebvre and Foucault are precursors of *Postmodern geographies*. In his view, "it is space not time that hides consequences from us" and hence, following their example, the concept of spatiality may be used to open up "a different way of seeing the world in which human geography ... provides the most revealing critical perspective."[56] But when Jameson is added to the diagram complications set in.

---

[54] Deleuze, *Foucault*, pp. 52–60, 80–81; John Rajchman, *Michel Foucault: The freedom of philosophy* (New York: Columbia University Press, 1985) pp. 50–67; idem, "Foucault's art of seeing," *October* 44 (1988) pp. 88–117.

[55] Soja, *Postmodern geographies*, pp. 122, 124–25; Henri Lefebvre, *The production of space* (Oxford: Blackwell Publishers, 1991) pp. 27–30; this was first published in French in 1974.

[56] Soja, *Postmodern geographies*, p. 23.

Mapping Jameson   There can be no doubt about Jameson's commitment to some form of spatial analysis – although a number of objections can be registered against his accounts of "postmodern hyperspace" – and his recent writings have been unwavering in their commitment to a cultural politics informed by a deep sense of spatiality. Indeed, he has repeatedly applauded Lefebvre's work and once complained, in an acerbic aside, that Lefebvre's "conception of space as the fundamental category of politics and of the dialectic itself – the one great prophetic vision of these last years of discouragement and renunciation – has yet to be grasped in all its pathbreaking implications."[57] Jameson's attitude to Foucault is more ambivalent, but he is evidently sympathetic to a reading of his work (and an appropriation of it) "in terms of the *cognitive mapping* of power, the construction of spatial figures."[58] But at present I am less interested in Jameson's view of space than his attachment to historicism, because Soja insists that the agreeable prospect of a postmodern geography – of which Jameson too is supposed to be a precursor – is blocked only by its continued salience.

Jameson begins his excavation of *The political unconscious* with the unambiguous injunction: "Always historicize!"[59] In doing so he reveals a horizon of meaning which can, I think, be traced throughout his work, including the later essays which deal more directly with questions of spatiality.[60] In those later writings Jameson certainly recognizes that one of the constitutive features of the postmodern is what he calls a "weakening of historicity," to such a degree that "our daily life, our psychic experience, our cultural languages are today dominated by categories of space rather than by categories of time." In consequence, he argues, the past has been turned into a vast collection of images. All that remains is a "libidinal historicism": a playful collage in which the archive is ransacked and its fragments pasted

[57] Fredric Jameson, "Architecture and the critique of ideology," in Joan Ockman, Deborah Berke, and Mary McLeod (eds.), *Architecture, criticism, ideology* (Princeton: Princeton Architectural Press, 1985) pp. 51–87; the quotation is from p. 53. Jameson's acclamation is not unexpected, since both he and Lefebvre share a more or less Hegelian reading of Marx.

[58] Anders Stephanson, "Regarding postmodernism: A conversation with Fredric Jameson," in Douglas Kellner (ed.), *Postmodernism / Jameson / Critique* (Washington, D.C.: Maisonneuve Press, 1989) pp. 43–74; the quotation is from p. 47.

[59] Fredric Jameson, *The political unconscious: Narrative as a socially symbolic act* (Ithaca: Cornell University Press) p. 9. Jameson goes on to represent this as "the one absolute and we may even say 'transhistorical' imperative of all dialectical thought." See Michael Sprinker, "The part and the whole: Jameson's historicism," in his *Imaginary relations: Aesthetics and ideology in the theory of historical materialism* (London: Verso, 1987) pp. 153–76 and Hayden White, "Getting out of history: Jameson's redemption of narrative," in his *The content of the form: Narrative discourse and historical representation* (Baltimore: Johns Hopkins University Press, 1987) pp. 142–68.

[60] The remainder of this paragraph is primarily based on a reading of Jameson, "Cultural logic," and *idem*, "Cognitive mapping," in Cary Nelson and Lawrence Grossberg (eds.), *Marxism and the interpretation of culture* (Urbana, Ill.: University of Illinois Press, 1988) pp. 347–60.

into glittering, shimmering, "overstimulating ensembles." Jameson invokes Lacan to represent this "random cannibalization of all the styles of the past" as a break in the signifying chain. "If we are unable to unify the past, present and future of our sentences," he laments, "we are similarly unable to unify the past, present and future of our own biographical experience." As the historical imagination is suppressed, critical depth and distance are erased.[61] Postmodern hyperspace has "finally succeeded in transcending the capacities of the individual human body to locate itself, to organize its immediate surroundings perceptually and cognitively to map its position in a mappable external world." When Jameson calls for a new "cognitive mapping," therefore, his purpose is not to privilege geography over history but rather to *restore* a (highly particular) sense of historicity.[62] As Preziosi shrewdly remarks,

> For Jameson, his version of Marxism is a place coextensive with the space of History. To arrive in that space, it is necessary to "pass through" texts, *and above all the texts and hyperspaces of postmodernism*, in order to grasp the latter's "absent causes": their History.[63]

Jameson's relation to any "postmodern geography" is thus doubly coded. He engages with poststructuralism and postmodernism (far more assiduously than most of their critics) not to celebrate them but to subvert them. This is the source of most of the misreadings of his project – at any rate, of those which have assimilated him to postmodernism *tout court* – and his description of E. L. Doctorow's *Ragtime* could, I think, be applied with

---

[61] This is exactly what concerned C. Wright Mills in his celebration of *The sociological imagination* (New York: Oxford University Press, 1959). Soja describes this book as "paradigmatic" (Soja, *Postmodern geographies*, p. 13) but this is not especially helpful. It may have been paradigmatic of historicism (at least as Soja understands it) but it was patently *not* paradigmatic of social theory as a whole. In failing to make this distinction Soja blunts the force of Mills's critique, which was directed against the ruling orthodoxy of Parsonian sociology. Parsons's unrelenting functionalism suppressed the very sense of historicity that Mills (and subsequently Jameson) sought to recover. The political implications of this point ought to be obvious and I sharpen them below.

[62] Hence his confession that "cognitive mapping" is "nothing but a code word for 'class consciousness'": Jameson, *Postmodernism*, p. 418. It is for this reason, I imagine, that Preziosi concludes that "the chief problem with what he construes as postmodernism is its 'unmappability' for the traditional instruments of Jameson's nostalgically Lukácsian perspective": Donald Preziosi, "*La vi(ll)e en rose:* reading Jameson mapping space," *Strategies* 1 (1988) pp. 82–99; the quotation is from p. 87. In general I agree, but with one important caveat. Jameson's admiration for Lukács (and Bloch) is well known, but it is misleading to imply that he wants to reinstate their problematic – he categorically rejects the possibility of a "return" of any kind – and his perspective is thus hardly nostalgic.

[63] *Ibid.*, p. 91; my emphasis. For other purposes one might also emphasize the capitalized "History" since this bears directly on Jameson's critique of postmodernism and its "enfeeblement" of a specifically Marxist sense of historicity: see Linda Hutcheon, *The politics of postmodernism* (New York: Methuen, 1989) pp. 113–14.

equal cogency to his own writings. Doctorow, he reminds us, "has had to
elaborate his work by way of that very cultural logic of the postmodern
which is itself the mark and symptom of his dilemma."[64] Jameson's own
dilemma is double-sided, however, and in drawing attention to postmodern
space and spatiality he is seeking to inculcate, through those very figures,
an enlarged sense of historicity. He wants to reclaim the possibility of
"making history" more or less as Marx understood it. If one wants to add
"making geography" to that, as Soja clearly wants to do (and as Jameson
seems prepared to do), then Jameson's own project – whatever inflection
Soja might want to give it – remains firmly within the horizon of historical
materialism. When Jameson writes about "the essential *mystery* of the
cultural past" – the emphasis is his but I retain it for good reason – he
is adamant that this can be resolved "only if the human adventure is one":

These matters can recover their original urgency for us only if they are retold within
the unity of a single great collective story; only if, in however disguised and symbolic
a form, they are seen as sharing a single fundamental theme – for Marxism, the
collective struggle to wrest a realm of Freedom from a realm of Necessity; only if
they are grasped as vital episodes in a single, vast unfinished plot.[65]

And that is precisely how Jameson sees them. Throughout his writings he
constantly calls upon Marxism to provide the master-narrative that makes
his work possible. That narrative is not unchanged or unchanging and
Jameson is not proposing a return to any of its classical versions. But if
his attempt to reconstruct (or "map") the totality of late capitalism does
not involve him in any return to modernism – something that he repeatedly
if not altogether persuasively disavows – then his continued commitment

[64] Jameson, "Cultural logic," pp. 70–71. Jameson himself describes his project as an attempt "to
undo postmodernism homeopathically by the methods of postmodernism: to work at dissolving
the pastiche by using all the instruments of the pastiche itself, to reconquer some genuine historical
sense by using the instruments of what I have called substitutes for history": Stephanson,
"Regarding postmodernism," p. 59. For a fuller discussion see Steven Best, "Jameson, totality and
the poststructuralist critique," in Kellner (ed.), *Postmodernism/Jameson/Critique*, pp. 333–68.

[65] Jameson, *Political unconscious*, pp. 19–20. The sense of incompletion established in the last
clause (history as an "unfinished plot") is of the utmost importance since Jameson's historicism
is directed toward the *future*. As Eagleton remarks in his discussion of Jameson's style – something
by no means incidental to his project as a whole – Marxism draws its poetry "from a future to
which it is simultaneously deferred. There is no historical conjuncture except from the standpoint
of a desirable future." See Terry Eagleton, *Against the Grain: Essays 1975–1985* (London: Verso,
1986) p. 68.

It is for this reason that Jameson is so preoccupied with the Utopian impulse – the anticipatory
moment – of critical theory. He increasingly expresses this (I think vital) impulse in spatial terms,
through the anticipation of "properly spatial Utopias, in which the transformation of social relations
and political institutions is projected onto the vision of place and landscape": Jameson, *Postmodernism*,
p. 160. But I see this as what Jameson would himself call a "transcoding" of historicity rather
than its transcendence by spatiality.

to the concept of totality plainly distances him from any prospective *post* modernism too. As Eagleton puts it, what appears wrong with the world from Jameson's perspective

is not so much this or that phenomenon but the fact that we cannot see all of these phenomena together and see them whole – the fact that they are isolated, fragmented, compartmentalized.... What is wrong with capitalism on this reckoning is its power to disconnect; and Marxism will accordingly become a hermeneutics ...deciphering the submerged grammar of bourgeois society, explaining how all of these disconnected phenomena are secretly totalized.[66]

It then makes perfectly good sense for someone like Harvey to invoke Jameson in his historico-geographical analysis of *The condition of postmodernity*. Harvey makes no secret of his own hostility to postmodernism and, in particular, its licence to "revel in the fragmentations and the cacophony of voices ... of the modern world" while "denying that kind of meta-theory which can grasp the political-economic processes ... that are becoming ever more universalizing in their depth, intensity, reach and power over daily life."[67] But to the extent that Soja seeks to enlist Jameson under the sign of *Postmodern geographies* he must render his own project deeply problematic.

Ironically, Jameson's totalization is so complete and the unity of his History – that "single, great collective story" – so intense that his desire to breathe new life into the redemptive historiography of modern Marxism ends in a paralysis of the *differential temporalities* of postmodernity. Davis captures the consequences with a particular precision:

Jameson's [account of] postmodernism tends to homogenize the details of the contemporary landscape, to subsume under a master concept too many contradictory phenomena which, *though undoubtedly visible in the same chronological moment*, are nonetheless separated in their true temporalities.[68]

---

[66] Eagleton, *Against the grain*, pp. 76–77. Compare Jameson's remarks on totality and totalization in his *Postmodernism*, pp. 400–406.

[67] Harvey, *Condition of postmodernity*, pp. 116–17. The parallels are repeated across a number of other levels. Both Jameson and Harvey treat postmodernism as the "cultural logic" or the "cultural dominant" of late capitalism (though in Harvey's case this is translated into a regime of "flexible accumulation") and there are clear resonances between Jameson's call for a new "cognitive mapping" and Harvey's much earlier celebration of the "geographical imagination": see his *Social justice and the city* (London: Edward Arnold, 1973) pp. 23–24. Those resonances become more pronounced as Harvey moves deeper into Marxism, and he eventually suggests – in a deliberate echo of Jameson – that the existing horizon of the geographical imagination may not be able to encompass the time-space compressions of postmodernity: "There is an omnipresent danger that our mental maps will not match current realities." See Harvey, *Condition of postmodernity*, pp. 304–305.

[68] Mike Davis, "Urban renaissance and the spirit of postmodernism," *New Left Review* 151 (1985) pp. 106–13; the quotation is from p. 107 (my emphasis). Davis is referring specifically to Jameson, "Cultural logic," but other critics would generalize his objection. For a response more sympathetic to Jameson, see Best, "Jameson, totality and the poststructuralist critique."

I have accentuated that qualifying clause because it is a compelling reminder of the dark side of spatiality: of the danger of being blinded by simultaneity. And for this reason one ought to welcome the very specific sense in which Soja claims to object to historicism. As he uses the term, it is intended to connote

an overdeveloped historical contextualization of social life and social theory that actively submerges and peripheralizes the geographical or spatial imagination. This definition does not deny the extraordinary power and importance of historiography as a mode of emancipatory insight, but identifies historicism with the creation of a critical silence, an implicit subordination of space to time that obscures geographical interpretations of the changeability of the social world and intrudes upon every level of theoretical discourse, from the most abstract ontological concepts of being to the most detailed explanations of empirical events.... It is the dominance of a historicism of critical thought that is being challenged, not the importance of history.[69]

The corollary of all this is equally unacceptable, however, and I am by no means convinced that Soja avoids it. If, as he claims, his intention is "not to erase the historical hermeneutic but to open up and recompose the territory of the historical imagination through a critical spatialization," this is seriously compromised by the way in which he pursues his history of the present. His single-minded determination to expose one "hidden narrative" – the subordination of space to time – runs the risk of inviting a simple reversal in response: the subordination of time to space. There are several passages in which Soja appears to accept such an invitation, but it can be forestalled, I think, by moving against the grain of his intellectual history to expose another "hidden narrative."[70]

The domain of the dead   I am referring to the history of Western geography and, more particularly, to its modern – and, I think, its *characteristically* modern – obsession with space. Although Soja claims that the instrumentality of space was "lost from view" at the close of the nineteenth century, this was surely a highly selective myopia. For the fin de siècle coincided with the institutionalization of modern geography, and its admission to the academy (often against considerable opposition) was secured in part through

---

[69] Soja, *Postmodern geographies*, pp. 15, 24.

[70] *Ibid.*, p. 12. Soja's reference to a "hidden narrative" and his intention to deconstruct the conventional history of social theory is derived from Eagleton, *Against the grain*, p. 80: "To 'deconstruct,' then, is to reinscribe and resituate meanings, events and objects within broader movements and structures; it is, so to speak, to reverse the imposing tapestry in order to expose in all its glamorously disheveled tangle the threads constituting the well-heeled image it presents to the world." My concern, obviously, is that Soja's deconstruction weaves its threads back into a narrative as coherent as the one it seeks to replace.

a recognition of its political salience in a world of European rivalries and imperialist adventures. Nation and empire, territory and frontier were as much the common currency of the discipline as the resource inventories of its self-styled "commercial" geographies. After the Franco-Prussian war of 1870–71 most of the major European powers recognized the strategic importance of a knowledge of geography – not least for the conduct of military campaigns and occupations – and in the following decades Ratzel's *Politische Geographie*, Vidal de la Blache's *France de l'Est* and Mackinder's *Democratic ideals and reality* all accentuated, in ostensibly more intellectual but no less ideological terms, the geopolitical significance of space and spatiality. This was reinforced, in more mundane but perhaps even more practical ways, by the imbrications of geographical knowledge and colonial power. Soja is aware of these developments but he makes remarkably little of them, and the omission is not without consequence.[71]

From one point of view, Soja's disinterest seems perfectly proper. After all, his focus is on the trajectory of *critical theory* (in the most general of senses) and both of these terms were foreign to most of human geography until the second half of the twentieth century. Certainly, the installation of geography within the academy had little or no impact on the formation of a radical political culture or on the course of Western Marxism. Those geographers working on the left were marginalized – for various reasons – and none of the three authors listed above could (or would want to) claim any such affiliation.[72] Their writings had few theoretical pretensions, and those who followed them into positions of disciplinary power established geography as a resolutely empiricist science estranged from the theoretical engagements of even the bourgeois social sciences.[73] By the end of World War I, as Soja argues, geography had become an accomplice in

---

[71] Soja does recognize the involvement of modern geography within the state apparatus – in particular in the three arenas of "military intelligence, economic planning and imperial administration" (Soja, *Postmodern geographies*, p. 37) – but he offers no systematic discussion of the structural ties between the two. This is unfortunate because it confines his interpretative analytics (at this point only: Soja's theorization of spatiality escapes these objections) to the actions of individuals; he is unable to say very much about the structures that made their actions possible and gave them such salience. For an analysis of the constitution of *post*-war geography closer to my own inclinations – and, I think, more conformable with Soja's project – see Allen Scott, "The meaning and social origins of discourse on the spatial foundations of society," in Peter Gould and Gunnar Olsson (eds.), *A search for common ground* (London: Pion, 1982) pp. 141–56.

[72] Even so, Soja's argument is more general. He claims that during the fin de siècle "the politics and ideology embedded in the social construction of human geographies and the crucially important role the manipulation of these geographies played in the late nineteenth-century restructuring and early twentieth-century expansion of capitalism seemed to become either invisible or increasingly mystified, left, *right and centre*" (Soja, *Postmodern geographies*, p. 34); my emphasis.

[73] I describe them as "bourgeois" not only to distinguish the social sciences from the developing discourse of historical materialism but also to draw attention to the class interests embedded in their constitution: see Göran Therborn, *Science, class and society: On the formation of sociology and historical materialism* (London: New Left Books, 1976).

its own exile. It built a wall around itself and retreated inside, shielded from the intellectual debates that raged beyond its confines. Soja borrows a remark of Foucault's to complain that this "cocooned" geography – "entombed" would perhaps be more accurate – "treated space as the domain of the dead, the fixed, the undialectical, the immobile – a world of passivity and measurement rather than action and meaning." Content to describe "outcomes deriving from processes whose deeper theorization was left to others," he continues, by midcentury "the discipline of Modern Geography was theoretically asleep."[74]

Now Soja's disaffection with this "infinite regression of geographies upon geographies" is one I share and, in this sense, his disinclination to explore it in any depth is understandable. But from another point of view the decision is much less satisfactory. If I change the emphasis of the previous paragraph to say that Soja focuses on the *trajectory* of critical theory, then one can see that this "tunnel-vision," if I may so call it, is out of place in a text whose construction is designed to "sidetrack the sequential flow" in order to "take coincident account of simultaneities." This is more than a matter of textual consistency because a more rigorous deconstruction of modern geography would surely show that its vision of space as "the domain of the dead" was in large measure precisely the product of its *antihistoricist* gaze.

Hartshorne's *The nature of geography*, published on the eve of World War II, could not have been plainer. Summoning support from Kant (and more immediately from Hettner), Hartshorne insisted that it was essential "to distinguish clearly between the two points of view – that which studies the associations of phenomena in terms of time [history] and that which studies their associations in terms of place [geography]."[75] Hartshorne's project was an exercise in policing disciplinary boundaries, of course, fully symptomatic of – and in part constitutive of – that "involution" to which Soja so forcefully objects. It cast a long shadow. In the postwar years the emergence of spatial science, so far from overthrowing the Hartshornian orthodoxy (as its protagonists claimed), continued and even intensified the marginalization of history and historical geography. Its minimalist geometries of points, lines, and hexagons moved the center of geography still further away from the crowded and complicated canvas of history. The intellectual itinerary that underwrote these maneuvers was itself profoundly antihistorical. As Smith shows in an exceptionally careful commentary, Hartshorne's inquiry was predicated on the unshakeable conviction that "Geography is today simply what it has always been."[76] In effect, he turned

[74] Soja, *Postmodern geographies*, pp. 37–38.

[75] Richard Hartshorne, *The nature of geography* (Lancaster, Penn: Association of American Geographers, 1939) pp. 175–88; the quotation is from p. 184.

[76] Neil Smith, "Geography as museum: Private history and conservative idealism in *The Nature of Geography*," in J. Nicholas Entrikin and Stanley D. Brunn (eds.), *Reflections on Richard Hartshorne's*

the discourse of geography into a museum and appointed himself curator. "The intellectual artifacts [were] arranged in historical sequence, room by room, for the purpose of glorifying the present."[77] Spatial science simply opened an additional room in order to reinstate what Haggett called "the geometrical tradition [that] was basic to the original Greek conception of the subject."[78] In other versions it closed the museum altogether. Bunge considered it absurd to pay any attention to the writings of previous generations because their authors would have held different views had they been alive today. One of his closing questions radicalized the hostility of spatial science (and of modern geography more generally) to historical inquiry through a revealing play on words. "Do we need the concept of time in geography?" he asked. "Perhaps movement and space are the more convenient primitives for geography. Space will tell."[79]

Now I do not mean to imply that Soja's "reassertion of space" inculpates him in any of these spatial fetishisms; it clearly does not. In contrast to the exceptionalist position of the writers in the previous paragraph, Soja locates his project within critical social theory and, as he would be the first to remind me, the context makes a difference. Neither do I pretend that the relation between history and critique is symmetrical. There is nothing in the historical imagination that automatically provides a critical perspective: Those who seek to enlarge it may be (and often are) as conservative as those who seek to suppress it. But turn this around and it is immensely difficult to envisage a radical politics that is not informed by a deep sense of historicity. Critical social theory is predicated on precisely this assumption and Soja's reassertive incorporation of spatiality must surely retain that historical sensibility if it is not to render space as "the domain of the dead" and thereby diminish its own practical intent.[80]

The Nature of Geography, Occasional Publications of the Association of American Geographers 1 (1989) pp. 91–120; the quotation is from p. 110. Smith's remark makes sense only if one remembers that Hartshorne's was a highly selective historiography. Geography was not locked in a perpetual struggle with history, and Ratzel, Vidal de la Blache, and Mackinder all recognized the importance of an historical perspective in their work. But Hartshorne paid little attention to their substantive writings – he virtually ignored the last two altogether – and, like most disciplinary policemen, he was preoccupied with taking down methodological statements.

[77] *Ibid.* Hartshorne's glorification of the present was not an endorsement of modernism; quite the opposite. Smith describes *The nature of geography* as "the reveille of antimodernism."

[78] Peter Haggett's *Locational analysis in human geography* (London: Edward Arnold, 1965) p. 15. Haggett is obviously right if one limits discussion to Eratosthenes and Ptolemy, but Van Paassen shows that the importance of "mathematical geography" was by no means a constant within the classical tradition and that what he calls "the genealogical genre" continued to occupy a central place down to Strabo and beyond: Christian van Paassen, *The classical tradition of geography* (Groningen, Netherlands: J. B. Wolters, 1957).

[79] William Bunge, *Theoretical geography* (Lund, Sweden: Gleerup, 1962) pp. 1 and 248.

[80] Cf. Felix Driver, "The historicity of human geography," *Progress in human geography* 12 (1988) pp. 497–506.

America and the end of history Antihistoricism is not confined to geography, however, and this more general subordination of history has distinctive coordinates. More than 30 years ago Mills – who Soja lambastes for his historicism – had already proposed that "the modern age [was] being succeeded by a post-modern period" for which "historical explanations [may be] less relevant than for many other periods." Mills did not relish the prospect and he insisted that "such a retreat from history makes it impossible . . . to understand precisely the most contemporary features of this one society."[81] That one society was, of course, the United States, and there may be something distinctively American about all this. Perhaps the difference between us is that Soja writes as an American whereas I am – still – a European?[82] At any rate, I want to pursue the possibility a little further, not because I think it will provide a satisfactory answer but because it may provide a more precise location for Soja's project.

In his "Foreword to historical geography," Sauer drew attention to what he saw as "a peculiarity of our American geographical tradition," namely "its lack of interest in historical processes and sequences, even their outright rejection." He held that this strongly reflected the Midwest origins of American geography:

In the simple dynamism of the Midwest of the early twentieth century, the complex calculus of historical growth or loss did not seem particularly real or important. . . . In this brief moment of fulfillment and ease, it seemed that there must be a strict logic of the relationship of site and satisfaction, something approaching the validity of natural order. . . . Actors in the last scenes of a play that had begun in the early nineteenth century, the authors of these studies were largely unaware that they were a part of a great historic drama. They came to think that human geography and history were really quite different subjects.[83]

Sauer was writing from California and his department at Berkeley enjoyed a particular salience within the intellectual landscape of American geography: a sort of ex-centric centricity. In a very real sense, I think, Soja seeks to recenter American geography in southern California and to present a nascent "Los Angeles School" as a (postmodern?) replacement for Sauer's (anti-modern?) Berkeley School. I hope this is not as facile as it appears; the one is hardly equivalent to the other. But their juxtaposition is more than a *jeu*

---

[81] Mills, *Sociological imagination*, pp. 173–74 and 184.

[82] Nigel Thrift has made the interesting suggestion that there is a considerable gulf between American and European thought, which in geography has been concealed by the fusions of "Anglo-American" geography: most decisively in the 1960s and, perhaps, most ineffectually in the 1990s.

[83] Carl Sauer, "Foreword to historical geography," *Annals of the Association of American Geographers* 31 (1941) pp. 1–24; the quotation is from pp. 3–4. This was published hard on Hartshorne's heels and was intended as a rebuttal of his views. To Sauer, Hartshorne was the voice of the Midwest sounding "The Great Retreat."

d'esprit and my parenthetical hesitations are far from innocent. There is no doubt that Sauer was a convinced historicist and I do not think it would be difficult to show that his cultural geography was, in its essentials, an antimodernist program. But the connection that Soja seeks to make between antihistoricism and postmodernism is an altogether more difficult question.

Let me try to indicate why this should be so. There are several senses in which postmodernism renders "history" problematic and some of them are of considerable interest to any radical politics. The critique of "total history," the accent on textuality and intertextuality, the processual character of representation – all of these things provide a serious challenge to empiricist and even some postempiricist histories.[84] I imagine that claims of this sort would be endorsed by Soja too. But the discursive space of postmodernism is a differentiated one and some of its voices have proclaimed "the end of history" in ways that simultaneously announce the beginning of – what shall I call it? Hardly geography. Geometry? Consider for a moment Baudrillard's cartoon contrast between twentieth-century America, which he describes as "the original version of modernity," and Europe, which is merely "the dubbed or subtitled version."

America ducks the question of origins; it cultivates no origin or mythical authenticity; it has no past and no founding truth. Having known no primitive accumulation of time, it lives in a perpetual present.[85]

If this is the "evaporation of history," as one commentator puts it, it is achieved most mesmerically in the heat of the desert.[86] As Baudrillard travels through America, the desert comes to occupy a central place in his geometric imaginary. The desert is "a space from which all substance has been removed" and all traces erased; its planar emptiness is projected almost cinematically onto the city, where one is supposedly "delivered from all depth" into "an outer hyperspace." Through these projections Baudrillard senses the power and the exhilaration of "pure open space" until, finally, he arrives at the very edge of what he calls "astral America," Soja's "quintessential postmodern place": Los Angeles.

There is nothing to match flying over Los Angeles by night. A sort of luminous, geometric, incandescent immensity, stretching as far as the eye can see, bursting

---

[84] Linda Hutcheon, *The politics of postmodernism* (London: Routledge, 1989) pp. 62–92.

[85] Jean Baudrillard, *America* (London: Verso, 1988) p. 76. When I call this a "cartoon contrast" I have in mind Kellner's description of Baudrillard as "the Walt Disney of contemporary metaphysics": Douglas Kellner, *Jean Baudrillard: From Marxism to postmodernism and beyond* (Cambridge, England: Polity Press, 1989) p. 179. Kellner intends this as a criticism, of course, but I am by no means sure that Baudrillard would read it as such.

[86] Sharon Willis, "Spectacular topographies: *Amérique*'s post-modern spaces," *Threshold* IV (1988) pp. 62–68; the quotation is from p. 63.

out from the cracks in the clouds. Only Hieronymus Bosch's hell can match this inferno effect. The muted fluorescence of all the diagonals: Willshire, Lincoln, Sunset, Santa Monica. Already, flying over San Fernando Valley, you come upon the horizontal infinite in every direction. But, once you are beyond the mountain, a city ten times larger hits you. You will never have encountered anything that stretches as far as this before. Even the sea cannot match it, since it is not divided up geometrically. The irregular, scattered flickering of European cities does not produce the same parallel lines, the same vanishing points, the same aerial perspectives either. They are medieval cities. This one condenses by night the entire future geometry of the networks of human relations, gleaming in their abstraction, luminous in their extension, astral in their reproduction of infinity. Mulholland Drive by night is an extraterrestrial's vantage point on earth, or conversely, an earth-dweller's vantage point on the Galactic metropolis.... [L.A.] is in love with its limitless horizontality, as New York may be with its verticality.[87]

Now contrast the almost lyrical exultation of this passage with what I take to be Jameson's more measured (and more modernist) response. I am aware that Jameson has learned much from Baudrillard's earlier writings, but when he distinguishes between America and Europe he emphatically does not valorize the first over the second.

Our relationship to our past as Americans must necessarily be very different and far more problematical than for Europeans whose national histories ... remain alive within their contemporary political and ideological struggles. I think a case could be made for the peculiar disappearance of the American past in general, which comes before us in unreal costumes and by way of the spurious images of nostalgia art.[88]

Here the erasure of history becomes problematic, an occasion for concern rather than celebration. Jameson wants to recover (and deepen) a critical sense of historicity: to find, as he once put it, an historical way of "systematizing something which is resolutely unhistorical."[89] The same is true of another author whom Soja regards with evident approval. In *All that is solid melts into air* Berman argues that "we find ourselves today in the midst of a modern age that has lost touch with the roots of its own modernity." His solution is an unequivocally modernist one:

We can learn a great deal from the first modernists, not so much about their age as about our own. We have lost our grip on the contradictions that they had to

[87] Baudrillard, *America*, pp. 51–52, 124–25.

[88] Leonard Green, Jonathan Culler, and Richard Klein, "Interview with Fredric Jameson," *Diacritics* 12 (1982) pp. 72–91; the quotation is from p. 74.

[89] Stephanson, "Regarding postmodernism," p. 74. The "something" is, of course, postmodernism.

grasp with all their strength, at every moment of their everyday lives, in order to live at all. Paradoxically, these first modernists may turn out to understand us – the modernization and modernism that constitute our lives – better than we understand ourselves. If we can make their visions our own, and use their perspectives to look at our own environments with fresh eyes, we will see that there is more depth in our lives than we thought.... It may turn out, then, that going back can be a way to go forward: that remembering the modernisms of the nineteenth century can give us the vision and courage to create the modernisms of the twenty-first.[90]

Where, then, does Soja stand? With Baudrillard or with Jameson and Berman? The question is a real one – I am not inventing differences where none exist – but the answer is far from straightforward. In one sense, this is hardly surprising since postmodernism is supposed to be doubly coded and to subsume both complicity and critique. Although the divide between the two may be less of a line and more of a margin – a zone of contestation and subversion – it is still possible, I think, to make a provisional mapping. In his latest writings it seems clear that Soja echoes many of Baudrillard's more recent preoccupations and endorses many of his more substantive claims.[91] But for the most part *Postmodern geographies* remains a resolutely modern project: in its textual strategy, in its spatial thematic, in its theoretical armature. Although Soja is skeptical (and at times astonishingly hostile) toward historical inquiry, in practice he offers an *historical geography* of modernity that is itself thoroughly *modernist*. I want to be quite clear about what I am saying here. I do not mean to imply (and Soja does not claim) that postmodernism involves a complete break with modernism. Nor do I intend this as a blanket criticism: many of the most compelling features of Soja's argument reside precisely in its modernism. But I also think that some of the most serious weaknesses in Soja's analysis nevertheless flow from a failure to attend to the more radical implications of the postmodern critique. This double valency is particularly clear in his remarkable account of Los Angeles: "the place where 'it all comes together'" and which he then tries to "take apart again." Let me now do the same.

---

[90] Berman, *All that is solid melts into air*, pp. 17 and 36. In the original edition Berman castigated what he called "a mystique of postmodernism" which "strives to cultivate ignorance of modern history and culture" (p. 33) and in his preface to the second edition he invokes Habermas to reaffirm his own (political) vision of modernism against the postmodern critique (p. 10). Ten years later, his "Why modernism still matters," in Scott Lash and Jonathan Friedman (eds.), *Modernity and identity* (Oxford: Blackwell Publishers, 1992) pp. 33–58, includes an impassioned assault on the aestheticism of some (Francophone) versions of postmodernism: "If modernism had found both its fulfilment and its ruin in the streets, this postmodernism saved its devotees the trouble of ever having to go out at all" (p. 44).

[91] Edward Soja, "Inside exopolis: Scenes from Orange County," in Michael Sorkin (ed.), *Variations on a theme park: The new American city and the end of public space* (New York: Noonday Press, 1992), pp. 94–122.

Learning from Los Angeles

I want to build on the preceding discussion by considering some of the most important conceptual categories that Soja uses in his deconstruction of late twentieth-century Los Angeles. In particular, I will focus on the way in which two key words – "modernization" and "landscape" – are mobilized to sustain an insistently modern reading of this supposedly post-modern place.

Modernization and modernism

Soja's analysis of Los Angeles is framed by a particular view of the historical geography of modernity. Its basal logic is derived from Mandel's account of long waves of capitalist development that Soja translates into a model of *modernization.*

Modernization is, like all social processes, unevenly developed across time and space and this inscribes quite different historical geographies across different regional social formations. But on occasion, in the ever-accumulating past, it has become systematically synchronic, affecting all predominantly capitalist societies simultaneously. This synchronization has punctuated the historical geography of capitalism since at least the early nineteenth century with an increasingly recogniz-able macro-rhythm, a wave-like periodicity of societal crisis and restructuring.[92]

The translation is not innocent, however, and I want to signpost two of the theoretical maneuvers it accomplishes. The first concerns Berman's extended celebration of historical materialism, *All that is solid melts into air*, and the conjunction he proposes between Marx, modernism, and modernization. The second concerns Weber's concept of modernity and the connections he sought to establish between the triumph of specific modes of rationality, "the iron cage" of capitalism and the developmental history of the West. I will consider each in turn.

All that is solid melts into air The vocabulary of modernization enables Soja to bring Berman into the frame. Insofar as Berman advertises *All that is solid melts into air* as a study in the dialectics of modernization and modernism, then Soja can represent modernism as "responsive" to modernization, as "the cultural, ideological, reflective and . . . theory-forming response to modernization" and as a "reaction formation."[93] Although the

---

[92] Soja, *Postmodern geographies*, p. 27.

[93] *Ibid.*, pp. 26, 29. The concept of a reaction-formation is Freud's. Very roughly, "its supposed presence will generate precisely the opposite behavior to what would otherwise have been expected":

last of these concepts is derived from Freud, the lineage of classical Marxism and its base-superstructure couple is clearly not far below the conceptual surface. That Soja is uncomfortable about its proximity is evident from his summary account of Harvey's analysis of the landscape of late twentieth-century capitalism.

> The landscape has a textuality that we are just beginning to understand, for we have only recently been able to see it whole and "read" it with respect to its broader movements and inscribed events and meanings. Harvey's inaugural reading focuses on the hard logics of the landscape, its knife-edge paths, its points of perpetual struggle, its devastating architectonics, its insistent wholeness. Here, capital is the crude and restless *auteur*. . . . Capital is seen as two-facedly choreographing the chronic interplay of time and space, history and geography, first trying to annihilate with temporal efficiency the intransigent social physics of space that it seeks to transcend. This stressful ambivalence etches itself everywhere, organizing the landscape's material forms and configurations in an oxymoronic dance of destructive creativity. Nothing is wholly determined, but the plot is established, the main characters clearly defined and the tone of the narrative unshakably asserted.[94]

Soja regards Harvey's historico-geographical materialism as "an invitation to theoretical paralysis" – a judgment I find astonishing – and so he immediately seeks to distance himself from its supposedly "late modern" metanarrative.[95]

> The real text, of course, is much more subtly composed and filled with many different historical and geographical subtexts to be identified and interpreted. Capital, above all, is never alone in shaping the historical geography of the landscape and is certainly not the only author or authority.[96]

I doubt that Harvey would disagree: his essays on nineteenth-century Paris reveal a much more assured command of the complexities of historical geography than the usual reductionist caricatures of his work. But Soja insists that while his own postmodern geography "continues to draw

---

Russell Keat, *The politics of social theory: Habermas, Freud and the critique of positivism* (Oxford: Blackwell Publishers, 1981) p. 143. Soja invokes it here, I presume, to establish the critical location of modernism and to signal that it is not a direct "reflection" of modernization.

[94] Soja, *Postmodern geographies*, pp. 157–58.

[95] Soja's description of Harvey's project as "late modern" is clearly intended to signal its distance from his own ostensibly *post* modern geography. Suspending (for a moment) the success or otherwise of Soja's attempt to distance himself from modern Marxism, his attempt to do so does not mean that his text is left *without* a metanarrative. Soja has a particular story to tell and other voices are admitted on his terms: They are allowed to speak only through him. The distance from Lyotard is thus considerable since Lyotard, *Postmodern condition*, p. xxiv defines the postmodern as an "incredulity toward metanarratives."

[96] Soja, *Postmodern geographies*, p. 158.

inspiration from the emancipatory rationality of Western Marxism, [it] can no longer be confined within its contours." Even so, those contours continue to define the base camp, as it were, from which his expedition sets forth:

The landscape being described must be seen as a persistently capitalist landscape with its own distinctive historical geography, its own particularized time-space structuration. The initial mapping, at least, must therefore never lose sight of the hard contours of capitalism's "inner contradictions" and "laws of motion" no matter how blurry or softened history and human agency have made them. The plot has thickened but not enough to obliterate an enduring central theme.[97]

That theme reappears time and time again and in the same central place in Soja's deconstruction of the landscape of Los Angeles.

Underneath [its] semiotic blanket there remains an economic order, an instrumental nodal structure, an essentially exploitative spatial division of labor, and this spatially organized urban system has for the past half century been more continuously productive than almost any other in the world. But it has also been increasingly obscured from view, imaginatively mystified in an environment more specialized in the production of encompassing mystifications than practically any other.[98]

This is the modern problematic (and modern Marxism) writ large: it depends on one of the very "depth models" that Jameson claims postmodernism has discarded.[99] And in Soja's version of it, culture is rendered epiphenomenal, superstructural, and its collective representations made into so many veils to be drawn aside to reveal the basal contours of political economy.

Modernity, rationality, and local knowledge If Soja remains close to Marx, however, he is still a long way from Weber, and the vocabulary of modernization performs what Habermas calls "two abstractions on Weber's concept of modernity."

[It] dissociates "modernity" from its modern European origins and stylizes it into a spatio-temporally neutral model for processes of social development in general.

---

[97] *Ibid.*, pp. 73, 158. This perhaps explains Soja's hostility to post-Marxism. He insists that his usage of the prefix "post-" in postmodernism connotes " 'following upon' or 'after' but does not mean a complete replacement of the modified term" (*Ibid.*, p. 16). I agree; but he is not prepared to extend the same privilege to post-Marxism, which he dismisses as "naive and simplistic" (*Ibid.*, p. 170). It is easy to see why: he seeks to retain most of the features that post-Marxism rejects.

[98] *Ibid.*, p. 246.

[99] Jameson, "Cultural logic," p. 62. It is also, of course, the same depth model that Jameson himself uses. In view of Soja's characterizations of modernism (above), I might add that Jameson includes both Marx and Freud among the architects of these various depth models.

Furthermore, it breaks the internal connections between modernity and the historical context of Western rationalism, so that processes of modernization can no longer be conceived of as . . . the historical objectification of rational structures.[100]

It should be said at once that the charge of "spatio-temporal neutrality" applies more to Berman than to Soja. For all his finely textured portraits of particular cities (Paris, St Petersburg, New York), Berman uses them as a kind of spatial palette from which to compose an undifferentiated, almost timeless representation of modernity – what Anderson sees, from a different perspective, as "a fundamentally *planar* development"[101] – more or less indifferent to spatio-temporal context. It is to avoid precisely this danger, I take it, that Soja appeals to Mandel to provide an historical and geographical grid for the long waves of capitalist development: both a periodization and a regionalization of the modernization process. But the appeal is only partially successful. It certainly enables Soja to distinguish "four modernizations" that recomposed the "macro-geographical landscape" and to relay their convulsive rhythms to the reverberating spaces of the capitalist city.[102] If this means that Soja is able to provide a more discriminating *historical* matrix than Berman, however, his "summative mappings" of the North American city nevertheless delineate an evolutionary trajectory that (ironically) is quite unable to convey the variable *geographical* matrix within which urbanization has taken place. It is only when Soja turns away from these prototypical models toward an analysis of Los Angeles – only when he locates this city-region within the specific context of what he calls the "Californianization" and the globalization of capitalism – that he is able to bring the dynamics of capitalist spatialization into sharp focus.

If the reading that I am proposing here is to be sustained then, clearly, when Soja describes Los Angeles as "the quintessential postmodern place,"

---

[100] Habermas, *Philosophical discourse*, p. 2. Significantly, Habermas links this "dissociation" with both postmodernism and *posthistoire*.

[101] Anderson, "Modernity and revolution," p. 101.

[102] Soja's periodization does not repeat Jameson's misreading of Mandel, which wires postmodernism to the logic of "late capitalism": see Davis, "Urban renaissance." For Soja, late capitalism emerged some time around World War II and its mix of Fordism and Keynesianism sustained the progressive enlargement of a "state-managed urban system." It was *this* conjuncture that was so dramatically dislocated during the early 1970s – the period that so concerns Jameson – when the installation of a *new* regime of flexible accumulation ("post-Fordism") inaugurated a compound cycle of internationalization and global restructuring that has had the most profound consequences for the late twentieth-century city. In one sense, the correction makes little difference to Jameson's account because he relegates the economy to the margins of his argument, where it functions as an ever-present but largely unexamined horizon against which postmodernism is thrown into relief. In another sense, however, it does make a difference. I find Soja's analysis of the political economy of urbanization more persuasive precisely because it foregrounds those questions that are placed in the background of Jameson's account: but I do not see how it then departs in any substantial way from Harvey's analysis.

"a paradigmatic place," "the epitomizing world-city," he must understand these terms in ways far removed from the context-free generalizations offered by Berman. That he does in fact do so is suggested by this central passage:

Ignored for so long as aberrant, idiosyncratic, or bizarrely exceptional, Los Angeles, in another paradoxical twist, has, more than any other place, become the paradigmatic window through which to see the last half of the twentieth century. *I do not mean to suggest that the experience of Los Angeles will be duplicated elsewhere. But just the reverse may be true, that the particular experiences of urban development and change occurring elsewhere in the world are being duplicated in Los Angeles*, the place where it all seems to "come together."[103]

This is a genuinely provocative insight that brightly illuminates the construction of what Clifford (almost) calls historical geographies of "emergent differences." Like Soja, Clifford draws attention to the connective imperative between the local-global dialectic and the crisis of representation. As the experiences of "dwelling and travel [have become] less and less distinct," he argues, so "geopolitical questions must now be asked of every inventive poetics of reality."

An older topography of experience and travel is exploded. One no longer leaves home confident of finding something radically new, another time or space. Difference is encountered in the adjoining neighborhood, the familiar turns up at the ends of the earth.... Twentieth-century identities no longer presuppose continuous cultures or traditions. Everywhere individuals and groups improvise local performances from (re)collected pasts, drawing on foreign media, symbols and languages.[104]

The parallel seems exact; but Clifford is preoccupied with the predicament of what he calls *ethnographic modernity* and those two words dislocate any simple parallel between the two projects. "By the 1920s," Clifford remarks, "a truly global space of cultural connections and dissolutions has become imaginable: local authenticities meet and merge in transient urban and suburban settings." This new imaginary is being put in place during Soja's fin de siècle where it becomes constitutive of the experience of modernity: emphatically not postmodernity. Still more awkward (for Soja), Clifford

---

[103] Soja, *Postmodern geographies*, p. 221; my emphasis. That said, there *is* a generalizing impulse here too. Part of the purpose of Soja's deconstruction of Los Angeles is "to appreciate the specificity and uniqueness of a particularly restless geographical landscape *while simultaneously seeking to extract insights at higher levels of abstraction*" (p. 223; my emphasis). "Abstraction" is not quite the same as "generalization" but in Soja's discussion the two are very close.

[104] James Clifford, *The predicament of culture: Twentieth-century ethnography, literature and art* (Cambridge: Harvard University Press, 1988) pp. 6 and 14. Clifford actually says "histories of emergent differences" but I take the geographical to be required by the logic of his argument.

argues that these juxtapositions – a cultural *bricolage* – gave a sharp edge to the cross-cuts between the avant-garde and ethnography. What Clifford has in mind in his discussion of "ethnographic surrealism" in particular is not only the surrealists' interest in and borrowing from the (to them) palpably exotic – the cultures of Africa or the South Pacific, for example – but also, and perhaps more significantly, their interest in "making the familiar strange": in using modernism to shock, perplex and "de-familiarize" the taken-for-granted lifeworlds of the modern city.[105]

In one sense, this is Soja's objective too. His experimental geography – "taking Los Angeles apart" – derives much of its power from an arresting re-presentation of a series of commonplace models of the city (most obviously those developed by the Chicago School) in a radically new and aggressively contrary vocabulary. In another sense, however, Soja's concerns into a different vocabulary, of course, but Habermas's own concerns are very different. His parodic intimations of the world of Burgess and Hoyt function as a telling critique of the theoretical mainstream but largely fail to evoke the lifeworlds of mainstreet.[106] I do not think this is simply a consequence of the authors Soja chooses to parody. Whatever the sensibilities of other members of the Chicago School, it is certainly true that neither Burgess nor Hoyt displayed much interest in ethnography, and it is perhaps scarcely surprising that a pastiche should repeat their own priorities. But Soja's postmodern geography *as a whole* seems indifferent to the importance of ethnography and in particular to the experiments in "polyphony" which have so vigorously animated postmodern ethnography. I suggest that this is, in part, a direct consequence of the visual metaphoric through which Soja conceives of spatiality. Modern ethnography is itself no stranger to a visual metaphoric and for this very reason *part of Clifford's purpose is to displace it.*

---

[105] *Ibid.*, pp. 117–51. An example of obvious relevance to Soja's experimental geography is Louis Aragon's *Le paysan de Paris*. This was originally published in serial form in *La revue européene* between 1924 and 1925; it has since been translated as *Paris peasant* (London: Picador, 1980). Aragon described the text as a "mythology of the modern" but it is not only its radicalization of culture that distinguishes his approach from Soja's: Aragon also disavows the conventions of Soja's depth model. "The world exists in a state of unthinkable disorder," he remarks, and "the extraordinary thing about this is that [people] should have habitually sought beneath the surface appearance of disorder for some mysterious order" (p. 203).

[106] Several critics have charged Clifford with a similar scholasticism. "We have moved back from the tent in the Trobriands filled with natives to the writing desk in the campus library": Paul Rabinow, "Representations are social facts: Modernity and post-modernity in anthropology," in James Clifford and George Marcus (eds.), *Writing culture: The poetics and politics of ethnography* (Berkeley: University of California Press, 1986) pp. 234–61; the quotation is from p. 244. Rabinow's critique is undoubtedly important but I do not think it blunts the point I am making here. Clifford insists that ethnographers are participants in a conversation with other cultures and that their textual strategies should be formed through "a cultural poetics that is an interplay of voices": James Clifford, "Introduction: partial truths," in Clifford and Marcus, *Writing culture*, pp. 1–26; the quotation is from p. 12. It is precisely this dialogical dimension that is missing from Soja's text (though in practice it seems to be attenuated in Clifford's too).

Once cultures are no longer prefigured visually . . . it becomes possible to think of
a cultural poetics that is an interplay of voices, of positioned utterances. In a
discursive rather than a visual paradigm, the dominant metaphors for ethnography
shift away from the observing eye and toward expressive speech (and gesture). The
writer's "voice" pervades and situates the analysis, and objective, distancing rhetoric
is renounced.[107]

Expressed like that, the gap between Clifford and Soja seems to be
considerable; but this is more complex than it appears. Soja's "voice"
certainly pervades and situates *Postmodern geographies* – to such an extent
that he is at once composer and conductor of the orchestra – but the
performance remains his own (very much his own) creation. The problem
is not so much how "the writer" finds his or her voice, or at any rate not
only that, but how to *listen to* and *learn from* other voices: particularly those
outside the enclosures of theoretical discourse.

I can make the same point in a different way by turning to Habermas's
other objection. One of the more extraordinary lacunae in Soja's discussion
of modernization is any consideration of Weber's account of rationality
and modernity. Not only is it quite wrong to dismiss Weber's project as
a "methodological individualism," as Soja does – the structural readings of
his "developmental history" are among the most productive contributions
to modern social theory – but Weber's interrogation of reason and rationality
has been one of the pivots around which the contemporary debate over
modernism and postmodernism has been articulated.[108] The importance of
Weber to Habermas's theory of communicative action is well-known. To
be sure, Habermas says little enough about the spatiality of social life, but
it would not be difficult to rework his discussion of rationalization to show
how late twentieth-century instrumentalities of political and economic systems
have deformed ("colonized") the lifeworlds of people in cities like Los Angeles:
and, perhaps, *particularly* in cities like Los Angeles. This is to translate Soja's
concerns into a different vocabulary, of course, but Habermas's own concerns
are transformed too: exposing the connections between the systematic production of space – "systematic" in exactly the sense that Habermas would
understand the term – and the struggles for place which then play such an
important role in the formation of his "counter-public spheres."[109]

I realize that Habermas's hostility to postmodernism makes it difficult
for Soja to contemplate a strategy of this sort; but the question of rationality

[107] Clifford, "Introduction," p. 12.
[108] See, for example, Wolfgang Schluchter, *The rise of Western rationalism: Max Weber's developmental
history* (Berkeley: University of California Press, 1981) and Sam Whimster and Scott Lash (eds.),
Max Weber, rationality and modernity (London: Allen and Unwin, 1987).
[109] See Derek Gregory, "The crisis of modernity? Human geography and critical social theory,"
in Richard Peet and Nigel Thrift (eds.), *New models in geography: The political economy perspective* (London:
Unwin Hyman, 1989) pp. 348–85.

that Weber posed with such clarity and which Habermas elaborates with such sensitivity is no less central to the critique of modernity advanced in a strikingly different register by Foucault. Soja would obviously prefer to conduct an analysis of the city in these terms and when he "takes Los Angeles apart" he draws upon a number of Foucauldian themes. But in doing so he severs Foucault's analytics of space from the genealogy of the subject.[110] This not only narrows the scope of the argument: it also snaps the connective imperative that provides Foucault's project with much of its critical momentum.[111] As a result, Soja is unable to bring into focus those spaces in which – and in part through which – resistance to the disciplinary logic of rationalization takes place. Foucault was of course reluctant to invoke the progress of rationalization in general and spoke instead of "specific rationalities," of what Habermas would call "the historical objec-tification of rational structures" (though the correspondence is by no means exact).

Foucault's analysis of these specific rationalities underwrote a conception of resistance that depended in its turn on the reactivation of local knowledges and their enlistment in struggles against specific techniques of power. The discussion of "heterotopias" of which Soja makes so much is intimately involved with the discovery of these sites of contestation and, so it seems to me, is a constitutive dimension of any possible "politics of space."[112] If these struggles are localized, however, Foucault was neverthe-less able to identify a common objective: the subversion of a specifically modern technique of power.

This form of power applies itself to immediate everyday life which categorizes the individual, marks him by his own individuality, attaches him to his own identity,

---

[110] Foucault's problematic of the self – of *assujetissement* ("subjection" and "subjectification") – undergoes a series of major modifications. In his earliest writings, including *Madness and civilization*, Foucault addressed a number of existentialist theses. Much of the rest of his work, and most obviously *Discipline and punish*, was devoted to displacing the self as a centered subject altogether. But in the closing volumes of his *History of sexuality* Foucault seemed to move much closer to an historicist conception of self-formation. For illuminating discussions, see Peter Dews, *Logics of disintegration: Post-structuralist thought and the claims of critical theory* (London: Verso, 1987) pp. 144–70 and Mark Poster, *Critical theory and poststructuralism: In search of a context* (Ithaca: Cornell University Press, 1989) pp. 53–69. Soja's account of Foucault's "ambivalent spatiality" says nothing about these – or any other – changes in his project.

[111] At least one commentator has identified the singularity of Foucault's contribution to the social sciences as "an analysis of the forms of subjectification involved in modern constructed spaces": Paul Hirst, "Power/knowledge: constructed space and the subject," in Richard Fardon (ed.), *Power and knowledge: Anthropological and sociological approaches* (Edinburgh: Scottish Academic Press, 1985) pp. 171–91; the quotation is from p. 172.

[112] Michel Foucault, "Of other spaces," *Diacritics* 16 (1986) pp. 22–27; see also Georges Teyssot, "Heterotopias and the history of spaces," *Architecture & Urbanism* 121 (1980) pp. 79–100. Soja has since provided a more nuanced reading of heterotopias in his "Heterotopologies: a remembrance of other spaces in Citadel-LA," *Strategies* 3 (1990) pp. 6–39.

imposes a law of truth on him which he must recognize and which others have to recognize in him. It is a form of power which makes individuals subjects.... We have to promote new forms of subjectivity through the refusal of this kind of individuality which has been imposed on us for several centuries.[113]

Foucault's *cri de coeur* requires considerably more discussion than I can provide here, but even in its unelaborated version it surely shows that the relation between specific rationalities, local knowledges, and forms of subjectivity form a conceptual triangle that has to be central to any serious appropriation of his work. Soja's insistence on the spatiality of Foucault's project is not misplaced, but the mapping of sites of contestation cannot be conducted independently of this basic triangulation. To see what happens when one attempts to do so, it is only necessary to turn to Soja's deconstruction of the landscape of Los Angeles.

Landscapes and lifeworlds

Soja constantly represents Los Angeles as a "landscape" and the term carries a considerable conceptual baggage. Consistent with Soja's visual metaphoric, Cosgrove has shown that "landscape" emerged as a distinctive "way of seeing" – the term is, of course, Berger's – in the fifteenth and early sixteenth centuries. He contends that the formation and eventual hegemony of this "visual ideology" owed a great deal to Alberti's *Della pittura* (1435), which set out a geometric technique for establishing linear perspective within a painting. Perspective radiated far beyond the canvas, however, and Cosgrove is concerned to identify the ways in which it was implicated in a specifically capitalist constellation of power, property, and possession. Most important for my present purposes, he argues that its representation of three-dimensional space on a two-dimensional surface

directs the external world towards the individual located outside that space. It gives the eye absolute mastery over space.... Landscape is thus a way of seeing, a composition and structuring of the world so that it may be appropriated by a detached, individual spectator to whom an illusion of order and control is offered through the composition of space according to the certainties of geometry.[114]

113 Michel Foucault, "The subject and power," in Hubert Dreyfus and Paul Rabinow, *Michel Foucault: Beyond structuralism and hermeneutics* (Brighton, England: Harvester Press, 1982) pp. 208–26; the quotation is from pp. 212 and 216. In the light of what I wish to argue later, it is impossible for me to ignore the intrusive male pronoun in this passage. For feminist discussions of Foucault, see Irene Diamond and Lee Quinby (eds.), *Feminism and Foucault: Reflections on resistance* (Boston: Northeastern University Press, 1988); Jana Sawicki, *Disciplining Foucault: Feminism, power and the body* (New York: Routledge, 1991); Lois McNay; *Foucault and feminism* (Cambridge, England: Polity Press, 1992).

114 Denis Cosgrove, "Prospect, perspective and the evolution of the landscape idea," *Transactions of the Institute of British Geographers* 10 (1985) pp. 45–62; the quotation is from pp. 48 and 55.

I want to suggest that this is Soja's technique too: his representation of the landscape of Los Angeles is conducted from a series of almost Archimedian points from which he contemplates a series of abstract geometries.

The surveillant eye Soja introduces his most experimental essay on Los Angeles as an attempt to evoke a "spiraling tour" around the city that he made with Lefebvre and Jameson in 1984. He makes it plain before he sets out that "any totalizing description" is impossible and that what follows is of necessity "a succession of fragmentary glimpses" or, significantly, "field notes." All of this serves to establish Soja's credentials and to authorize his physical presence in the field: the "being-there" of modern ethnography.[115] And yet what follows is so obviously *not* a report of any field experience. Soja starts with "an imaginative cruise directly above the contemporary circumference of the Sixty-Mile Circle," hopping from military base to military base, "rampart" to "rampart" (figure 23). Then he darts inside the circle, "within the wall," where his surveillant eye captures what he claims conventional imagery misses. This is "the premier industrial growth pole of the twentieth century," a propulsive silicon landscape energized by state expenditures to such a degree that, in his view, Los Angeles has become "the prototypical Keynesian state-city."[116] He then hurtles outside the circle to fly through (or rather above) the "outer spaces." These production complexes ring the central city from LAX to Orange County and on some readings their orbits are made "to transform the metropolitan periphery into the core region of advanced industrial production." But Soja moves quickly to cancel any such paradoxical inversion. Although this is a postmodern landscape, he insists, it has a center: His reading of it will provide no "negation of nodality," no "Derridean deconstruction of all differences between the 'central' and the 'marginal.'" Instead, Soja soars back to the center – "an urban panopticon counterposed to the encirclement of watchful military ramparts and defensive outer cities" – where he climbs to its pinnacle. There, looking down and out from City

---

[115] Clifford, *Predicament of culture*, pp. 21–54 calls this "ethnographic authority." In a later essay Soja translates "being there" as "*être-LA*": in other words, being in Los Angeles; see his "Heterotopologies," p. 35.

[116] The invocation of Keynes is entirely appropriate since Keynes's original argument did not turn solely on the welfare function of the state but was also tied to its expenditure on warfare. Harvey provides a more general discussion of the Keynesian city but he also pays considerable attention to "the struggle for survival in the *post*-Keynesian transition" (my emphasis) that has been under way since the early 1970s. Although Soja makes nothing of this transition, several of the general strategies that Harvey identifies are captured with great clarity in the analysis of late twentieth-century Los Angeles provided by Davis, "Chinatown." See David Harvey, *The urbanization of capital* (Oxford: Blackwell Publishers, 1985) pp. 202–21 and *idem*, "From managerialism to entrepreneurialism: The transformation of urban governance in late capitalism," *Geografiska Annaler* 71B (1989) pp. 3–17.

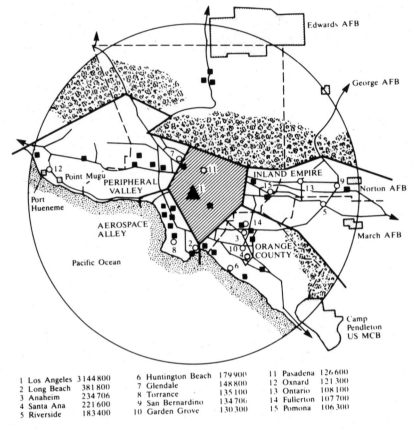

1 Los Angeles	3144800	6 Huntington Beach	179900	11 Pasadena	126600
2 Long Beach	381800	7 Glendale	148800	12 Oxnard	121300
3 Anaheim	234706	8 Torrance	135100	13 Ontario	108100
4 Santa Ana	221600	9 San Bernardino	134700	14 Fullerton	107700
5 Riverside	183400	10 Garden Grove	130300	15 Pomona	106300

Figure 23   Soja's flight over Los Angeles [map: Edward Soja, by permission].

Hall, he gazes upon the "heterotopic" politico-economic landscape of downtown, the "dazzling array of sites in this compartmentalized corona of the inner city," and its "lateral extensions."

I have stripped Soja's account of much of its detail in order to show how the entire excursion is conducted from a series of external viewpoints. Even the language, resonant with Foucauldian images of panopticism and surveillance, conveys a sense of separation between the spectacle and the spectator. "Only from the advantageous outlook of the center," Soja declares, "can the surveillant eye see everyone collectively, disembedded but interconnected."[117] And what does that eye – Soja's "I" – *see* from those viewpoints? Towers and freeways, sites and districts, zones and areas, enclaves and pockets, gradients and wedges: a landscape without figures.

[117] Soja, *Postmodern geographies*, p. 236.

Soja's essay becomes a morphology of landscape that, like Sauer's original, is rarely disturbed by the human form.[118] Los Angeles is represented as "a shattered metro-sea of fragmented yet homogenized communities ... the most extraordinary crazy quilt ... a dazzling, sprawling, patchwork mosaic": in short, a *geometry*. And it turns out that this was the destination Soja had in mind all along. "However much the formative space of Los Angeles may be global (or perhaps Mandelbrotian, constructed in zig-zagging nests of fractals)," he advised his fellow space-travelers, "*it must be reduced to a more familar and localized geometry to be seen.*"[119]

I make so much of all this because it makes Soja's essay so astonishingly univocal. We never hear the multiple voices of those who *live* in Los Angeles – other than Soja himself – and who presumably learn rather different things from it. At first sight, de Certeau's description of New York provides a remarkably close parallel:

To be lifted up to the summit of the World Trade Center is to be lifted out of the city's grasp.... An Icarus flying above these waters, he can ignore the devices of Daedalus in mobile and endless labyrinths far below. His elevation transfigures him into a voyeur. It puts him at a distance. It transforms the bewitching world by which one was "possessed" into a text that lies before one's eyes. It allows one to read it, to be a solar Eye, looking down like a god.... The 1370 foot high tower that serves as a prow for Manhattan continues to construct the fiction that creates readers, makes the complexity of the city readable, and immobilizes its opaque mobility on a transparent text.[120]

But de Certeau occupies this vantage-point *only to vacate it at the first opportunity*:

The ordinary practitioners of the city live "down below," below the threshold at which visibility begins. They walk – an elementary form of this experience of the city: they are walkers, *Wandersmänner*, whose bodies follow the thicks and thins of

---

[118] I would obviously not want to push this comparison too far – Soja's conception of geography registers an advance over Sauer's in so many ways – but neither do I think Soja has nothing to learn from Sauer. One of the major preoccupations of the Berkeley School was the ecological dimension of landscape and a similar concern would evidently not be out of place in the study of such an environmentally fragile region as southern California. To be sure, it appears here and there in *Postmodern geographies*, but when Soja starts "taking Los Angeles apart" comments of this kind are almost invariably confined to footnotes and marginalized from the main text. Cf. Margaret Fitzsimmons, "The matter of nature," *Antipode* 21 (1989) pp. 106–20. The politics of nature – indeed, a whole "culture of nature" – are also focal to Polanski's *Chinatown*.

[119] Soja, *Postmodern geographies*, p. 224; my emphasis. The reference to fractals is not gratuitous. Fractals are "self-similar": that is to say, any segment of the curve will contain the geometry of the whole curve within itself. This is presumably analogous to Soja's claim that "the particular experiences of urban development and change occurring elsewhere in the world are being duplicated in Los Angeles" (p. 221).

[120] Michel de Certeau, *The practice of everyday life* (Berkeley: University of California Press, 1984) p. 92. Here the male pronoun seems to me consistent with the masculinity of this gaze.

an urban "text" they write without being able to read it. These practitioners make use of spaces that cannot be seen; their knowledge of them is as blind as that of lovers in each other's arms. The paths that correspond in this intertwining, unrecognized poems in which each body is an element signed by many others, elude legibility. It is as though the practices organizing a bustling city were characterized by their blindness. . . .

Escaping the imaginary totalizations produced by the eye, the everyday has a certain strangeness that does not surface, or whose surface is only its upper limit, outlining itself against the visible. Within this ensemble, I shall try to locate the practices that are foreign to the "geometrical" . . . space of visual, panoptic . . . constructions.[121]

It would I think be wrong to conclude from this that de Certeau means to imply ignorance on the part of the *Wandersmänner*. Although he says they are "unable to read," their spaces "cannot be seen," their knowledge is "blind," these images are being used to register a series of oppositions to the identification of the visible – of what can be seen – with knowledge: what can be known. De Certeau is not dismissing what it is they know, it seems to me, but insisting that their knowledge escapes the solar, surveillant eye and cannot be illuminated by its rays precisely because it is not confined to the visible: it is incorporated in and inseparable from their everyday practices. What interests me here (and in de Certeau's work more generally) is not so much its hostility to the visual metaphoric – one which, as I have already shown, dominates *Postmodern geographies* too – as its implication of an altogether different analytics of power, which is both the consequence and the reciprocal of Foucault's.

While de Certeau certainly admires Foucault's cartography of power and the careful delineation of its "strategies," his own objective is to recover the innumerable actions through which those strategies are subverted: the domain of what he calls *tactics*. "The space of a tactic is the space of the other," he explains, never fixed, rarely visible, always constituted through the fleeting and unexpected appropriation of the spaces of established power. A tactic is clandestine, almost guerilla-like in the way in which it takes advantage of a situation, and revealed "not in its own products . . . but in an art of using those imposed on it." De Certeau summarizes the distinction between the two by suggesting that a strategy privileges space – at one point he even equates it with the "victory of space over time" – whereas a tactic privileges time: "always on the watch for opportunities that must be seized 'on the wing.'" If this opposition could be sustained, then Soja's militant antihistoricism would stand in the way of an analysis of the tactical. But de Certeau allows that tactics form a "complex *geography* of social ruses" (my emphasis) which he clearly regards as being in some substantial sense constitutive of "everyday life." Not surprisingly, therefore,

---

[121] *Ibid.*, p. 93.

he describes Lefebvre's writings as "an indispensable source" for the recovery of those practices – the French is better: *arts de faire* – that "are neither determined nor captured by the systems in which they develop."[122]

Soja's accent on Lefebvre's *espace vécu* would presumably license a similar analysis. Yet there is virtually no trace anywhere in *Postmodern geographies* of "the swarming of [those] procedures that, far from being regulated or eliminated by panoptic administration, have reinforced themselves in proliferating illegitimacy." These are not counterstrategies as such: de Certeau is not so much concerned with the formation of alliances, coalitions, or campaigns as the ordinary, everyday refusal to submit to the law of what he calls "the jungle of functionalist rationality."[123] But they are nonetheless effective in forging cultures of consolation and, on occasion, cultures of resistance.

The politics of everyday life   Let me try to make my objection more concrete. In three brilliant essays Davis has situated the cultural politics of Los Angeles within an analytical framework that is in considerable agreement with (and, indeed, derives in large measure from) Soja's own.[124] Like Soja, he accentuates the political economy of urban and regional restructuring and in particular its imbrication in the "hypertrophic" expansion of global circuits of financial capital: symptoms, so he suggests, of crisis rather than renewal. He captures the besieged nature of the urban landscape with a succession of images even more startling than Soja's: Thus the new Figueroa and Bunker Hill complexes are set in a "protective maze of freeways, moats, concrete parapets, and asphalt no-man's lands"; "the sumptuary towers of the speculators" are "vivariums for the upper middle classes, protected by astonishingly complex security systems"; and the aggressive iconography of "fortified skyscrapers" signals "the coercive intent of postmodernist architecture . . . to polarize [the city] into radically antagonistic spaces." He moves behind the carceral institutions that Soja sees from his central lookout to reveal the formidable apparatus of repression that has been put in place by local politicians and the police to turn "vast stretches of southern California's sumptuous playgrounds, beaches and entertainment centres . . . [into] no-go areas for young Blacks or Latinos."

Unlike Soja, however, Davis also illuminates the social struggles that are part of the production and consumption of these metropolitan social spaces. To be sure, many of them are far from heroic and Davis is not

---

[122] *Ibid.*, pp. xix, 22, 38, 205 n. 5.

[123] *Ibid.*, pp. xviii, 96.

[124] Davis, "Urban renaissance"; *idem*, "Chinatown"; *idem*, "Los Angeles: Civil liberties between the hammer and the rock," *New Left Review* 170 (1988) pp. 37–60. Davis draws on and commends Soja's work in each of these essays. See also his synoptic *City of Quartz: Excavating the future in Los Angeles* (London: Verso, 1990).

starry-eyed about them. He makes no secret of the rock-houses and gang topographies that mark the contours of the crack economy. But he also pays attention to the ways in which many of the Third World populations that have been drawn around or across the Pacific Rim as part of the "internationalization" of Los Angeles have insisted on creating their own (multiple) identities against the stereotypes of City Hall and the media. This "immigrant working class," so he argues, "does not simply submit to the city for the purposes of capital, it is not merely the collective victim of 'urban crisis'; it also strives to transform and create the city, its praxis is a material force, however unrecognized or invisible in most accounts of contemporary Los Angeles."

Through their skilled involvement in what de Certeau would call "the practice of everyday life," these people – "ordinary people" – have created a series of distinctive urban cultures. "Like the enclave Asian boomtowns in the region," Davis declares, the Spanish-speaking neighborhoods of Los Angeles

> are more than melting pots for eventual assimilation to some hyphenated ethnicity. Together with their integral worlds of work and intineraries of movement, these residential environments comprise a virtually parallel urban structure – a second city...Spanish-speaking Los Angeles – the second-largest Mexican, Guatemalan, and Salvadorean city in the hemisphere – has far in excess of the necessary critical mass of institutions and media to define its own distinctive urbanity: a different, more "classical" way of living in the city based on gregarious, communitarian uses of markets, boulevards, parks and so on. The great Latino shopping streets – Broadway in Downtown and Brooklyn in Boyle Heights – have more in common with the early twentieth-century city, with the culture of Ragtime, than they do with a deathwish "postmodernity."[125]

Taken together, these materials sharpen what I take to be one of de Certeau's central points: that it is impossible to recover human geographies from a contemplation of their abstract geometries. Soja would no doubt say the same and his condemnation of spatial science is, as I have already indicated, unimpeachable. Yet in practice he renders the landscape of Los Angeles as a still life. Ever faithful to the Enlightenment interdiction against "blurring the genres," Soja's avowedly most experimental – and his supposedly most postmodern – essay turns out to be a paradigmatic illustration of Lessing's division between the spatial and the temporal arts. Far from challenging the conventions of Enlightenment aesthetics, Soja's geography

---

[125] Davis, "Chinatown," pp. 77–78. The closest Soja comes to recognizing these everyday accomplishments is when he notes "the underground economy [that] thrives in the interstices of urban life, succoring ethnicities and providing the necessary niches for personal survival" (Soja, *Postmodern geographies*, p. 219). But the instrumentality of his language dulls any really sensitive reading of the production and reproduction of these cultural landscapes.

respects its restrictions on capturing the swirl of human action. In using Davis's essays as a counterpoint to this thematic, I do not mean to imply that he is a disciple of de Certeau. But I do think that his accounts are more vivid in their evocation of Los Angeles for reasons that at least intersect with de Certeau's critique.

Partly, it is a matter of ethnographic sensibility. Although Davis often relies on the *Los Angeles Times* to secure his accounts, it is impossible to read his *City of Quartz* without realizing how much he knows beyond the vicarious pages of the newspaper. Davis grew up in southern California and he opens the book with a marvelous quotation from Benjamin: "A native's book about his city will always be related to memoirs; the writer has not spent his childhood there in vain." Indeed not, and Davis will subsequently argue that "a radical structural analysis of the city (as represented by the 'Los Angeles School') can only acquire social force if it is embedded in an alternative *experiential* vision": which means attending to and learning from those everyday practices that evade the surveillant eye. But it is also is a matter of conveying those experiences through narrative forms since, as de Certeau puts it, "a theory of narration is indissociable from a theory of practices."[126] And here too Davis is on common ground with him, mobilizing "the art of speaking" – of telling stories – for tactical, parapolitical purposes. He has an astonishing capacity to make the city come alive: to illuminate the most local of detail in such a way that it reveals a far wider canvas. His writings dissolve the Enlightenment dualism that skewers Soja's text and in doing so they provide a pellucid demonstration of the ways in which social struggles are embedded in the production of space. In Soja's contrastive view, of course, Los Angeles is "difficult to grasp persuasively in a temporal narrative": but in consequence I think he seriously diminishes the possibilities for a politics of space.[127]

To be sure, inquiries like de Certeau's are insufficient by themselves. There is much that cannot be understood at street level and it is a commonplace of contemporary anthropology that many ethnographies suffer from a weakness opposite to the one I have been pressing here: namely, a chronic failure to register the larger politico-economic systems that enter into the constitution of the localized lifeworlds that are their primary concern. But that is precisely why Davis's essays are so illuminating. For they can be situated at the flash point between political economy and ethnography and in doing so a space is opened in which, in Marcus and Fischer's telling phrase, anthropology can be "repatriated as cultural critique."[128]

---

[126] Davis, *City of Quartz*, p. 87; my emphasis; de Certeau, *Everyday life*, p. 78.

[127] Soja, *Postmodern geographies*, p. 22.

[128] George Marcus and Michael Fischer, *Anthropology as cultural critique: An experimental moment in the human sciences* (Chicago: University of Chicago Press, 1986) p. 111 and *passim*. The point is not a narrowly disciplinary one, of course.

Such a possibility immediately rebounds on the practical intent of Soja's program, however, because a politics that foregrounds political economy and distances itself from ethnography (as Soja's seems to do) is likely to be a thoroughly traditional – which is to say a thoroughly instrumentalist – one.

In contradistinction to the assumptions of an older school of critical theory, most writers would now recognize that ordinary people often have a remarkably sophisticated awareness of the impingements and encroachments of abstract systems on their everyday lifeworlds. They are not dupes living in one-dimensional societies.[129] For this reason their insights and understandings have a central place in the formulation of a radical democratic imaginary and, according to some commentators, in the construction of a postmodern politics. A project of this kind breaks with what Laclau and Mouffe call the "logic of privileged points" – the identification of a single vantage point from which to map the social order and the enlistment of a unitary agent to redraw its contours – and substitutes a scrupulous respect for heterogeneity and difference. Such a dissolution of what Lefort once termed the "landmarks of certainty" is a part of the postmodern condition as both Harvey and Soja construe it, but Laclau and Mouffe spell out its implications for a radical democratic politics in a depth and a direction that neither of those authors contemplates. Their argument is complex (and contentious) but its outlines are clear enough.

Our societies are confronted with the proliferation of political spaces which are radically new and different and which demand that we abandon the idea of a unique constitutive space of... the political.... [W]e are in fact always multiple and contradictory subjects, inhabitants of a diversity of communities (as many, really, as the social relations in which we participate and the subject-positions they define), constructed by a variety of discourses and precariously and temporarily sutured at the intersection of those subject – positions.[130]

The most obvious sign of this proliferation of political spaces and political subjects is the emergence of new social movements seeking to

---

[129] Insofar as postmodernism effaces the boundary between high culture and popular culture – what Huyssen calls "the great divide" – then it has to be equally sensitive to arguments of this kind and their implications for cultural politics: see Andreas Huyssen, *After the great divide: Modernism, mass culture and postmodernism* (London: Macmillan, 1986) and Jim Collins, *Uncommon cultures: Popular culture and post-modernism* (New York: Routledge, 1989).

[130] Chantal Mouffe, "Radical democracy: Modern or postmodern?" in Andrew Ross (ed.), *Universal abandon? The politics of postmodernism* (Minneapolis: University of Minnesota Press, 1988) pp. 31–45; the quotation is from pp. 43–44. Mouffe herself refers to Claude Lefort, *Essais sur la politique* (Paris: Éditions du Seuil, 1986) p. 29. For a fuller discussion see Ernesto Laclau and Chantal Mouffe, *Hegemony and socialist strategy: Towards a radical democratic politics* (London: Verso, 1985).

challenge forms of domination and subordination that do not lie solely within (and cannot be reduced to) the capital-labor nexus. A recognition of their importance is not confined to postmodernism, of course, and Habermas's redemption of the project of modernity clearly accentuates the importance of those multiple struggles lying at the interface between "system" and "lifeworld." But movements of this kind are also vital moments in any progressive postmodern politics: Yet they have little or no place in Soja's pages. This is to say things too quickly, however, because Soja does acknowledge the importance of campaigns around housing and homelessness, and he also draws attention to the formation of coalitions between middle-class and inner-city community groups. Soja is clearly far from sanguine about their prospects of success but he is plainly more receptive to their concerns and more sympathetic to their territorial bases than Harvey. After making several concessions to the legitimate claims of many of these new social movements, Harvey objects that their localization and differentiation means they "become a part of the very fragmentation which a mobile capitalism and flexible accumulation can feed upon."[131]

Soja's prospectus for a "postmodern politics of resistance" shows a much greater willingness to entertain the radical possibilities contained within such struggles and to endorse their often explicitly spatial agendas. Hence he insists that his postmodern geography "be attuned to the emancipatory struggles of all those who are oppressed *by the existing geography of capitalism* (and existing socialism as well) – exploited workers, tyrannized peoples, dominated women." But, as the emphasis I have added indicates, Soja locates all these peripheries in relation to a single center; he convenes all their oppressions within the plenary geography of capitalism (or its dual). And in doing so, he refuses to abandon what Laclau and Mouffe call "the assumption of a sutured society."[132] It is in this sense that I think Soja occludes the significance of new social movements and for this reason that he detects a "numbing depoliticization" of social life in contemporary Los Angeles: If the politics of restructuring have been marked by a "cacophonous silence" perhaps one ought to listen elsewhere.[133] Organized labor may well have been outmaneuvered by capital but there are many other struggles that simply cannot be mapped by such a unitary conceptual apparatus.

I wrote the first version of this essay before the riots of late April and early May 1992, but Davis had already sounded a warning:

In Los Angeles there are too many signs of approaching helter-skelter: everywhere in the inner city, even in the forgotten poor-white boondocks with their zombie

---

[131] Harvey, *Condition of postmodernity*, p. 303.

[132] Soja, *Postmodern geographies*, p. 5, 74; Laclau and Mouffe, *Hegemony*, p. 179.

[133] Soja, *Postmodern geographies*, p. 219. Cf. Roger Keil, "The urban future revisited: Politics and restructuring in Los Angeles after Fordism," *Strategies* 3 (1990) pp. 105–29.

populations of speed-freaks, gangs are multiplying at a terrifying rate, cops are becoming more arrogant and trigger-happy, and a whole generation is being shunted towards some impossible Armageddon.[134]

It arrived eighteen months later. Four white police officers had been charged with assaulting Rodney King, an African-American motorist, and their actions had been captured on amateur videotape. Defense lawyers had the case moved out of the city (where the beating took place) to Simi Valley, a predominantly white middle-class suburb. The officers were acquitted, and this is how *Newsweek* described the aftermath:

After years of neglecting the pent-up misery of the inner cities, the country shuddered at the bloody wake-up call. Out of a city endlessly burning, out of the heart of Simi Valley and the soul of South-Central Los Angeles, a verdict seen as a miscarriage of justice induced a convulsion of violence that left 44 dead, 2,000 bleeding and $1 billion in charred ruins. The 56 videotaped blows administered by Los Angeles police to Rodney King last year had landed hard on everyone's mind. But they fell like feathers on a suburban jury that acquitted the cops of using excessive force. "That verdict was a message from America," said Fermin Moore, owner of an African artifacts shop near Inglewood. The reply from the inner city was a reciprocal "F- - - you." First South-Central blew. Then fires licked up to Hollywood, south to Long Beach, west to Culver City and north to the San Fernando Valley. The nation's second largest city began to disappear under billows of smoke.[135]

One would have to be a fool to deny the salience of class in all this; but it would be equally foolish to ignore the ways in which class relations were cross-cut with intricate ethnic, gender, and generational divisions. It is not my purpose to offer some simple analysis of what took place: my point is precisely that no simple analysis will be able to do justice (literally so) to the complexities of despair and discrimination in late twentieth-century Los Angeles. If the "hard edges of the capitalist, racist and patriarchal land-scape" still seem "to melt into air," as Soja once thought, then, as his own Bermanesque-Marxisant phrasing implies, they do so in part because he identifies modernity so completely with capitalism.[136]

I hope this will not be misunderstood. Capitalism is of course a constitutive dimension of modernity: but it is not the only one. Laclau and Mouffe draw upon many of the same sources as Harvey and Soja – most notably the work of Aglietta and the Regulation School – in seeking to make sense of the transformations of late twentieth-century capitalism and the installation of new regimes of accumulation. They therefore have no

---

[134] Davis, *City of Quartz*, p. 316.

[135] *Newsweek*, May 11, 1992, p. 30.

[136] Soja, *Postmodern geographies*, p. 246.

difficulty in recognizing that "there is practically no domain of individual or collective life which [now] escapes capitalist relations."[137] But they also cast their conceptual net much wider.

One cannot understand the present expansion of the field of social conflictuality and the consequent emergence of new political subjects without situating both in the context of the commodification and bureaucratization of social relations on the one hand and the reformulation of the liberal-democratic ideal – resulting from the expansion of struggles for equality – on the other.[138]

In this context perhaps the most startling absence from *Postmodern geographies* is any sustained engagement with feminism. This is perfectly consistent with the modernist cast of Soja's project as a whole, I suppose, and a number of writers have drawn attention to the masculinity of the modern gaze and its systematic devaluation of the experiences of women.[139] In this particular sense (a more restricted sense than that in which Soja uses the term) aesthetic modernism may indeed be a "reaction formation."[140] But the contemporary conjunction between postmodernism and feminism makes such an omission from a postmodern geography truly astonishing. This is a complex and fractured terrain and I do not wish to imply that post-modernism and feminism can be mapped onto the same discursive space. Different writers clearly have different views on the relations between the two, but I am persuaded by those who have identified a tense area of overlap in which the social construction of differences is a central concern.[141] I should also say that I do not wish to minimize the difficulties facing Soja (and me).[142] It is one thing for men to *attend* to the experiences of

---

[137] *Ibid.*, Laclau and Mouffe, *Hegemony*, p. 161.

[138] *Ibid.*, p. 163.

[139] See, for example, Griselda Pollock, "Modernity and the spaces of femininity," in her *Vision and difference: Feminity, feminism and histories of art* (London: Routledge, 1988) pp. 50–90 and Janet Wolff, "The invisible *flâneuse:* Women and the literature of modernity," *Theory, culture and society* 2 (3) (1985) pp. 37–46. These two essays are sharp reminders that the view from the street is equally susceptible to voyeurism. In the first version of this essay, I said I would leave it to others to reflect on the phallocentrism of Soja's landscape and his final contemplation of it from the "monumental erection" of City Hall (Soja, *Postmodern geographies*, p. 224). Since it was published two critics have done exactly that: see Rosalyn Deutsche, "Boys town," *Environment and Planning D: Society and Space* 9 (1991) pp. 5–30 and Doreen Massey, "Flexible sexism," *loc. cit.*, pp. 31–57.

[140] Andreas Huyssen, "Mass culture as woman: Modernism's other," in his *Great Divide*, pp. 44–62; see especially the discussion of the modernist aesthetic as a reaction formation on pp. 53–55.

[141] Nancy Fraser and Linda Nicholson, "Social criticism without philosophy: An encounter between feminism and postmodernism" and Jane Flax, "Postmodernism and gender relations in social theory," both in Linda Nicholson (ed.), *Feminism/Postmodernism* (London: Routledge, 1990) pp. 19–38 and pp. 39–62.

[142] Cf. Stephen Heath, "Male feminism," in Alice Jardine and Paul Smith (eds.), *Men in feminism* (New York: Methuen, 1987) pp. 1–32.

women but quite another for them to *represent* those experiences. Any act of translation is an intervention in and within a matrix of power and different languages (and, by extension, different lifeworlds) are made to submit to what Asad calls "forcible transformation" and what Spivak calls "epistemic violence" in the act of translation itself.[143]

Here I sense – though I cannot fully feel – the force of the language of "landscape" that Soja routinely deploys. Within its pictorial lexicon, so Barrell suggests, it is impossible to capture "the simultaneous *presence* of someone within the center of knowledge ... and his *absence* from it, in a position from which he observes but does not participate."[144] Put more plainly, while Soja and I are plainly involved in patriarchy, we are equally removed from women's experience of the oppression that accompanies it: and doubly so if it is represented in this way. But that does not absolve either of us from a confrontation with its barbarities and I am surprised that his concern with textual strategy has not sensitized him to the politics of its representation.[145]

These are not easy issues to summarize and since they raise such searching questions about Occidentalism and postcolonialism it would probably be wrong to attempt to do so. But I do want to insist that the tension that I have been trying to describe in the previous paragraphs is not a superficial but a structural feature of *Postmodern geographies* and that it has extremely serious consequences for the construction of a critical theory sensitive to the intrinsic spatiality of social life. One last vignette might secure my argument. In his closing essay Soja insists that "there is ... always room for resistance, rejection and redirection ... creating an active politics of spatiality, struggles for place, space and position within the regionalized and nodalized urban landscape."[146]

So I would hope. Yet just ten pages later Soja is prepared to portray Los Angeles as "a lifespace comprised of Disneyworlds":

Like the original "Happiest Place on Earth" the enclosed spaces are subtly but tightly controlled by invisible overseers despite the open appearance of fantastic freedoms of choice.[147]

---

[143] Talal Asad, "The concept of social translation in British social anthropology," in Clifford and Marcus (eds.), *Writing culture*, pp. 140–64; Gayatri Chakravorty Spivak, "Can the subaltern speak?" in Grossberg and Nelson (eds.), *Marxism and culture*, pp. 271–313. I cite these two essays for a particular reason: neither of them is exclusively concerned with feminism and both of them problematize the more general ontological conceit of the Western subject. Their arguments therefore have a wider purchase on *Postmodern geographies* (and postmodernism) than I am able to discuss here.

[144] John Barrell, "Geographies of Hardy's Wessex," *Journal of historical geography* 8 (1982) pp. 347–61; the quotation is from p. 358. Here too the male pronoun is apposite.

[145] This is of course more than a matter of style: see Chapter 2, this volume.

[146] Soja, *Postmodern geographies*, p. 235.

[147] *Ibid.*, p. 246.

If this sentence virtually erases the space for resistance, it should now be clear that it does so in large measure because *those "overseers" include Soja himself.* Although he claims to "have been looking at Los Angeles from many different points of view" – more accurately, I should say, from many different viewpoints – these do not provide different "ways of seeing," as he seems to think: rather, they all provide the *same* way of seeing. For, as Cosgrove remarks of the concept of landscape more generally,

[P]erspective was employed to control space and to direct it towards the external spectator.... The people who occupy the landscape... do not themselves participate as subjects responsible for their world; they are puppets controlled by the artist. ...The experience of the insider, the landscape as subject, and the collective life within it are all implicitly denied. Subjectivity is rendered the property of the artist and the viewer – those who control the landscape – not those who belong to it.[148]

I do not of course mean to deny the significance of those circuits of cultural capital that are materially inscribed in the "Disneyworlds" of contemporary Los Angeles.[149] Neither do I seek to confine politics to the conceptual field. The substantive sources of exploitation and oppression cannot be reduced to a series of discursive strategies. But considerations of this kind indelibly shape the construction of any critical human geography and, as Soja would himself urge, the reassertion of space in critical social theory cannot be accomplished without scrupulous attention to its discursive constitution. Commenting on a painting of the city of Florence, executed soon after the rediscovery of linear perspective, one writer has recently drawn attention to the artist shown at the edge of the canvas.

Seated there as he is above the city, he incarnates at its birth a new ideal of knowledge according to which the further we remove ourselves from the world the better we can know it. It is an ideal, however, which by definition means a knowledge of the world which is increasingly disincarnate. On the hill above the city only his eyes remain "in touch" with the world observed below. But at that distance such eyes... can no longer know the words of anger or love uttered by those living in the city.[150]

This image continues to haunt my reading of *Postmodern geographies.* For surely postmodernism calls that "ideal of knowledge" into question? And yet doesn't Soja's visual ideology offer much the same perspective on late twentieth-century Los Angeles?

---

[148] Denis Cosgrove, *Social formation and symbolic landscape* (London: Croom Helm, 1984) pp. 25–26.

[149] For all that, Soja offers no systematic analysis of "Disneyfication": cf. Sharon Zukin, *Landscapes of power: From Detroit to Disneyworld* (Berkeley: University of California Press, 1991).

[150] Robert Romanyshyn, *Technology as symptom and dream* (London: Routledge, 1989) p. 38.

Watching the detectives

Let me now draw the threads of my discussion together. Like Soja, I accept that postmodernism can have an insistently critical edge: but I wish to affirm that its edge remains sharp *only to the extent that it remains in contact with modernism*. Any attempt to sunder the one from the other – to treat them as wholly separate or antithetical – seems to me to dull the critical sensibilities that need to be embedded in our ways of thinking about the contemporary condition.[151] Unlike Soja, however, I believe that the same is also true of post-Marxism: that it makes no sense to see its various formulations as an unremitting deconstruction of historical materialism *tout court*. Since I have shown that Soja's thinking remains imbricated in both modernism and Marxism then, to that extent, I clearly regard his text as a considerable intellectual achievement. I do not write commentaries of this length on books that I think unimportant or uninteresting. But I am uneasy about Soja's unwillingness to explore some of the lines of inquiry that transcend these two (for him) foundational problematics. In the first place, his "reassertion" of space in critical social theory is on occasion conducted in such aggressive terms that it turns into an unrelieved anti-historicism. And in the second place, his master-narrative is sometimes so authoritarian that it drowns out the voices of other people engaged in making their own human geographies. Both concerns seem to me to rebound on the practical intent of Soja's project; both are clarified by an engagement with postmodernism and post-Marxism.

These remarks allow me to return, at last, to my fictional detective. Kracauer was a close contemporary of Simmel and in his unpublished account of *The detective novel* he suggested that the modern genre typically represents purely formal relationships. Its characters "spread themselves out in an unbounded spatial desert and are never together even when in close proximity in the metropolis." One might even say that the revelation of reality takes place through the power of abstract reason embodied in the figure of the detective.[152] And pursuing these inquiries under the cover of historical materialism, Mandel suggests that

There is nothing so astonishing about the fact that literate people should be obsessed with mystery stories. After all, as Ernst Bloch once pointed out: isn't the whole of bourgeois society operating like a big mystery anyway?[153]

---

[151] Here I am particularly indebted to Andreas Huyssen, "Mapping the postmodern," in his *Great Divide*, pp. 178–221.

[152] Frisby, *Fragments*, p. 128.

[153] Ernest Mandel, *Delightful murder: A social history of the crime story* (London: Pluto Press) p. 72.

If so, my worry is probably that Soja's way of working is too much like Maigret's. But then, unlike some other writers on postmodernism, one has to be grateful that it isn't Clouseau's....

PART III

Between Two Continents

Introduction

Children, do you remember when we did the French Revolution? That great landmark, that great watershed of history. How I explained to you the implications of that word "revolution"? A turning round, a completing of a cycle. How I told you that though the popular notion of revolution is that of categorical change, transformation — a progressive leap into the future — yet almost every revolution contains within it an opposite if less obvious tendency: the idea of a return. A redemption; a restoration. A reaffirmation of what is pure and fundamental against what is decadent and false. A return to a new beginning.

Graham Swift, *Waterland*

Yet how well did we, whose business it was to do so, pick out these critical processes at the time of their happening, or link them with the changes derived from them? And why did we miss them, if not because we were unaccustomed to thinking in terms of processes?

Carl Sauer, *Foreword to historical geography*

These essays have their origins in the intellectual and political project embedded within David Harvey's *The condition of postmodernity*.[1] There are two senses in which I think of them as moving between two continents. In the first place, debates over postmodernism have so often been turned into arguments over post-Marxism that it is scarcely surprising that so many of the most sustained challenges to postmodernism should have been launched from the platform of historical materialism. I should say at once that I find myself caught in the middle: still very much interested in the development of historical materialism; suspicious of claims that it provides the single master key to unlock human history and geography; yet skeptical

---

[1] David Harvey, *The condition of postmodernity: An enquiry into the origins of cultural change* (Oxford: Blackwell Publishers, 1989).

of some of the assumptions and implications of postmodernism. In thinking about these questions I have learned a great deal from the writings of Callinicos, Eagleton, and Habermas (among others). None of these three defends an unreconstructed Marxism and each of them has made a series of creative contributions to the development of various traditions of Western Marxism, but they have all registered a series of uncompromising objections to postmodernism and its satellite discourses.[2]

This is not an entirely random selection, however, because all three are also *European* writers. To be sure, they work within a wider horizon of meaning, but a number of other commentators on the Left have suggested that the postmodern has a distinctively and contrarily *American* genealogy. I make this point not to minimize the contributions of Baudrillard, Derrida, Lyotard, Vattimo or any of the other European thinkers who have shaped our sense of the postmodern (and in somewhat more affirmative ways), but because – in the second place – Harvey's diagnosis of the condition of postmodernity seems to me to derive from his experience of *both* America *and* Europe. Born in Britain, he studied at the University of Cambridge and then taught for several years at the University of Bristol; in 1969 he moved across the Atlantic to Johns Hopkins University in Baltimore; and in 1987 he returned to England to the Halford Mackinder Chair' of Geography at the University of Oxford. It will soon become obvious that I mean this in more than a biographical sense, however, and it will help to make sense of the discussions that follow if, in the first section of this introduction, I sketch the outlines of the "American" argument. This intersects with critical traditions that have developed in and around historical materialism, but in the second section I will try to reverse tack and consider those responses to Harvey's project which, while they are by no means hostile to Marxism, resist being subsumed within its totalizing framework.

American dream

Andreas Huyssen and Fredric Jameson, two of America's most insightful cultural critics, have put forward particularly thoughtful versions of the American argument. Huyssen is probably the more forceful. The postmodern accrued its emphatic connotations in the United States, he declares, "not Europe." In his view, the prehistory of the postmodern – by which he

---

[2] Alex Callinicos, *Against postmodernism: A Marxist critique* (Cambridge, England: Polity Press, 1989); Terry Eagleton, "Capitalism, modernism and postmodernism," in his *Against the grain: Selected essays 1975–1985* (London: Verso, 1986) pp. 131–47; Jürgen Habermas, *The philosophical discourse of modernity: Twelve lectures* (Cambridge, England: Polity Press, 1987).

means the turbulent period of the 1960s – was distinguished by a determined assault on the austere culture of high modernism. There is nothing remarkable about an observation of this kind, but Huyssen uses it to advance a further and more striking claim: that postmodernism was an attempt "to revitalize the heritage of the European avant-garde and to give it an American form."[3] In Europe itself, he argues, the movement (such as it was) collapsed. The events of 1968 marked not so much a breakthrough as the "replayed end" of the avant-garde.

Whether in the German protest movement or in May '68 in France, the illusion that cultural revolution was imminent foundered on the hard realities of the status quo. Art was not reintegrated into everyday life. The imagination did not come to power.[4]

In America, however, proto-postmodernism (if that is not too ridiculous a way of putting it) was much more resilient and its exuberance was largely undiminished. Where Europeans might well react with a sense of *déjà vu*, Huyssen believes that "Americans could legitimately sustain a sense of novelty, excitement and breakthrough." This is not the same as saying that the imagination somehow "came to power" on that side of the Atlantic, whatever such a phrase might turn out to mean, but Huyssen has no doubt that postmodernism in America did much to enlarge what he calls the "temporal imagination": that it helped to promote and even secure that quintessentially American and altogether "unshaken confidence of being at the edge of history."[5]

I should add two riders to these observations. First, there is no doubt that the confidence that Huyssen identified *has* been shaken in recent years, not least by the shudderings of the American economy and the rise of other supereconomies around the Pacific Rim, which has provoked an aggressive new "othering" of Japan in particular. But American confidence has also been remade and repackaged by events elsewhere in the world: most

---

[3] Andreas Huyssen, "The search for tradition: Avant-garde and postmodernism in the 1970s" and "Mapping the postmodern," both in his *After the Great Divide: Modernism, mass culture and postmodernism* (London: Macmillan, 1986) pp. 160–77, 178–221; the quotations are from pp. 188, 190, 195.

[4] *Ibid.*, p. 166.

[5] *Ibid.*, pp. 166–167. When Huyssen speaks of the temporal imagination rather than the *historical* imagination he is choosing his words with care. To other commentators, postmodernism came to represent a threat to the historical imagination. " 'Historicism' effaces history" is how Jameson puts it. On his reading, history is transformed into "a vast collection of images" to be selected and recombined at will, and "the past as 'referent' finds itself gradually bracketed, and then effaced altogether": Fredric Jameson, "Postmodernism, or the cultural logic of late capitalism," *New Left Review* 146 (1984) pp. 53–92; the quotations are from pp. 65–66. At the limit, and most obviously in the work of Edward Soja, postmodernism becomes identified with a tenaciously *geographical* imagination: see Chapter 4, this volume.

obviously by the collapse of Communist regimes in Eastern Europe and the cacophonous dissolution of the Soviet Union – which overlapped with Fukuyama's Hegelian proclamation of "the end of history," the political and economic triumph of liberal democracy announced from within the U.S. State Department – and by the defeat of Iraq in the Gulf War, which was supposed to confirm the inauguration of a "new world order" in which the United States would do the ordering. But these changes do not really alter the substance of Huyssen's argument, because there are elements of the postmodern in the construction and mediatization of both episodes.

Second, following directly from these remarks, it would be quite wrong to think of Huyssen as a protagonist for any unreconstructed postmodernism; but it would be equally misleading to read him as crudely dismissive. He declares flatly that he does *not* believe "that a cultural criticism indebted to the tradition of Western Marxism is bankrupt or obsolete today"; but he also wants to salvage the postmodern from any collusion with neoconservatism and to open up its "productive contradictions," many of which (so he seems to say) are contained within its appropriations of modernist strategies and their reinscription in altogether new constellations.[6]

Jameson's account of postmodernism is situated within a similar complexity of terrain and, not surprisingly, his reading of the sixties is not so very different either: but it is markedly less celebratory. Two examples will perhaps sharpen the point. First, he draws attention to the progressive redefinition of "the edge of history." During the sixties, Jameson argues that countless people were admitted to the discourse of history for the first time, and he observes that this took place "internally as well as externally: those inner colonized of the First World – 'minorities,' marginals, women – fully as much as its external subjects and official 'natives.' "[7] History could no longer be represented as a coherent totality sutured around a single subject-position. The predicament of culture became (and very largely remains) the predicament of recognizing our own otherness. If this is to insist that the dynamic of Third World history entertains "some privileged relationship of influence" on the unfolding of First World history, however, as Jameson claims, it is emphatically not to imply that this relationship has sustained any emancipatory shift in the political register. On the contrary, he makes it plain that new relations of dependence and penetration were simultaneously installed by a global capitalism "in full and innovative expansion."[8] Since then, of course, capitalist expansion

---

[6] Huyssen, *Great Divide*, pp. xii, 200, 218.

[7] Fredric Jameson, "Periodizing the '60s," in his *The ideologies of theory: Essays 1971–1986*. Vol. 2: *Syntax of history* (Minneapolis: University of Minnesota Press, 1988) pp. 178–208; the quotation is from p. 181.

[8] *Ibid.*, pp. 184, 185–86. Be that as it may, Jameson's own appropriations of "the Third World" are not beyond reproach: see Aijaz Ahmad, "Jameson's rhetoric of otherness and the 'national allegory,' " *Social text* 17 (1987) pp. 3–25.

has been jagged and uneven, but the ascendance of market triumphalism in Eastern Europe speaks to much the same point; so too does American intervention to protect its oil interests in the Gulf.[9]

Second, Jameson agrees that postmodernism (then) was a way of "making creative space" for those artists who were oppressed by the conventions of high modernism and who wanted to establish new connections with popular culture. But he also argues that postmodernism (even then), so far from being adversarial or oppositional, "constitutes the very dominant or hegemonic aesthetic of consumer society itself and significantly serves the latter's commodity production as a virtual laboratory of new forms and fashions."[10] This too is supposed to bear an American trademark. According to Jameson, postmodernism expresses the "inner truth" of the emergent social order of late capitalism or, as he put it in perhaps his most seminal essay, postmodernism is "the cultural logic of late capitalism."[11] There is a strong implication, although it is never made quite explicit, that the internationalization of capital that has supposedly animated the extraordinary motility of "postmodern hyperspace" has been driven by American powers and interests.[12]

For the most part, I suspect that Harvey would share the broad analytical sense of these readings. He would undoubtedly want to make a number of additions and qualifications, particularly in relation to the periodization of "late capitalism," but he nonetheless draws upon both Huyssen and Jameson to characterize the cultural politics of the sixties, and in particular the formation of the New Left and its critique of everyday life, as moments in a "global movement of resistance to the hegemony of high modernist culture." I think it important to take this conjuncture with all possible seriousness and to pay the closest attention to Harvey's suggestion (following Huyssen and Jameson) that this movement was "the cultural and

---

[9] Cf. Fredric Jameson, "Conversations on the New World Order," in Robin Blackburn (ed.), *After the fall: The failure of communism and the future of socialism* (London: Verso, 1991) pp. 255–68; idem, "Thoughts on the late war," *Social Text* 28 (1991) pp. 142–46.

[10] Jameson, "Periodizing," p. 196.

[11] *Idem*, "Postmodernism and consumer society," in Hal Foster (ed.), *The anti-aesthetic: Essays on postmodern culture* (Port Townsend, Wash.: Bay Press, 1983) pp. 111–25; the quotation is from p. 113; idem, "Cultural logic."

[12] So, for example, Jameson appears to assent to the suggestion that he ties postmodernism "to the rise of American capital on a global scale." "Once modernism broke down," he remarks, "the absence of traditional forms of culture in the United States opened up a field for a whole new cultural production across the board. . . . [A] *system* of culture could only emerge from this American possibility. The moment American power begins to be questioned, a new cultural apparatus becomes necessary to reinforce it." This is (I hope) a counter-factual rather than a purely functionalist argument. See Anders Stephanson, "Regarding postmodernism – a conversation with Fredric Jameson," in Andrew Ross (ed.), *Universal abandon? The politics of postmodernism* (Minneapolis: University of Minnesota Press, 1988) pp. 3–30; the quotation is from p. 8.

political harbinger of the subsequent turn to postmodernism."[13] Much of what follows is an attempt to do exactly that. But I should say at once that I think Harvey is quite wrong to say that the New Left disabled itself "in insisting that it was culture and politics that mattered, and that it was neither reasonable nor proper to invoke economic determination even in the last instance" – the inclusion of the Althusserian phrase is almost enough to jar the comment on its own – and that an interest in cultural politics was (and presumably is) somehow inimical to an understanding of class relations and struggles. This is a caricature of the New Left; whatever its failings, they did not include an unwillingness to attend to problems of political economy, and the address of cultural politics was not (and is not) a distracting sideshow to the main event.[14] But this is not the place to argue about the New Left.

What I think is more important, in the context of Harvey's critique of postmodernism, is a nuanced understanding of the distinctively American cultural geography of these developments and a consideration of the distinctively European genealogy of the theoretical dispositions which he seeks to make about them. In Chapter 5, therefore, I return to an earlier modernism and explore two iconographic traditions that formed a template for what came to be called the American Dream: "liberty" and "power."[15] As I try to show there, these two thematics are combined in a particularly arresting way in *The condition of postmodernity*. In Chapter 6 I explore this constellation of power and liberty in a different direction by tracing its connections with a distinctively French "politics of space." In order to do so, however, I must also clarify the limits of Harvey's critical practice, because the conception of liberty – of emancipation – which it provides has been criticized for its own closures and exclusions.

"Dreams of unity"

The most important critical responses to *The condition of postmodernity* have opened two main avenues of inquiry. The first of them has exposed Harvey's marginalization of feminism. *The condition of postmodernity* is announced as "an enquiry into the origins of cultural change," but according to Rosalyn Deutsche it soon becomes obvious that Harvey has set his sights on "a general theory of contemporary culture."

What, then, does it mean that in the book he altogether ignores one of the most significant cultural developments of the past twenty years – the emergence of

---

[13] Harvey, *Condition of postmodernity*, p. 38.

[14] *Ibid.*, p. 354.

[15] According to *Webster's*, the phrase "American Dream" dates from 1933.

practices in art, film, literature, and criticism that are informed by specific kinds of feminism?[16]

Deutsche argues that it is *precisely* Harvey's claims to completeness and comprehensiveness that distinguish his work from feminism. His theorizing turns on a series of visual metaphors derived from a modernist optics: "establishing a binary opposition between subject and object, it renders the subject transcendental, the object inert, and so underpins an entire regime of knowledge as mastery." And yet, Deutsche continues, "such knowledge, like vision, is highly mediated by fantasy, denial and desire": by "dreams of unity" (the phrase is Bruno's) in which all differences are reconciled within some plenary totality.[17] She identifies two consequences that she thinks follow from this.

In the first place, Harvey's visual metaphoric for theory construction is in tension with his ambivalence toward, even mistrust of, *other* visual images: in them, he finds "only insufficiency, absence, fragmentation." But he does so, Deutsche contends, because he is unable to recognize that meaning – *all* meaning, including ostensibly "theoretical" meaning – is culturally and conventionally produced, and clings instead to a belief in "fundamental and unambiguous meanings" whose provenance is authenticated by his own supposedly objective theoretical optics.[18]

In the second place, the gender of the epistemological noun – "mastery" – is deliberate. For the would-be "universal" subject is in fact a *masculine* subject, whose very claim to occupy a privileged ground – an autonomous, sovereign position – betrays his own fear at occupying what Deutsche calls "a position of threatened wholeness in a relation of difference." Hence her conclusion: Harvey "may not consider feminism worth knowing about," she writes, "but feminism, although it hardly knows everything, knows something about him."[19]

---

[16] Rosalyn Deutsche, "Boys town," *Environment and Planning D: Society and Space* 9 (1991) pp. 5–30; the quotation is from p. 6.

[17] *Ibid.*, pp. 10–11.

[18] *Ibid.*, pp. 15, 22. Deutsche uses Harvey's (brief) reading of Cindy Sherman's photography as an index of the difference between them. Harvey suggests that Sherman's photographs "focus on masks without commenting directly on social meanings other than the activity of masking itself" and that, in consequence, they are indifferent "towards underlying social meanings" and indeed complicitous with fetishism (*Condition of postmodernity*, p. 101). Deutsche claims that Sherman's work "locates fetishism precisely in this belief in fundamental and unambiguous meanings" ("Boys town," p. 28). Deutsche's point is a sharp one, but it may be too sharp for this particular example. Laura Mulvey shows in some considerable detail how Sherman's work traces the failure of the fetish through images of the feminine; but she also concedes the risk involved: the possibility that Sherman may herself be "stuck in the topographic double bind of the fetish and its collapse." See Laura Mulvey, "A phantasmagoria of the female body: The work of Cindy Sherman," *New Left Review* 188 (1991) pp. 136–50.

[19] Deutsche, "Boys town," pp. 12, 29. Substantially the same argument could be made by juxtaposing Harvey's "playing the *flâneur*" with Wilson's reading of the same figure as "a shifting

Similarly, Doreen Massey claims that Harvey's Marxism-modernism is constructed around a masculine subject-position whose particularity is never recognized. She too focuses on Harvey's visual obsessions to secure her argument. Like Deutsche, she objects that his interpretation of *Blade Runner*, which draws upon Giuliana Bruno's work, refuses to yield ground to Bruno's feminist criticism; he is only able to advance upon his fictive "universal" through a series of slippages and evasions that reverse or even erase the tenor of Bruno's argument.[20] Massey draws particular attention to – and takes particular exception to – the sequence of illustrations that Harvey uses to compare modernism and postmodernism: Titian's *Venus*, which was reworked as Manet's (modernist) *Olympia* and incorporated into Rauschenberg's (postmodernist) *Persimmon*. As Massey points out, all of them depict a naked woman and assume a masculine subject-position on the part of the viewer.[21] Although Harvey's primary purpose was to demonstrate what he took from Crimp to be the characteristic postmodernist movement from production to reproduction, "the fiction of the creating subject [giving] way to frank confiscation, quotation, excerption, accumulation and repetition of already existing images," he has subsequently claimed that the conscious subtext – "the additional point I sought to make," but one which seemed so obvious to him that he thought it needed no comment – was that the subordination of women could "expect no particular relief by an appeal to postmodernism."[22] Maybe not; but Deutsche and Massey make it plain, as Harvey does not, that there *are* versions of postmodernism that provide stringent criticisms of the authoritarianism of the male gaze.

These feminist objections thus intersect with, though they cannot be folded into, critical forms of postmodernism: the relation between the two is one of creative tension. But neither can they be cleanly separated from historical materialism. Both Deutsche and Massey resist the subsumption of feminism into Marxism, and in particular Harvey's repeated description of feminist struggles against patriarchy as "local" struggles that are to be located within the "global" class struggle against capitalism: but they do

---

projection of angst rather than a solid embodiment of male bourgeois power": Elizabeth Wilson, "The invisible *flâneur*," *New Left Review* 191 (1992) pp. 90–110. But I have to say that I am uneasy about the transcendental cast of some of these psychoanalytical readings; that is no doubt precisely the point of them, but they seem (in *their* turn) insufficiently attentive to differences in context and constituency.

[20] Doreen Massey, "Flexible sexism," *Environment and Planning D: Society and Space* 9 (1991) pp. 31–57; Deutsche, "Boys town," pp. 6–7; Giuliana Bruno, "Ramble city: Postmodernism and *Blade Runner*," *October* 41 (1987) pp. 61–74.

[21] Massey, "Flexible sexism," pp. 44–50.

[22] Harvey, *Condition of postmodernity*, pp. 54–57; Douglas Crimp, "On the museum's ruins," in Foster (ed.), *The anti-aesthetic*, pp. 43–56; David Harvey, "Postmodern morality plays," *Antipode* 24 (1992) pp. 300–26; the quotation is from p. 318. The same qualification appears as an addendum in reprinted copies of *The condition of postmodernity* (1992).

not minimize the importance of political economy, as some of Harvey's more obdurate critics seem to have done.

Some "postmodern geographers," if I may so call them, are more hostile to historical materialism (or at least to Harvey's version of it). They also object to what Michael Dear characterizes as Harvey's "unrelieved antipathy" toward postmodern thought. "He seems incapable of tolerating differences," so Dear concludes, and the question of "otherness," of listening to and learning from other voices, other discourses is raised only to be muted and marginalized.

Harvey has driven inexorably through the maze of postmodernity primarily to dissolve differences. But why on earth would anyone want to do that? Why should we sacrifice our hard-won openness to "thick description" in favor of some new/old closure? Why pillory those who reject totalizing discourse? Criticism need not mean betrayal. And just because one dislikes fragmentation doesn't mean that it can be wished away; nor can one concede to difference on condition that it stays in its proper place.

Precisely because of his inability to deal with difference, Harvey's response to the challenges of postmodernism and deconstruction has been to reconstruct a thoroughly modernist (Marxian) rationality.... [I]n taking an axe to postmodernism, Harvey leaves his own historical materialism almost totally unexamined.

Dear claims that it is precisely this (absent) reflexivity which is so fully present in postmodernism, constantly opening and inspecting its theoretical baggage, and which has also made possible the transcendence of classical and Western Marxism by post-Marxism. The consequence of not releasing those exclusionary catches, Dear continues, is an extraordinarily confining set of assumptions. In Harvey's case these are particularly burdensome. For how is it, he asks, that Harvey can seek to redeem the project of modernity, while recognizing *its* deformations and pathologies, and yet concern himself so exclusively with the negativities of postmodernism that its (different) productive capacities are simply denied? Dear concludes that his critique founders on this contradictory practice, its credibility exploded by such a devastating "political paradox."[23]

I will comment on these readings (and Harvey's reply) in Chapter 5. For now, I simply wish to emphasize that they do not exhaust the terrain of critical response and that I think it also important to explore what both Deutsche and Dear consider to be the most valuable and original contribution

---

[23] Michael Dear, review of *The condition of postmodernity* in the *Annals of the Association of American Geographers* 81 (1991) pp. 533–39; the quotation is from p. 537. The same argument is repeated in his "The premature demise of postmodern urbanism," *Cultural anthropology* 6 (1992) pp. 535–49. Dear's appeal to post-Marxism is specifically to Ernesto Laclau and Chantal Mouffe, *Hegemony and socialist strategy: Towards a radical democratic politics* (London: Verso, 1985).

of *The condition of postmodernity*: Harvey's dovetailed discussion of the production and politics of space, in which he identifies some – though only some – vitally important intersections between capitalism, modernity, and spatiality.[24] As I try to show in Chapter 6, however, a serious appreciation of Harvey's contribution requires a consideration of his politico-intellectual project as a whole. But there I am also (and probably more) concerned to explore its filiations with and its differences from the work of Henri Lefebvre. There are several reasons for doing so, but I hope that such a tactic might also open up a more positive and productive dialogue between Harvey and his critics. Deutsche has been influenced by Lefebvre in a number of her other writings, for example, and has in fact commended his interventions in the strongest of terms:

[Lefebvre's] analysis of the spatial exercise of power as a construction and conquest of difference, although it is thoroughly grounded in Marxist thought, rejects economism and opens up possibilities for advancing analysis of spatial politics into realms of feminist and anti-colonial discourse [and into the theorization of radical democracy].[25]

It is those possibilities that interest me too, although I do not want to minimize the importance of a critical analysis of class formations within capitalist space-economies. I am not sure what Harvey's other critics will make of Lefebvre: He figures prominently in Soja's *Postmodern geographies*, of course, but since that is such a profoundly modern text it is difficult to know how his ideas will be received by those whose postmodernism is folded into poststructuralism. Lefebvre was an implacable opponent of structuralism and structural Marxism, but he was also sharply critical of Derrida and Foucault too. Yet many of his substantive preoccupations – with the decorporealization of space and the banalization of the everyday – which he shared with the situationists must now be shared with postmodernists too. I hope to show that the rest of us need not be left out either.

---

[24] Deutsche, "Boys town," p. 17; Dear, "Review," p. 535.

[25] Rosalyn Deutsche, "Uneven development: Public Art in New York City," *October* 47 (1988) pp. 3–52; the quotation is from p. 29. I have added the appeal to "radical democracy" – an appeal to Laclau and Mouffe's Post-Marxism – from the revised version of this essay, which appears under the same title in Russell Ferguson, Martha Gever, Trinh T. Minh-ha and Cornel West (eds.), *Out there: Marginalization and contemporary cultures* (Cambridge: MIT Press, 1990) pp. 107–32.

5

Dream of Liberty?

I hear the ruin of all space, shattered glass and toppling masonry, and time one livid final flame.

James Joyce, *Ulysses*

And in moments of despair or exaltation, who among us can refrain from invoking the time of fate, of myth, of the Gods?

David Harvey, *The condition of postmodernity*

Cover version

The injunction not to judge a book by its cover is usually a sensible one. Nevertheless, I want to try to show that the illustration used for the cover of David Harvey's *The condition of postmodernity* can be made to illuminate its political and intellectual genealogy in a number of different ways.[1] The image in question is Madelan Vriesendorp's *Dream of Liberty* (figure 24), and in what follows I will suggest that its juxtaposition with Harvey's text constitutes a provocative site at which meanings are produced. I hope it will be allowed that such a proposal does not turn on recovering the artist's or author's intentions. It is a commonplace of critical practice that texts escape their authors' intentions; so too do images. But I am aware that this does not license a riot of interpretations, and for those who might be uncomfortable at my suggestions and wonder about Harvey's own reading of *Dream of Liberty*, here it is:

Postmodernity, it is said, is about fiction rather than function. But it is about a certain kind of fiction, in which quite disparate worlds collide and intermingle, in which time and space collapse in on each other, to produce a flat landscape in

---

[1] David Harvey, *The condition of postmodernity: An enquiry into the origins of cultural change* (Oxford: Blackwell Publishers, 1989).

Figure 24 Dream of Liberty *[Madelan Vriesendorp, 1974; Deutsches Architektur-museum, Frankfurt].*

which anything goes and all voices are treated as equal. This painting, out of the postmodernist stable, illustrates these theories with clarity and precision.[2]

[2] David Harvey, internal memorandum, Basil Blackwell, Inc., June 1989. I am grateful to John Davey for providing me with a copy.

It might be helpful to keep this in mind as I proceed. I begin with the two images that occupy the foreground of *Dream of Liberty* – the Statue of Liberty and the Chrysler Building – and then move out into the landscape in which Vriesendorp places them.

Imagining liberty

In 1856 Frédéric-Auguste Bartholdi, a young French sculptor, was in Egypt seeking inspiration for a *grand projet*. He was entranced by the pyramids, but most of all by the Sphinx at Giza. Half-human, half-lion, it had crouched near the great pyramid of Chephren for more than 4,000 years. Bartholdi began to dream of a work on a comparable scale. Eventually, he proposed the construction of a lighthouse in the shape of a beautiful woman, draped in a robe with her right arm holding a torch aloft. The statue would be raised at the Mediterranean entrance to the Suez Canal as a symbol of the nineteenth-century expansion of Europe. Its proposed title, *Egypt carrying light to Asia*, was intended to assert (as its construction was intended to confirm) the historical mission of the West – with a colonized Egypt acting as its handmaiden – to bring "enlightenment" to the East.[3] The insensitivity of the image is arresting to our contemporary sensibilities, but Napoleon's expedition to Egypt (1798–1801), although a military and political failure, had succeeded in annexing the land and its people to the discourse of what Edward Said calls "Orientalism." Napoleon's armies had been accompanied by a team of over 100 scholars and the results of their labors were published between 1809 and 1828 as the *Description de l'Égypte*. The 23 volumes of this extraordinary text reveal the ambitions of Orientalism to have been almost boundless. Said's indictment of the project includes the following:

To restore a region from its present barbarism to its former classical greatness; to instruct (for its own benefit) the Orient in the ways of the modern West; to subordinate or underplay military power in order to aggrandize the project of glorious knowledge acquired in the process of political domination of the Orient; to formulate the Orient, to give it shape, identity, definition with full recognition of its place in memory, its importance to imperial strategy, and its "natural" role as an appendage to Europe; to dignify all the knowledge collected during colonial occupation with the title "contribution to modern learning" when the natives had neither been consulted nor treated as anything except as pretexts for a text whose

---

[3] This account is drawn from Oscar Handlin, *Statue of Liberty* (New York: Newsweek, 1971); Pierre Provoyeur and June Hargrove (eds.), *Liberty: The French-American statue in art and history* (New York: Harper and Row, 1986); Mary Shapiro, *Gateway to Liberty* (New York: Vintage, 1986); and Marvin Trachtenberg, *The Statue of Liberty* (New York: Penguin, 1986).

usefulness was not to the natives; to feel oneself as a European in command, almost at will, of Oriental history, time and geography.[4]

This was just as powerful as the physical ligatures put in place by the completion of the Suez Canal itself: in its way, perhaps more so. Said argues that Foucault's account of the connective imperative between "power" and "knowledge" and his genealogy of *assujetissement* – in the twin sense of "subjection" and "subjectification" – ought not to be confined to the continent of Europe.[5] Certainly, the discourse of Orientalism continued to shape the Eurocentric construction of "the other" throughout the nineteenth and into the twentieth centuries. In France, Harvey reminds us, Michelet and the Saint-Simonians used the imagery of Orientalism "to justify the penetration of the Orient by railroads, canals and commerce, and the domination of an irrational Orient in the name of a superior Enlightenment rationality."[6] Seen from such a perspective there is nothing unusual about the symbolism of Bartholdi's dream.

He spent two years making plans and models for his statue, but in 1869 – the year the Canal opened – the Khedive withdrew his support and it was not until 1871 that Bartholdi was able to reactivate his scheme. By then, in the wake of the Paris Commune, the project had been transformed and relocated. One of his patrons, Édouard de Laboulaye, suggested that a monument be raised on the shores of the New World to symbolize *Liberty Enlightening the World*. The representation of Liberty as a woman derived from classical antiquity but this "whole allegorical apparatus," as Maurice Agulhon calls it, had been codified in France in the late seventeenth century. When the Revolution occurred, Liberty already had an established iconographical status and, indeed, a decree of 1792 adopted her as the seal of the republic: "the image of France in the guise of a woman, dressed in the style of Antiquity, standing upright, her right hand holding a pike surrounded by a Phrygian cap or cap of liberty."[7] By the opening

[4] Edward Said, *Orientalism* (Harmondsworth, England: Penguin, 1985) pp. 79–88; the quotation is from p. 86; cf. Anne Godlewska, "The Napoleonic Survey of Egypt," *Cartographica* 25 (1–2) Monograph 38–39 (1988).

[5] For a particularly powerful study of nineteenth-century Egypt in these terms, see Timothy Mitchell, *Colonizing Egypt* (Cambridge: Cambridge University Press, 1988). I discuss these matters in more detail in Chapters 1 and 2, this volume.

[6] David Harvey, *Consciousness and the urban experience* (Oxford: Blackwell Publishers, 1985) p. 201. Harvey also draws attention to the gendering of the discourse of Orientalism. "The Orient was seen ... as the locus of irrational and erotic femininity" which, as he says himself, was to be "penetrated" by the essentially masculine rationality of the West.

[7] Maurice Agulhon, *Marianne into battle: Republican imagery and symbolism in France, 1789–1880* (Cambridge: Cambridge University Press, 1981) pp. 11–13, 16. The Phrygian cap was placed on the head of newly released Roman slaves to symbolize their freedom. See also Lynn Hunt, *Politics, culture and class in the French Revolution* (Berkeley: University of California Press, 1984) pp. 60–66, 93–94, 115–19 and the catalogue accompanying the exhibition, *La France: Images of woman and ideas of nation, 1789–1989* (London: South Bank Centre, 1989).

decades of the nineteenth century the use of "Marianne" as a symbol of both Liberty and the Republic had become a commonplace. This was true in the most literal of senses. "Where was this woman to be seen?" asks Agulhon. The answer: "all over the place." Paris had two statues of her – in the Place de la Concorde and the Place Vendôme – and many other towns had their own effigies. In 1848 she appeared on the second seal of the Republic, wearing a diadem of corn with seven rays of the sun encircling her head in a spiked halo.[8] The resemblance to the head of the American statue is striking, but a sunburst was also the Bartholdi family emblem and, still more significantly, it was intimately associated with the reign of Louis XIV, the Sun King. To adorn Liberty with a sunburst was thus in a sense "to 'crown' her," Kaja Silverman argues, "and thereby to align her with a tradition of stable and conservative government."[9] Two new statues were hastily constructed, one on the Champ de Mars and the other in the Place de la Concorde (where it was flanked by an obelisk brought from Luxor in Egypt).[10]

That Liberty should be represented by a woman was not without irony. In practice, Joan Landes remarks,

the assault on patriarchalism was limited both by force ... and by the redirection of women's public and sentimental existence into a new allegory of republican, virtuous family life. Liberty herself is a profoundly ironic symbol, a public representation of a polity that sanctioned a limited domestic role for women.... If Liberty represented woman, surely it was as an abstract emblem of male power and authority.[11]

The power of patriarchy was reasserted still more forcefully after the fall of the Commune with the triumph of the bourgeois republic and its cult of respectability. As one historian puts it, "the official *Mariannes* who adorned town halls by the 1880s wore a halo of flowers and the motto *Concorde*, moving towards that anodyne statue which France sent to her fellow capitalist republic as the Statue of Liberty."[12] Anodyne indeed:

---

[8] Agulhon, *Marianne*, p. 86; illustration p. 90.

[9] Kaja Silverman, "Liberty, maternity, commodification," *New Formations* 5 (1988) pp. 69–89; the quotation is from p. 73.

[10] Agulhon, *Marianne*, p. 65.

[11] Joan Landes, *Women and the public sphere in the age of the French Revolution* (Ithaca: Cornell University Press, 1988) pp. 159, 161; see also Margaret Iversen, "Imagining the Republic: The sign and sexual politics in France," in Peter Hulme and Ludmilla Jordanova (eds.), *The Enlightenment and its shadows* (London: Routledge, 1990) pp. 121–39.

[12] Roger Magraw, *France 1815–1814: The bourgeois century* (London: Fontana, 1983) p. 212. The representation of women in fin-de-siècle painting was much less anodyne, of course, but the fundamental point remains: "Women's bodies [were portrayed] as the territory across which men artists [could] claim their modernity." See Griselda Pollock, "Modernity and the spaces of femininity," in her *Vision and difference: Femininity, feminism and the histories of art* (London: Routledge 1988) pp. 50–90; the quotation is from p. 54.

Silverman argues that Bartholdi virtually erased the corporeality of the body. Thus he "completely buries the female form beneath her classic drapery" and "any thought that a body might nevertheless lurk beneath those folds is abruptly put to flight by the possibility of entering the statue and climbing up inside it." Silverman is well aware of the sexual connotations of such a reading and moves quickly to foreclose them. "Liberty is precisely an extension of the desire to 'return' to the inside of the fantasmatic mother's body," she proposes, *"without having to confront her sexuality in any way."* Viewed in this light, therefore, Liberty is rendered nonthreatening and even "safe."[13]

It was thus from within a many-layered iconographical tradition that Laboulaye's proposal was made. He was Professor of Comparative Law at the Collège de France and although he had never crossed the Atlantic he was nonetheless regarded as France's greatest expert on the United States.[14] Like many other republicans at the time, he regarded the United States as a model of the ideal society and he and his companions were convinced that a Statue of Liberty, given by France to America, would symbolize their most cherished principles.[15] For this reason Bartholdi was urged to ensure that the statue "not be liberty in a red cap, striding across corpses with her pike at the port" – a reference to Delacroix's famous *Liberty guiding the people to the barricades* – but "the American liberty whose torch is held high not to inflame but to enlighten." Bartholdi agreed. "Revolutionary Liberty cannot evoke American Liberty," he declared, "which, after a hundred years of uninterrupted existence, should appear not as an intrepid young girl but as a woman of mature years, calm, advancing with the light but sure step of progress."[16]

For all the enthusiasm of the project's initiators, however, public subscriptions were slow – even Gounod conducting *La Liberté éclairant le monde* at the Paris Opéra brought in a mere 8,000 francs – and in America they were slower still. Bartholdi made a show of offering the statue to Philadelphia and Boston; other American cities submitted bids until at last prominent subscribers in New York were goaded into action. In 1875 Bartholdi started work in his Paris *atelier*. He soon realized the magnitude of the task and invited Gustave Eiffel to design the wrought-iron bracing

---

[13] Silverman, "Liberty," pp. 74, 82 (my emphasis).

[14] Theodore Zeldin, *France 1848–1945*; vol. 2: *Intellect, taste and anxiety* (Oxford: Clarendon Press, 1977) p. 129.

[15] For much the same reason, a number of radical Republicans subsequently proposed building a colossal statue of *Liberty* on the summit of Montmartre in Paris. There it would effectively obscure the basilica of Sacré-Coeur, which they regarded as a symbol of "the intolerance and fanaticism of the right": see Harvey, *Consciousness*, pp. 221–49.

[16] Silverman, "Liberty," p. 75; Iversen, "Imagining," p. 136. These sensibilities are memorialized in the inscription on the base of the statue, which commemorates the example of the American Revolution and the sympathy offered to France by the United States during the "difficult" period of the Franco-Prussian War and the Paris Commune.

Figure 25 The Statue of Liberty on the Rue de Chazelles, Paris [Victor Dargaud, 1884].

needed to support the copper sheets that would form the outer skin of the sculpture. It took several years to complete the disembodied sections of the statue, but by the spring of 1883 Bartholdi was at last ready to assemble them. By the end of the year, as Dargaud's canvas shows, the statue, still surrounded in scaffolding, was looming above the Rue de Chazelles (figure 25). The government of France presented the Statue of

Liberty to the United States of America on the Fourth of July, 1884. Five
months later it was dismantled, shipped across the Atlantic and reassembled
on Bedloe's Island in New York harbor. When the inauguration ceremony
was held in October 1886, all women were barred, except for the wives
of the French delegation; American suffragists held their own simultaneous
ceremony. Bartholdi led the French delegation, alongside Ferdinand de
Lesseps, the architect of the Suez Canal, and when he returned to France
he declared triumphantly that he had no doubt that "the monument will
last as long as those built by the Egyptians."[17]

Representing power

Already other powerful icons were soaring into the sky over Manhattan.
The skyscraper made its first appearance in Chicago in the closing decades
of the nineteenth century, and it was one of the most vigorous symbols
of the density of offices and corporations in so-called "organized capitalism." It depended on a series of advances in engineering, most obviously
on the elevator and the steel frame, and it presumably owed something
(though exactly how much is a moot point) to the upward pressure on
land rents in the center of the modern metropolis. But the skyscraper was
more than the architecture of the possible, more even than the architecture
of necessity. To one observer, contemplating the New York skyline in the
closing stages of World War I, "a skyscraper, shouldering itself aloft" was
"a symbol, written large against the sky, of the will-to-power."[18]

This Nietzschean imagery received its most dramatic expression during
the following decade. The skyscrapers built in the 1920s were doubly coded.
On the one side, they were "clothed in a fantastic scenography" and seen
as "the totems of American capitalism."[19] No building reveled in these
associations more obviously than the Chrysler Building, whose Art Deco
façade celebrated the unrestrained iconography of the automobile – "the
frieze at the thirtieth floor represented a long line of Chrysler hubcaps
speeding towards the corners, where monstrously scaled radiator caps took

---

[17] Silverman, "Liberty," p. 86; Handlin, *Statue*, p. 165.

[18] Claude Bragdon, *Architecture and democracy* (New York: Alfred Knopf, 1918) pp. 5–6. Much
more recently Jencks has endorsed a parallel reading. "Profit and power are tied by architects to
a Nietzschean symbolism": Charles Jencks, *Skyscrapers – skyprickers – skycities* (New York: Rizzoli,
1980) p. 15.

[19] Robert Stern, Gregory Gilmartin, and Thomas Mellins, *New York 1930: Architecture and urbanism
between the two world wars* (New York: Rizzoli, 1987) p. 589. According to Harvey, "it is no accident
that . . . in the age of monopoly capitalism it is the Chrysler Building and the Chase Manhattan
Bank which brood over Manhattan Island": David Harvey, *Social justice and the city* (London: Edward
Arnold, 1973) p. 32.

wing as twentieth-century gargoyles"[20] – and whose height expressed the vaunting ambition of its parent corporation and its architect, William Van Alen (figure 26). When it was completed in August 1930 the Chrysler Building was the tallest office building in the world. But it was soon challenged by the Empire State Building, finished less than a year later and one of the few construction projects to survive the stock market crash, "rising from the ruins," as Scott Fitzgerald put it, "lonely and inexplicable as the Sphinx."[21] Fitzgerald's Egyptian imagery was by no means exceptional, and neither was his astonishment at the building's survival. In fact, it took so long to rent that for a time it was known as the Empty State Building.

On the other side, most commentators agreed that the skyscraper also reflected an aggressive masculinity. To Claude Bragdon, the skyscraper was "a symbol of the American spirit in its more obvious aspect – that ruthless, tireless, assured *energism*, delightedly proclaiming 'What a great boy am I!' "[22] The appeal to the masculine was not accidental and it contained within it a latent contest of virilities. When H. Craig Severance's Bank of Manhattan Building was completed, for example, it topped the Chrysler Building by two feet. But Van Alen had concealed a steel mast inside the dome of the Chrysler Building which he promptly raised to tower more than 100 feet above his rival. "During this procedure," van Leeuwen remarks, "the building did what otherwise could only have happened in the imagination of the beholder: it grew visibly."[23] The sexual imagery implicit in the contest is captured with an astonishing exactness in this description of the dome:

Van Alen's design was a sort of cruciform groin sliced in seven concentric segments that mounted up one behind the other. The whole complex seemed animated by a mysterious vertical thrust: the vaults swelled upward toward the center.[24]

If this seems excessive, it shrivels into insignificance beside Dali's view of the New York skyline as what Conrad calls "a map of priapic rivalry."

---

[20] Stern, Gilmartin, and Mellins, *New York*, p. 608.

[21] *Ibid.*, p. 615.

[22] Claude Bragdon, *The frozen fountain* (New York: Alfred Knopf, 1932) p. 25.

[23] Thomas van Leeuwen, "The myth of natural growth II," in his *The skyward trend of thought: The metaphysics of the American skyscraper* (Cambridge: MIT Press, 1988) pp. 115–51; the quotation is from p. 150, which also contains a cut-away diagram.

[24] Stern, Gilmartin, and Mellins, *New York*, p. 608. In a reply to his critics Harvey concedes that skyscrapers "can in part be explained as phallic erections celebratory of a dominant masculine power"; but he also insists that "politico-economic will" also plays a role, and – by implication – the leading role. Otherwise, he asks, "What do we say about the Chrysler Building?" What indeed? See David Harvey, "Postmodern morality plays," *Antipode* 24 (1992) pp. 300–326; the quotation is from p. 313.

Figure 26 Beaux-Arts Ball, New York, 1931. Van Alen is in the center dressed as the Chrysler Building. [Source: Pencil Points, February 1931].

[Dali] sees Manhattan as the rearing multiple erection of some couchant, many-phallused monster. After sunset lights burst on in the skyscrapers as if in response to a libidinal massage, which caresses the prongs until, Dali says, they void themselves in the sky's vagina.[25]

These may be nothing more than the scabrous fantasies of male critics and commentators, of course, but I do not think it difficult to accept Conrad's suggestion that these buildings "did battle on their owners' behalf" and sought to "symbolize power over mortality and over the human body which, for all their wealth, they themselves [did not] possess."[26] The conjunction between corporate capitalism and phallocratic power was thus profoundly worldly and riven with both antagonism and anxiety.

Still, Bragdon confessed that the skyscraper could also be seen as a "most august and significant symbol" of human achievement: on one condition. "We [have to] look at these megathena of commerce with something of the same detachment and young-eyed wonder we bring to the contemplation of the pyramids – *to which they are related as life to death.*"[27] That sense of sublimity was by no means unusual and its evocation typically involved just this sort of transhistorical imagery. Sometimes New York was likened to Babylon; when Mesopotamia had been plundered for its icons, the Mayan civilizations of pre-Colombian America were pressed into service.[28] At other times, as here, commentators turned to Egypt, although few of them did so with the fervor of Dali:

New York, you are an Egypt! But an Egypt turned inside out. For she erected pyramids of slavery to death, and you erect pyramids of democracy with the vertical organ-pipes of your skyscrapers all meeting at the point of infinity of liberty![29]

This is surely even more perverse than his priapic celebration of Manhattan.

Dream of Liberty

Since the last fin de siècle both the Statue of Liberty and the skyscraper have been used time and time again not only to celebrate the values of modern America – which was (in part) their purpose – but also to serve

[25] Peter Conrad, *The art of the city: Views and versions of New York* (Oxford: Oxford University Press, 1984) pp. 146, 208.

[26] *Ibid.*, p. 209.

[27] Bragdon, *Fountain*, p. 35; my emphasis.

[28] Stern, Gilmartin, and Mellins, *New York*, pp. 511–13; Van Leeuwen, "Apokatastasis or the return of the skyscraper," in his *Skyward trend*, pp. 39–55.

[29] Salvador Dali, *The secret life of Salvador Dali* (New York: Dial Press, 1942) pp. 331–32.

as the occasion for an indictment of them. When Christopher Isherwood first visited New York, for example, he wrote of the skyscrapers as overblown "Father-fixations." The Statue of Liberty fared no better. She became "'the Giantess': an overbearing and unnurturing mother." The famous WeeGee photograph of the Statue blurred the distinctions even further. Now assimilated to the ideals of consumer society, what Conrad calls "the new litheness of her body" was likened to the shape of a Coca-Cola bottle and her torch to a hot dog.[30]

One of the most dramatic juxtapositions of these two symbols is to be found in Vriesendorp's *Dream of Liberty*. The appropriateness of this image for the cover of *The condition of postmodernity* is, I think, suggested by some astonishingly prescient remarks of Fredric Jameson. Referring to post-modernism's "eclipse of culture as an autonomous space," he writes:

> The break-up of the sign in mid-air determines a fall back into a now absolutely fragmented and anarchic social reality; the broken pieces of language (the pure signifiers) now fall again into the world, as many more pieces of material junk among all the other rusting and superannuated apparatuses and buildings that litter the commodity landscape and that strew the "collage city," the "delirious New York" of a postmodernist late capitalism in full crisis.[31]

Harvey would not, I think, identify postmodernism with late capitalism; but with that single reservation, this passage expresses his concerns exact-ly.[32] And the last two or three lines speak more or less directly to Vriesendorp's work: She was a member of the Office of Metropolitan Architecture (OMA) and worked closely with her partner, Rem Koolhaas, in the preparation of his *Delirious New York*.[33] Harvey identifies both OMA as a group and Vriesendorp as an individual with postmodernism and, following Klotz, claims that their projects are characterized by "the collage of fragments of reality and splinters of experience enriched by historical references."[34] In fact, Klotz is even more specific than that and proffers a reading of *Dream of Liberty* itself. The toppled towers of the skyscrapers supposedly evoke "the death agony of the metropolis," its despairing gloom

---

[30] Conrad, *Art of the city*, pp. 215, 219–20. In Lagarrigue's *Liberty-Cola* the head and draperies actually become a Coca-Cola bottle: see Silverman, "Liberty," p. 87.

[31] Fredric Jameson, "Periodizing the sixties," in his *The ideologies of theory: Essays 1971–1986*. Vol 2: *Syntax of history* (Minneapolis: University of Minnesota Press, 1988) pp. 178–208; the quotation is from p. 201. Compare this with Harvey's reading (above, pp. 327–28).

[32] Jameson was drawing on Ernest Mandel, *Late capitalism* (London: New Left Books, 1975), but Mandel's "late capitalism" coincides with what Harvey identifies as "Fordism" not the subsequent regime of flexible accumulation that Harvey sees as the counterpart to postmodernism.

[33] Rem Koolhaas, *Delirious New York: A retroactive manifesto for Manhattan* (New York: Oxford University Press, 1978). Jameson does not cite this text but the reference is unambiguous.

[34] Harvey, *Condition of postmodernity*, pp. 82–83.

relieved only by the Statue of Liberty, "like a new Prometheus . . . attempting to hold up the 'flame of liberty' and free herself from a civilization that has fallen into ruin."[35] But this is far from being an exhaustive or particularly sensitive interpretation, and I want to draw attention to some other aspects that can be situated within the same general frame of meaning but that are given a distinctive inflection by their proximity to Harvey's text. Taken together, I suggest, they speak even more directly to the political and intellectual project of *The condition of postmodernity*.

There are five points I wish to make. Firstly, there is no doubt that the metropolis whose ruins occupy the foreground is a modern, even a modernist city. The skyscrapers shown one behind the other are the Chrysler Building, the Empire State Building and the RCA Building. Their razed towers are set off from the mortuary architecture of premodern civilizations – the City of the Dead, the Sphinx and the line of pyramids – by an abrupt precipice. This is the conventional iconography of modernity, of course: the discontinuity that unequivocally divides the modern from the premodern, canceling and annulling the stasis of traditional societies and substituting the turbulence, the transience, and what Aragon called the "vertigo" of the modern. As the preceding discussion has shown, however, this caesura is transgressed by several powerful iconic continuities. Neither the Statue of Liberty nor the skyline of Manhattan erase the past and, indeed, it is clear that in both cases it is possible to identify particular, complex, and many-layered associations with ancient Egypt. More figuratively, but nonetheless startling in its associations in this context, Gauny suggests that "It is in the desert that seditious thought ferments, but it is in the city that such thought erupts. Liberty likes extreme crowds or absolute solitude."[36]

These complex filiations suggest the possibility, even the necessity, of reconsidering the historical geography of modernity, of displacing the metaphor of discontinuity and exploring instead the multiple meanings carried forward and embedded within the project of modernity. This is made all the more urgent by the continuities that can be posited in the opposite direction, mirrored in the backward glance of modernity.[37] Liberty may be

---

[35] Heinrich Klotz, *Revision of the modern* (New York: Architectural Design, 1985) p. 50. For the full discussion of OMA, see *idem, Revision der Moderne: Postmoderne Architektur 1960–1980* (Munich: Prestel, 1984) pp. 206–17.

[36] Louis Gabriel Gauny, *Le philosophe plébien* (Paris: La Découverte, 1983) p. 115; cf. Kristin Ross, *The emergence of social space: Rimbaud and the Paris Commune* (Minneapolis: University of Minnesota Press, 1988) p. 21.

[37] Two other members of OMA play with these continuities in their later work. Zoe and Elia Zenghelis's *Hotel Sphinx* (1975–1976) depicts a hotel in the shape of the Sphinx; its head contains a planetarium and a swimming pool. And in their *Planetarium with swimming pool*, the view from the hotel gives on to the distant (which is to say past) tower of the Empire State Building. These drawings are reproduced without comment in Heinrich Klotz, *The history of postmodern architecture* (Cambridge: MIT Press, 1988) pp. 314–15.

escaping from the ruins of the modernist city but, as I now want to propose, she is not thereby abandoning the project of modernity.

Secondly, then, Vriesendorp shows Liberty freed not from Bartholdi's scaffolding but from Van Alen's Chrysler Building: emerging, that is to say, from *within* the collapse of the modern metropolis, not apart from it. Although the plasticity of the statue's form contrasts with the angular and brittle geometries of the ruined cities that lie behind her, she is rendered in brick (unlike the original). If Liberty has turned her back on previous urban forms, one might suggest, she has not abandoned the urban *tout court*. This is also consistent with the ambitions of modernism. Not only was modernism an art of the city, but the modernist city was conceived by Le Corbusier and the *Congrès Internationaux d'Architecture Moderne* as a "city of salvation," as a means of resolving the contradictions of capitalism through the construction of an urban society. *La ville radieuse* – the "radiant city" that was to "succeed the reigning darkness" of the 1920s and 1930s – began with an affirmation of what Le Corbusier regarded as "an inalienable, unquestionable truth that is fundamental to all plans for social organization: individual liberty." To be sure, it was a strange sort of liberty that saw French colonialism as a *force moral*, offered unqualified praise of the Ford factory in Detroit ("everything is collaboration"), and sought an accommodation with the Vichy government.[38] It is scarcely surprising that Le Corbusier should have been so enthusiastic about "Authority" (and in strictly patriarchal terms: He once declared that "France needs a Father") and that some commentators have detected a strong Nietzschean undercurrent in his work.[39] For all his importance, it would be wrong to cast Le Corbusier as paradigmatic of modernism; the movement was never monolithic. But his emphasis on what Tafuri terms "the whole anthropogeographic landscape" – on the construction of urban society as the central bridge between architecture and utopia – was undoubtedly a vital moment in the trajectory of the modernist project.[40]

Thirdly, *Dream of Liberty* endorses what Martin Jay calls the "ocular-centrism" of modernity and the privilege characteristically accorded to vision

[38] My discussion of Le Corbusier and the modernist city is derived from Robert Fishman, "From the radiant city to Vichy: Le Corbusier's plans and politics, 1928–1942," in Russell Wadden (ed.), *The open hand: Essays on Le Corbusier* (Cambridge: MIT Press, 1977) pp. 244–83; James Holston, *The modernist city* (Chicago: University of Chicago Press, 1989) pp. 41–58; Stanislaus von Moos, *Le Corbusier: Elements of a synthesis* (Cambridge: MIT Press, 1979).

[39] See, for example, Charles Jencks, *Le Corbusier and the tragic view of architecture* (Cambridge: Harvard University Press, 1973) pp. 25–29, 181–82.

[40] Manfredo Tafuri, *Architecture and Utopia: Design and capitalist development* (Cambridge: MIT Press, 1976) p. 126. For a more qualified reading of Le Corbusier, which suggests that his textual strategies effectively undermine the modernist temper of his arguments, see Dennis Crow, "Le Corbusier's postmodern plan," in *idem* (ed.), *Philosophical streets: New approaches to urbanism* (Washington: Maisonneuve Press, 1990) pp. 71–92.

as "the master sense of the modern era."[41] The forked lightning, the sun (or moon) beyond the horizon, the brilliant light illuminating the composition from outside the frame: all of these suggest the transcendent power of vision. And this is true, crucially, of Liberty herself. I do not mean to dwell on the obvious point that the Bartholdi statue can be placed within the obsessively visual thematic of nineteenth-century French culture. The torch, "Egypt carrying light to Asia," "Liberty enlightening the world" are all unequivocal in their scopophilic implications. What is more interesting, I think, is Jay's insistence that twentieth-century French culture has become much less affirmative, more critical in its attitude to the visible, even (though I think this is to claim too much) scopophobic.[42] It is surely not necessary to accept this argument in its entirety to follow Jay's thesis until it intersects with Marina Warner's reading of Liberty:

The Statue of Liberty engulfs us because she inspires one of the great sensual pleasures of the eye, dependent not upon aesthetic delight but upon the psychology of vision, inherent in the layered meaning of the very word. She gluts the eye with a sense of power, springing from the sensation of seeing the future.[43]

When we look across from Liberty toward Manhattan, Warner continues, "we are looking at a future that has happened."[44] *Dream of Liberty* provides no comparable concretization, only an anticipation, but it still constitutes an unequivocally modernist icon of visibility. As such (and here I agree with Jay) it participates in a withdrawal of emotion: a Cartesian gaze that is distanced from the everyday lives of ordinary men and women. This "optical knowledge," as de Certeau calls it, has long been associated with urban templates of power, with the city as a field of programmed and regulated operations. "Perspective vision and prospective vision constitute the twofold projection of an opaque past and an uncertain future onto a surface that can be dealt with."[45] Does Vriesendorp's Liberty really subvert the system of signs that created her in the first place?

I do not know, but I suspect that – fourthly – Vriesendorp's emancipatory motif is more complicated than it appears. Liberty is shown being freed from the phallic towers of Manhattan and for this reason Klotz

---

[41] Martin Jay, "Scopic regimes of modernity," in Hal Foster (ed.), *Vision and visuality* (Washington: Bay Press, 1988) pp. 3–23; the quotation is from p. 3.

[42] *Idem*, "In the empire of the gaze: Foucault and the denigration of vision in twentieth-century French thought," in David Couzens Hoy (ed.), *Foucault: A critical reader* (Oxford: Blackwell Publishers, 1986) pp. 175–204.

[43] Marina Warner, *Monuments and maidens: The allegory of the female form* (London: Weidenfeld and Nicolson, 1985) p. 13.

[44] *Ibid.*, p. 13. "The same scopic drive haunts users of architectural productions by materializing today the utopia that yesterday was only painted": Michel de Certeau, *The practice of everyday life* (Berkeley: University of California Press, 1984) p. 92.

[45] *Ibid.*, pp. 93–95.

assumes that the painting represents, among other things, the emancipation of women.[46] Certainly, the sexual imagery is unmistakable when Vriesendorp's work is considered as a whole. Indeed, one critic has suggested that her drawings invoke "Dali's anthropomorphic dreams of New York."[47] So they do; but her rendering of the Chrysler Building departs significantly from his usual preoccupations. In *Après l'amour*, for example, Vriesendorp shows the Chrysler and Empire State Buildings lying on a bed, a deflated blimp between them (the mast on top of the Empire State Building was intended to function as a mooring for dirigibles). In her *Flagrant délit* the two buildings are surprised by the abrupt entry of the RCA Building into the bedroom. In both cases it is far from clear that the Chrysler Building has been coded in masculine terms, and the same ambiguity presumably characterizes *Dream of Liberty* since it is fully continuous with the imagery of these other paintings.[48] Perhaps this makes the Chrysler Building a more appropriate carapace for Liberty herself: but Liberty is hardly an unambiguous icon either. As I have already suggested, there is nothing paradoxical about a female figure being used to represent liberty in a patriarchal society. Warner argues that in such circumstances it is often the case that "the recognition of a difference between the symbolic order, inhabited by ideal, allegorical figures, and the actual order ... depends on the unlikelihood of women practicing the concepts they represent."[49] One might therefore wish to claim that the continued use of Liberty is in some sense compromised by these associations. I am not sure.

Beyond this one could also argue – fifthly – that in Bartholdi's original conception the statue represented another regime of domination, namely the discourse of Orientalism, though this is, of course, connected to patriarchal power in complex ways. Seen from this perspective, the erection by Chinese students in May 1989 of a styrofoam and plaster "Goddess of Democracy" in Tiananmen Square in Beijing, copied from a similar statue in Shanghai, directly facing the disfigured portrait of Mao Tse-Tung, seems both deeply courageous and profoundly ironic. But it differed from the Franco-American original in two ways. First, it was not, strictly speaking, a copy at all. Wu Hung suggests that it was a borrowed symbol that was modified into a new image: that "no matter how much the Goddess owed its form and concept to the Statue of Liberty," it was a distinctly Chinese image. Second, the Goddess of Democracy achieved its effect through its impermanence: It was a "monument that was *intended* to be destroyed [it was crushed by a tank], because its monumentality would derive from such

---

[46] Klotz, *Revision*, p. 50.

[47] Richard Pommer, "Review of *Delirious New York*," *Art in America* 67 (3) (1979) p. 19.

[48] These two drawings are reproduced without comment in Koolhaas, *Delirious New York*, pp. 66–67 and Pommer, "Review," p. 19.

[49] Warner, *Monuments*, p. xx.

self-sacrifice."[50] It may be possible to displace this reading, however, and return it to the United States to confront Orientalism and its successor projects directly. I have in mind Dean MacCannell's provocative essay in which he records his dismay at the restoration of the original statue because the enterprise, which was carried out in the mid-1980s, tacitly reaffirmed an (exclusive) connection between European immigration and the United States, which had become out of joint with the times:

> A mighty woman, standing with a torch in a harbor, cannot serve as the symbol of the arrival of a person who has sneaked into the country in the false bottom of a truck. As it always has for African slaves, it will eventually come to stand for the difference between the current arrival of Asians and Latinos, and the earlier arrival of "true" Americans. The only possible way to signify continuity between the old European and the new immigrations would be to enshrine the Statue of Liberty as a *ruin*.[51]

The movement to restore the statue was led by the son of European immigrants, Lee Iacocca. He was also head of the Chrysler Corporation.

The condition of postmodernity

I have spent so much time establishing the pretexts for these readings (and I am acutely aware of how provisional and partial they are) for two reasons. In the first place, they indicate in what I hope is a clear and concrete way some of the most important questions at stake in the debates over modernity and postmodernity. The connections and separations between the premodern, the modern, and the postmodern; the power of the visible, of Foucault's "eye of power"; the embeddedness of the project of modernity in the city; the sensitivity of modernism and postmodernism to the construction of difference: All of these matters are of central importance and I shall have more to say about them in the next chapter. In the second place, these readings bring into focus some of Harvey's deepest concerns together with those of his critics, and I want to hold this double focus for a moment because I think each has something to show the other. Following the same sequence as my discussion of *Dream of Liberty*, therefore, let me offer the following preliminary characterizations of the debate over *The condition of postmodernity*.

First, Harvey is plainly interested in interrogating the meanings carried forward and embedded within the project of modernity, and he does so

---

[50] Wu Hung, "Tiananmen Square: A political history of monuments," *Representations* 35 (1991) pp. 84–117; the quotations are from pp. 110, 113.

[51] Dean MacCannell, "The Liberty restoration project," in his *Empty meeting grounds: The tourist papers* (London: Routledge, 1992) pp. 147–57; the quotation is from p. 155.

as the very condition of its reaffirmation. He has long been interested in
the historical geographies of capitalist modernity, but these are now used
to inform his construction of the present in a much more direct fashion
and, in particular, to displace any rigid caesura between the modern and
the postmodern. That he does so by convening them within the contradictory unity of capitalism as totality is, I agree, problematic.[52] But a starkly
discontinuist history seems to me a much more disturbing prospect, not
least because it threatens to erase from critical inquiry the continued
presence of the past in the present: except, perhaps, as a designer's palette
for some postmodern pastiche. To say this is not to make the opposite
mistake and endorse an evolutionary history in which moments clip
together like magnets. Although Harvey is properly suspicious of unexamined notions of "progress," there is also a *telos* within historical materialism
that sometimes involves him in a "progressivist" conception of history.
That too needs to be scrutinized, because critical historiography is defined
neither by revolutions and displacements, nor by continuities and incre-
ments.

Second, Harvey seeks to inscribe the project of modernity within urbanism (rather than the purely physical form of the city) and to place his
hopes for its reconstruction and redemption on what he calls elsewhere
the "urbanization of revolution."

If the urbanization of capital and of consciousness is so central to the perpetuation
and experience of capitalism, and if it is through these channels that the inner
contradictions of capitalism are now primarily expressed, then we have no option
but to put the urbanization of revolution at the center of our political strategies.[53]

I have already outlined some of the ways in which modernism can be
understood as an art of cities, and many discussions of postmodernism are
equally preoccupied with architecture, built form, and a "generalized urban."
Where many modernisms were undertaken as interventions in the urban,
however, as attempts to construct bridges between architecture and utopia,
postmodernism typically uses the urban to fashion its own self-image. I
am troubled by these "complicities," as Gillian Rose calls them, not least
because of the closures they tacitly put in place.[54] Harvey's commentary in
The condition of postmodernity trades on both tactics, but in doing so he relies
on a particular conjunction of Marxism and modernism, which installs an

---

[52] Harvey, *Condition of postmodernity*, pp. 338–42.

[53] Harvey, *Consciousness*, p. 276.

[54] Gillian Rose, "Architecture to philosophy – the postmodern complicity," *Theory, culture and
society* 5 (1988) pp. 357–71. Since I do not want to dissolve differences, I should explain that this
Gillian Rose is *not* the Gillian Rose who has criticized *The condition of postmodernity* so forcefully:
cf. the latter's "Review," *Journal of historical geography* 17 (1991) pp. 118–21.

equally particular conception of critical theory. Its anticipatory-utopian moment – "the urbanization of revolution" – is linked to and made possible by an explanatory-diagnostic moment that valorizes class in such a way that other, equally salient modalities of power are made to wait their turn. I doubt that the "urbanization of revolution" in these singular terms will be enough to free Liberty from the discriminations, oppressions, and terrors of the contemporary city.

Third, Harvey uses a series of visual images to describe his mode of theorizing. He gazes on the city from high vantage points, constructs theory by triangulation from different viewpoints, and examines the world through different windows. But he also insists that none of this privileges sight; as he says himself, his project turns on the point that the internal relations of capitalism *cannot* be seen and, more specifically, that postmodernism is "the mirror of mirrors."[55] This does not mean that these hidden processes cannot be visualized, however, and there is no need to convert the critique of the gaze into a recoil from vision. What is required, as Donna Haraway suggests, is a recognition of the *embodied* nature of *all* vision, so that one can "reclaim the sensory system that has been used to signify a leap out of the marked body and into a conquering gaze from nowhere."[56] Harvey is aware of Haraway's argument, but I rehearse it here because he tries to answer his critics by saying that his visual metaphors signal a *cognitive*, not a perceptual, mode of theorizing, and that cognition – unlike perception – "always involves the transcendence of any phenomenological sense of situatedness through putting oneself in the other's place."[57] That may well be so, but his reply overlooks the fact that it is precisely those claims of transcendence that make many of his critics so uncomfortable. "Putting oneself in the other's place" is about more than empathy, imagination and sensitivity; the very language invokes just that "leap out of the marked body" which troubles Haraway, and at the very least raises awkward questions about the politics of representation and authorization which Harvey simply ignores. I suspect that his critics would want to examine the *construction* of "the other" – the production of that "place" – and to explore less invasive, less potentially colonizing ways of listening to and indeed learning from other voices. This does not exempt them from the moral responsibility of discriminating between those voices where it is necessary to do so, of course, and Harvey is right to remind his critics of their obligations. But, as they would surely reply, those discriminations still have to be made from

[55] Harvey, *Condition of postmodernity*, p. 336.

[56] Donna Haraway, "Situated knowledges: The science question in feminism and the privilege of partial perspective," in her *Simians, cyborgs and women: The reinvention of nature* (New York: Routledge, 1991) pp. 183–201; the quotation is from p. 188.

[57] Harvey, "Postmodern morality plays," pp. 302, 312; Harvey is responding specifically to Rosalyn Deutsche, "Boys town," *Environment and Planning D: Society and Space* 9 (1991) pp. 5–30.

somewhere in particular and not from some imaginary Archimedian point of leverage.

Fourth, although Harvey tells some of his feminist critics that he was "early influenced by writers such as Haraway, Hartsock and Martin" and that the incorporation of their work into his argument would have "immeasurably strengthened" his case, the fact remains that *The condition of postmodernity* is markedly insensitive to feminism. Harvey now concedes that he did not pay sufficient attention to feminist writings, but he also says that he has "yet to encounter criticisms which would change anything other than the mode of appearance of my argument rather than its fundamentals."[58] This is an odd claim to make in the middle of a debate over representation and authorization and, as I have just indicated, Haraway's work – though by no means hostile to historical materialism – would seem to qualify Harvey's confident assertions. In fact his critics' central point, put most bluntly by Gillian Rose, is that the omission of gender (mis)informs the *ontology* of his text: that it is a strategic absence entering fully into the architecture of *The condition of postmodernity*. If that is so, then it will not be enough to add (still less to "integrate") gender into an argument that is otherwise allowed to remain unaltered. Acceptance of this neither requires nor implies acceptance of Rose's identification of modernism/masculine and postmodernism/feminine, which seems both totalizing and essentializing, and, as Harvey says himself, it is perfectly possible to identify a cogent feminist critique of postmodernism.[59] But it is also the case that a consideration of feminist theory, whatever its position on postmodernism, would impact on the construction of some of Harvey's most basic concepts, including capitalism, power, modernity and spatiality. In his post-*Condition* commentary on these matters, however, I am made uneasy by his attempt to turn differences among feminists into divisions between them, seemingly in order to exploit those disagreements and defend his original argument.

Finally, *The condition of postmodernity* is profoundly ethnocentric. I do not for one moment think that Harvey perpetuates any new Orientalism, let me say, and in his other writings he draws on Said's powerful critique to great effect. And yet for all its vaunted internationalism, its praise of the "heroic modernism" that set its face against parochialism and nationalism, its search for communalities in the human condition that can articulate differences, *The condition of postmodernity* is overwhelmingly centered on what Harvey calls the "post-Renaissance Western world." Reconstructing the political economy of capitalist modernity, Harvey pays attention to changes

---

[58] Harvey, "Postmodern morality plays," p. 305.

[59] Rose, "Review," p. 120; see also Deutsche, "Boys town" and Doreen Massey, "Flexible sexism," *Environment and Planning D: Society and Space* 9 (1991) pp. 31–57. For a particularly vigorous feminist critique of postmodernism, see Somer Brodribb, *Nothing mat(t)ers: A feminist critique of postmodernism* (North Melbourne: Spinifex Press, 1992).

in the international division of labor and inflows of capital, commodities, and information, but he does so almost invariably from the perspective and the position of the West; his account not only is framed by but also focuses on *its* geopolitics, *its* monetary systems, *its* consumption patterns. For an inquiry into cultural change, however, the limitations of such a perspective are even more startling. The Enlightenment project constructed a particular conception of the other in time and space. Beyond Europe was, in a sense, before Europe; the colonial conquest of space made it possible for "imperial adventures and conflicts" all over the globe to be reported on the pages of a single Western newspaper; as the project of modernity was carried forward into the twentieth century in multiple projects of "modernization," so these provoked a series of "Third World discontents."[60] Harvey says all this; but that is virtually all he says. It is as though those episodes are marooned on the periphery of his argument, never reconstructed in the places where they *took* place but always brought back to the discursive center for display in the world-as-exhibition. His discussions of modernism and postmodernism as cultural movements are equally metropolitan, and yet *both* of them have been shaped, in culturally important and politically resonant ways, by their distinctive constructions of other peoples and other places. And neither of them can be confined to Europe and North America. Although Harvey cites one commentator, Andreas Huyssen, who "particularly castigates the imperialism of an enlightened modernity that presumed to speak for others," including "colonized peoples," he has already sadly blunted the point with his own warning about the dangers of being "seduced" by postmodernism's concern with "otherness."[61] In moving between his two continents, it seems that Harvey barely has time to register the presence of the others.

[60] Harvey, *Condition of postmodernity*, pp. 139, 250–52, 264, 327.

[61] *Ibid.*, p. 47; Andreas Huyssen, "Mapping the postmodern," in his *After the Great Divide: Modernism, mass culture and postmodernism* (London: Macmillan, 1986) pp. 178–221.

6

Modernity and the Production of Space

Then as they came out into the Place Vendôme and looked up at the column, Monsieur Madinier... suggested going up to the top to see the view of Paris....

Round them the grey immensity of Paris, with its deep chasms and rolling billows of roofs, stretched away to the bluish horizon; the whole of the Right Bank lay in the shadows cast by a great ragged sheet of coppery cloud, from the gold-fringed edge of which a broad ray of sunlight came down and lit up the myriad windows of the Left Bank, striking from them cascades of sparks and making this part of the city stand out bright against a sky washed very clean and clear by the storm.

Émile Zola, *L'Assommoir*

We can take as an obvious and graphic example of the attack on verticality the Communards' demolition of the Vendôme Column, built to glorify the exploits of Napoléon's Grand Army.... For Mendès the destruction of the column abolishes history – makes room for a timeless present, an annihilated past, and an uncertain future.... [whereas] for the Communards, the existence of the column freezes time: "a permanent insult," "a perpetual assault."... Whose time is it?

Time, said Feuerbach, is the privileged category of the dialectician, because it excludes and subordinates where space tolerates and coordinates. Our tendency is to think of space as an abstract, metaphysical context, as the container for our lives rather than the structures we help create. The difficulty is also one of vocabulary, for while words like "historical" and "political" convey a dynamic of intentionality, vitality and human motivation, "spatial," on the other hand, connotes stasis, neutrality and passivity. But the analysis of social space, far from being reactionary or technocratic, is rather a symptom of strategic thought ... that poses space as the terrain of political practice. An awareness of social space, as the example of the Vendôme Column makes clear, always entails an encounter with history – or better, a choice of histories.

Kristin Ross, *The emergence of social space*

May 68 and Harvey 69

In this essay I continue my reading of *The condition of postmodernity* by "repatriating" Liberty to France – though I blanch at the verb – in order to consider the ways in which David Harvey's interest in the production and politics of space can be articulated with the work of the French philosopher Henri Lefebvre.[1]

I have always thought it ironic that Harvey should have been at work on *Explanation in geography*, at once the touchstone and tombstone of spatial science, during the tumultuous events of May 1968. The radical student movement was then at its height on both sides of the Atlantic and in that fateful month in Paris a serious and sustained challenge was mounted to the legitimacy of state power. To be sure, the reaction of the Left to those struggles was by no means unequivocal, but it was partly in response to them that Harvey almost immediately broke away from the intellectual and political framework of spatial science and set about constructing what he would eventually come to call "historico-geographical materialism."[2] His first attempts were sketched out in *Social justice and the city*, which opened from an avowedly liberal perspective and then moved, abruptly but exuberantly, toward a series of explicitly "socialist formulations." Harvey predicted that "the emergence of Marx's analysis as a guide to enquiry" would mean that he was "categorized as a 'Marxist' of sorts": but at that time it was very much "of sorts." His interpretive essay on urbanism, in many ways the climax of the book, owed as much to Polanyi's anthropology as it did to Marx's critique of political economy and, not surprisingly, Harvey faced considerable difficulties in fashioning a coherent project out of such disparate materials.[3] Since then, however, he has moved away from the substantivist paradigm of exchange – though he still uses its vocabulary of reciprocity, redistribution, and market exchange from time to time – and for the past 20 years or so his intention has been to develop Marx's "paradigm of production" in such a way that the production of space can be brought within its horizon as an essential moment in the production and reproduction of social life.[4]

---

[1] David Harvey, *The condition of postmodernity: An enquiry into the origins of cultural change* (Oxford: Blackwell Publishers, 1989).

[2] *Idem, Explanation in geography* (London: Edward Arnold, 1969). The manuscript was completed in June 1968. By March of the following year he had "changed several opinions" and could "already identify errors and shortcomings in the analysis" (p. viii). On the move from spatial science to historico-geographical materialism, see his "From Models to Marx: Notes on the project to 'remodel' contemporary geography," in Bill Macmillan (ed.), *Remodelling geography* (Oxford: Blackwell Publishers, 1989) pp. 211–16.

[3] *Idem, Social justice and the city* (London: Edward Arnold, 1973) p. 17. Harvey's account of urbanism, "Urbanism and the city – an interpretive essay," will be found on pp. 195–284.

[4] See David Harvey, *The limits to capital* (Oxford: Blackwell Publishers, 1982) and his two-volume

But this journey into Marxism has its own irony, for one of the most common responses to the events of May 68 in France was a retreat from, even a repudiation of, historical materialism. An example of particular relevance to my discussion is that of Jean-François Lyotard. Like many of his *confrères*, the lesson he learned from May 68 was about the imbrications of power and desire, and the need to reconcile and ultimately to transcend Marx's political economy and Freud's libidinal economy.[5] Over the next ten years Lyotard's drift away from these two thinkers accelerated, signaled in the essays collected in *Dérive à partir de Marx et de Freud*, and announced with an extreme vigor – so extreme, indeed, that it approached a brutality – in his *Économie libidinale*, where he turned each against the other and Nietzsche against both of them.[6] The result was what Dews describes as a displacement of critique by affirmation; but Dews also suggests that Lyotard's next projects were attempts to make amends for the "amoralization" of that text. Whatever changes Lyotard made, however, his hostility toward totalizing thought remained a constant:

History, like politics, seems to have need of a unique point of perspective, a place of synthesis, a head or eye enveloping the diversity of movements in the unification of a single volume: a synthesizing eye, but also an evil eye which strikes dead everything which does not enter its field of vision.[7]

This passage, which echoes Nietzsche's realization of the connections between visual metaphors and abstract thought, has a direct bearing on Harvey's project; a number of critics have drawn attention to the affinities between Harvey's visual metaphoric and his construction of metatheory.[8] Lyotard cast these sentiments in perhaps their most polemical form in *La condition postmoderne*, a book that evidently attracted Harvey's ire (and which Lyotard himself once described as the very worst of his books). Harvey did not dissent from the explanation that Lyotard put forward for the postmodern condition, which depended upon a series of changes in the technical, economic and social conditions of communication that would not have

Studies in the history and theory of capitalist urbanization, published as *Consciousness and the urban experience* and *The urbanization of capital* (Oxford: Blackwell Publishers, 1985).

[5] Vincent Descombes, "The end of time," in his *Modern French philosophy* (Cambridge: Cambridge University Press, 1980) pp. 168–90.

[6] Jean-François Lyotard, *Dérive à partir de Marx et de Freud* (Paris: Union Générale d'Editions, 1973); *idem, Économie libidinale* (Paris: Minuit, 1974). See also Steven Best and Douglas Kellner, "Lyotard and postmodern gaming," in their *Postmodern theory: Critical interrogations* (New York: Guilford Press, 1991) pp. 146–80.

[7] Jean-François Lyotard, "Futilité en revolution," in his *Rudiments paiens: Genres dissertatif* (Paris: Union Générale d'Editions, 1977) p. 164; cited in Peter Dews, *Logics of disintegration: Poststructuralist thought and the claims of critical theory* (London: Verso, 1987) p. 212.

[8] See, for example, Rosalyn Deutsche, "Boys town," *Environment and Planning D: Society and Space* 9 (1991) pp. 5–30.

been out of place in a conventionally materialist account; but he did object strenuously to Lyotard's endorsement of postmodernism as an "incredulity toward metanarratives," including those of traditional social science and historical materialism. I do not think that Lyotard was as uncritical of postmodernism as Harvey implies, and certainly some of his subsequent essays have sharpened his concerns, but in an essay published three years later and reprinted as an appendix to the English translation his opposition to totalization was, if anything, even more provocative: "Let us wage a war on totality; let us be witnesses to the unpresentable; let us activate the differences and save the honor of the name." Harvey's own diagnosis of *The condition of postmodernity* was plainly a response to these assaults on totalizing theory, and I assume that his title was intended to signal his distance from Lyotard.[9]

But this was not, I think, the only provocation. Running right through Lyotard's writings is a concerted attempt to use aesthetics "politically," as it were, to undo the closures of theory. This tactic is particularly clear in his early work, most obviously so in *Discours, figure*, but it also appears in his later writings where he often uses Kant's *Critique of judgement* and the Kantian sublime to figure a (particular) postmodern sensibility, a radical heterogeneity which cannot be represented by the categories of analytic thought. What the aesthetic and the political have in common, so Lyotard believes, is a ceaseless *experimentation* that constantly questions the foundations of aesthetics and politics. He sees them both as domains "where the absence of a general rule produces a continual questioning of, and experimentation with, rules and forms, with no ultimate resolution possible."[10] Any attempt to force a resolution, to make a closure, only confirms capitalism as a space of representation that will tolerate no differences that cannot be domesticated and brought within its own horizon of meaning. Lyotard's objective is therefore not to have socialism displace capitalism – to occupy the same space – but to deconstruct the space of representation altogether. In *Le différend* he hails Kant's final critique as a prologue to what he calls "an honorable postmodernity" and tries to develop a concept of justice that can articulate differences rather than suppress them beneath some common ("totalizing") rule or convene them within a singular space of representation.[11] Let me emphasize again that these explorations are not

---

[9] Jean-François Lyotard, *The postmodern condition: A report on knowledge* (Minneapolis: University of Minnesota Press, 1984) pp. xxiv, 82; the main text was prepared for the Conseil des Universités of the government of Québec and was published in French in 1979. For Harvey's principal objections to Lyotard, see his *Condition of postmodernity*, pp. 44–47, 52, 117.

[10] Jean-François Lyotard, *Discours, figure* (Paris: Éditions Klincksieck, 1971); David Carroll, *Paraesthetics: Foucault, Lyotard, Derrida* (New York: Methuen, 1987) p. 168. See also Bill Readings, *Introducing Lyotard: Art and politics* (New York: Routledge, 1992).

[11] See Jean-François Lyotard, *Le différend* (Paris: Éditions de Minuit, 1983); see also David Ingram, "Legitimacy and the postmodern condition: the political thought of Jean-François Lyotard," *Praxis*

bound by the conventions of social theory or moral philosophy. "Capital is not an economic and social phenomenon," Lyotard declares, but "the shadow cast by the principle of reason on human relations." This makes him particularly anxious to resist the colonization of the expressive by the cognitive:

That "cognitive" discourse has conquered hegemony over other genres, that in ordinary language the pragmatic and inter-relational aspect comes to the fore, whilst "the poetic" appears to deserve less and less attention – all these features of the contemporary language-condition cannot be understood as effects of a simple modality of exchange, i.e. the one called "capitalism" by economic and historical science.[12]

These claims rebound on Harvey's project too, and indeed they are more subversive of it than any move away from Marx. In geography, the only comparable projects that I can think of are Olsson's astonishing linguistic experiments in which the limits of scientifically constrained languages are constantly transgressed, their theoretical logics endlessly interrupted, by the irruptions of what is (in his case) a sensibility informed by a modernist avant-garde.[13] Harvey is prepared to learn from aesthetics only so long as its discourses are kept within bounds, however, and his central objection to postmodernism is precisely that it transgresses those limits by "submerging" ethics beneath aesthetics. Interestingly, his assessment of the postmodern condition is also made in (cautiously) Kantian terms.

The aesthetic responses to conditions of time-space compression are important and have been so ever since the eighteenth-century separation of scientific knowledge ["pure reason"] from moral judgement ["practical reason"] opened up a distinctive role for them. The confidence of an era can be assessed by the width of the gap between scientific and moral reasoning. In periods of confusion and uncertainty, the turn to aesthetics (of whatever form) becomes more pronounced. Since phases of time-space compression are disruptive, we can expect the turn to aesthetics and to the forces of culture as both explanations and *loci* of active struggle to be particularly acute at such moments.[14]

international 7 (1987/9) pp. 286–305. For a personal discussion of Lyotard's own *différend* with Marxism, see his "A memorial of Marxism: For Pierre Souyri," in Jean-François Lyotard, *Peregrinations: Law, form, event* (New York: Columbia University Press, 1988) pp. 45–75.

[12] *Idem, The inhuman: Reflections on time* (Stanford: Stanford University Press, 1991) p. 69; this was first published in French in 1988.

[13] See the later essays in Gunnar Olsson, *Birds in Egg/Eggs in Bird* (London: Pion, 1980) and especially his *Lines of power/Limits of language* (Minneapolis: University of Minnesota Press, 1991).

[14] Harvey, *Condition of postmodernity*, p. 327. I should add that this passage is hedged around with qualifications: Harvey does not "necessarily" accept Kant's tripartite distinctions between cognitive-instrumental rationality ("science"), moral-practical rationality ("morality") and aesthetic rationality ("art"), and neither does he concede the force of Kant's arguments about the beautiful (nor, I presume, the sublime).

This does not uniquely define the postmodern condition, of course; one of the most important discussions of the aestheticization of politics is still Benjamin's account of early twentieth-century fascism and its valorization of submission and spectacle: what Herf terms a "reactionary modernism."[15] But Harvey's objections to aestheticism are given a new intensity by the postmodern collision between aesthetics and ethics. The claim is not a fantasy of his own making, and neither is it a purely critical one. More affirmatively, for example, Vattimo argues that "the postmodern experience of truth is an aesthetic and rhetorical one," and then connects this to a "de-historicization of experience" in which "everything tends to flatten into the level of contemporaneity and simultaneity": a phrase that lucidly mirrors Harvey's own concerns about the contemporary response to time-space compression.[16]

These outline sketches not only show what Harvey thinks he is writing against (I put it like that because his assessment of Lyotard is by no means unassailable); they also indicate, more positively, what Harvey hopes to find through his engagement with an altogether different tradition of French social theory developed by Henri Lefebvre. As I will show presently, Lefebvre was closely involved in the events of May 68 and his commitment to Marxism never wavered (though it was one which he shaped in his own way); and the comparison between him and Lyotard is instructive. As will soon become obvious, Lefebvre is interested in the conjunction of political economy and libidinal economy too, though in rather different forms, and he is by no means antagonistic to Nietzsche. He also has much to say about capitalism as a space of representations. But more important is that he remains not only committed to historical materialism *but to a particularly Hegelian version of it* which is thoroughly inimical to Lyotard's postmodern politics. Lefebvre once described Hegel as "a sort of Place de l'Étoile, with a monument to politics and philosophy at its center." From that vantage point,

historical time gives birth to that space which the state occupies and rules over. . . . Time is thus solidified and fixed within the rationality immanent to space. . . .

[15] See Russell Berman, "The aestheticization of politics: Walter Benjamin on fascism and the avant-garde," in his *Modern culture and critical theory: Art, politics and the legacy of the Frankfurt School* (Madison: University of Wisconsin Press, 1989) pp. 27–41; Jeffrey Herf, *Reactionary modernism: Technology, culture and politics in Weimar and the Third Reich* (Cambridge: Cambridge University Press, 1984).

[16] Gianni Vattimo, *The end of modernity: Nihilism and hermeneutics in postmodern culture* (Cambridge, England: Polity Press, 1988) pp. 6–12. See also Richard Shusterman, " 'Ethics and aesthetics are one': Postmodernism's ethics of taste," in Gary Shapiro (ed.), *After the future: Postmodern times and places* (Albany: State University of New York Press, 1990) pp. 115–34. Even here it is difficult to see how aestheticism can announce a distinctively postmodern break from modernism. Shusterman's title – "ethics and aesthetics are one" – is a quotation from Wittgenstein's *Tractatus*. A more discriminating analysis would thus have to fasten not on aestheticism as such but on different modes of aestheticism.

What disappears is history, which is transformed from action to memory, from production to contemplation. As for time, dominated by repetition and circularity, overwhelmed by the establishment of an immobile space which is the locus and environment of realized Reason, it loses all meaning.[17]

Like the Communards confronting the Vendôme Column, Lyotard wants to demolish this monument: to interrupt and displace the reconciliations which Hegelianism constantly seeks to install. Lefebvre wants to demolish the same monument – which is why he seeks to bring Marx and Nietzsche into explosive confrontation with Hegel[18] – but he does so in a radically different way. His purpose is to prise open the sutures between "immobilized space" and "realized Reason" by bringing the production of space into human history and disclosing the social processes through which "abstract space" has been historically superimposed over "lived space." And yet in doing so there is no doubt (in my mind) that Lefebvre's ultimate objective, like that of Harvey's historico-geographical materialism, stands in the shadows of the totalizing drive of Hegelian Marxism.

Hegel's ghost

Lefebvre is a puzzle to most commentators on Western Marxism. In Perry Anderson's *Considerations on Western Marxism*, for example, he is portrayed as the lonely philosopher; read by the exiled Benjamin in Paris, he was himself "an international isolate at the close of the thirties; within France itself, his example was a solitary one." In the course of his second voyage *In the tracks of historical materialism*, Anderson discovers a topographical break, a shift in the locus of intellectual production from continental Europe to "traditionally the most backward zones" of Marxism's cultural geography, Britain and North America; there he charts "the rise of historiography to its long overdue salience within the landscape of socialist thought as a whole," the emergence of an "historically centered Marxist culture"; and yet he still records Lefebvre, like some Ancient Mariner, as "the oldest living survivor" of his previous voyage, "continuing to produce imperturbable and original work on subjects typically ignored by much of the Left." A note reveals that Anderson has in mind Lefebvre's *La production de l'espace*: but apart from this bibliographical marker-buoy, submerged at the bottom of the page, those subjects are ignored by Anderson too.[19]

---

[17] Henri Lefebvre, *The production of space* (Oxford: Blackwell Publishers, 1991) p. 21; this was originally published in France as *La production de l'espace* (Paris: Anthropos, 1974; 3rd ed., 1986).

[18] *Ibid.*, pp. 21–24.

[19] Perry Anderson, *Considerations on western Marxism* (London: Verso, 1979) p. 37; *idem*, *In the tracks of historical materialism* (London: Verso, 1983) pp. 24, 30.

But it is a significant marker nonetheless, because most Anglophone commentators on Lefebvre have fastened on his involvement in existential Marxism and his critique of Althusser; they usually pass over his interest in space and spatiality with a cursory nod or even blank incomprehension.[20] The most detailed (and certainly the most sympathetic) English-language discussion of Lefebvre's work in the 1960s and of his involvement in the debates which were raging at the time over humanism and existentialism is virtually silent about his problematic of spatiality; and yet Lefebvre had signaled his concerns long before the publication of *La production de l'espace*. Indeed, he later insisted that his interest in space and spatiality had always been latent within his work as a whole.[21] Discussions of his critical response to Althusser's reading of Marx, particularly to the essays collected in *Pour Marx* and the papers prepared for the seminar which Althusser and Balibar convened at the École Normale Supérieure in 1965 and published as *Lire le Capital*, similarly fail to elucidate the connections between Lefebvre's critique of the structural moment and his involvement in the project which formed around the journal *Espaces et sociétés* at the end of the sixties. It is now something of a commonplace to say that structuralism and structural Marxism were drawn around spatial metaphors, but their spaces were almost Platonic in their stillness. In Lefebvre's view, this "so-called philosophical thinking," as he referred to structuralism and post-structuralism more generally, was guilty of "the basic sophistry whereby the philosophico-epistemological notion of space is fetishized and the mental realm comes to envelop the social and physical one."[22] Its "spaces," "continents" and "regions" were just so many empty abstractions. Lefebvre was not referring specifically to Althusser, but this passage from *Lire le Capital* conveys exactly what he had in mind:

Not only is the economic a structured region occupying its peculiar place in the global structure of the social whole, but even in its own site, in its (relative) regional autonomy, it functions as a regional structure and as such determines its elements.[23]

---

[20] For a combination of the two, see Edith Kurzweil, "Henri Lefebvre: A Marxist against structuralism," in her *The age of structuralism* (New York: Columbia University Press, 1980) pp. 57–85.

[21] Mark Poster, *Existential marxism in postwar France: From Sartre to Althusser* (Princeton: Princeton University Press, 1975) pp. 238–60; Henri Lefebvre, "L'espace en miettes," in his *Le temps des méprises* (Paris: Stock, 1975) pp. 217–47.

[22] Lefebvre, *Production*, pp. 5–6.

[23] Louis Althusser and Étienne Balibar, *Reading Capital* (London: New Left Books, 1970) p. 180. There were, of course, several attempts to project Althusser's problematic onto the terrain of a more concrete spatial analysis: see, for example, Manuel Castells, *La question urbaine* (Paris: Maspero, 1972; 1976), which was translated into English as *The urban question: A Marxist approach* (London: Edward Arnold, 1977), and Alain Lipietz, *Le capital et son espace* (Paris: La découverte/Maspero, 1977). Spatial metaphors were of central importance to Althusser's later epistemology too, with its three great "continents" of science (mathematics, physics, history) and the disciplinary "regions" which fall within them.

What Lefebvre urged in the conceptual place of these metaphors was a turbulent *spatiality* – a way of registering, within the same figure, the articulations of "mental space" and "material space" – which he believed would restore geography to history, history to geography, and human subjects to both of them. In doing so, Lefebvre relied upon a critical appropriation of Hegel, which established a considerable distance between his project and that of Althusser.

In his otherwise illuminating discussion of French Marxism, however, Kelly simply passes over these matters. It is perfectly true that Lefebvre had been instrumental in the introduction of a Hegelian tradition into French Marxism in the 1930s; it is also true, as Kelly shows, that this prepared the ground for the postwar confrontation with Althusser. In the intervening decades Stalin had purged "official" Marxism of its Hegelianism, but when Khruschev denounced Stalin in 1956 the French Communist Party began a process of de-Stalinization that was at the same time a rehabilitation of Hegel. Althusser's central point was that the one did not entail the other: in other words, and in contradistinction to the charges leveled by Thompson and others, Althusser endorsed the repudiation of Stalinism but opposed the reinstatement of Hegelianism. He made three claims that are important to my own discussion because they run counter to Lefebvre's own proposals.[24]

1 Althusser insisted that whatever Hegelian residues might be present in Marx's early writings, the materialism that he constructed in his later ("mature") texts was not a mirror image of Hegel's idealism. Had he merely inverted Hegel, Marx would have remained trapped within the logical structure of Hegelian thought; this would have committed him to an *essentialism*, and historical materialism would have collapsed into an economism.

2 Althusser castigated Hegel as an architect of *historicism*, a model of history in which potentialities present at the beginning of the historical process are supposed to prefigure and set in motion a sequence of events whose end result is self-realization. Instead, Althusser sought to construct a Marxism that would disrupt the homogeneity of such an historiography. He proposed a foliated model of capitalist society, articulated around distinct economic, political and cultural-ideological levels, and rather than conceive their development in the "same" historical time, he assigned each level its own temporality.

3 Althusser objected to Hegel's hypostatization of human subjectivity being carried over into Marxism. This is a complicated question, but in outline Hegel distinguished modernity from previous historical formations by its

---

[24] Michael Kelly, *Modern French Marxism* (Oxford: Blackwell Publishers, 1982); see also Gregory Elliott, *Althusser: The detour of theory* (London: Verso, 1987).

solution to the problem of normativity. Constitutionally unable to appeal to myth, tradition, or religion as a source of authority, modernity established its own ground of legitimation in a self-contained subjectivity. Althusser argued that Hegel's acceptance of this supposed freedom and unity of the human subject – a bourgeois fiction, if you like – implicated him in an ideology of *humanism* that was completely antithetical to any scientific Marxism.[25]

Each of these claims bears directly not only on Lefebvre's Hegelianism, as Kelly indicates, *but also on his vision of the production and politics of space.* My own reading of Lefebvre is not an Althusserian one, and I am not suggesting that his work can be collapsed into a simple Hegelianism. But I do think that his response to these questions – essentialism, historicism, humanism – *within the problematic of spatiality* considerably clarifies his contribution to Western Marxism. The same three questions also illuminate the intersections between his project and Harvey's historico-geographical materialism, and for this reason I want to tease out three corresponding threads that will run through the discussion that follows.

Essentialism and the space of the commodity

Lefebvre, like Harvey, is drawn to Marx's writings on the commodity, and particularly to the sketches set out in the *Grundrisse*. This text consists of Marx's notebooks on political economy, which he compiled in 1857–58 and which are usually regarded as preparations for the writing of *Capital* itself. The reasons Lefebvre gives for his preference are revealing:

Whereas *Capital* stresses a homogenizing rationality, founded on the quasi-"pure" form, that of (exchange) value, the *Grundrisse* insists at all levels on difference.... Less rigour, less emphasis on logical consistency, and hence a less elaborate formalization of axiomatization – all leave the door open to more concrete themes.[26]

The contrast with Althusser's reading of Marx could not be greater, since he virtually ignored the *Grundrisse*. Several critics have concluded that he did so deliberately.

The *Grundrisse* is a protracted challenge to Althusser's periodization of historical materialism, within whose logic it is quite anomalous. Terminologically and methodologically the most Hegelian of Marx's later writings, it displays a Marx in transit

---

[25] Althusser and Balibar, *Reading Capital*; see also Ted Benton, *The rise and fall of structural Marxism: Althusser and his influence* (London: Macmillan, 1984).

[26] Lefebvre, *Production*, p. 102.

to something rather different from the Althusserian vision of *Capital*. For ... the *Grundrisse* witnesses to a reappearance of what – on Althusser's account – Marx had exorcized: the spectre of Hegel.[27]

As I have repeatedly accentuated, Harvey has never shown any interest in Althusser's work. *Social justice and the city* was sharply criticized for its violation of the protocols of structural Marxism, and since then Harvey has continued to ignore them; he certainly has no sympathy with the break Althusser sought to install between the young Marx and the mature Marx, and his own understanding of historical materialism still owes as much to the *Grundrisse* as it does to *Capital*.[28] That said, Harvey is probably more concerned with "logical consistency" and "rigor" (one of his watchwords, in fact) than Lefebvre. At least some of the marks of spatial science are still present in his work, and his writings are tighter, his political economy of space more systematic than Lefebvre's sometimes poetic figurations. But they both recognize that such a political economy cannot be confined to any Althusserian "region." As Lefebvre remarks,

The chain of commodities parallels the circuits and networks of exchange. There is a language and a world of the commodity.... Linking commodities together in virtually infinite numbers, the commodity world brings in its wake certain attitudes toward space, certain actions upon space, even a certain concept of space.[29]

This prefigures Harvey's trajectory from *The limits to capital* to *The condition of postmodernity* with precision. I would say that even before he formalized his interest in historical materialism, Harvey's discussions of "created space" respected the overdetermination of the sign-process and the multiple ways in which spaces could be coded and decoded. His geographical imagination was not limited to the space-economy, and he drew attention to the power of painting, sculpture and film to shape spatial consciousness. So even then, in the first bloom of his interest in Marx and Marxism, I doubt that Harvey would have objected to these cautionary remarks subsequently made by an art historian who was himself acutely aware of the salience of historical materialism.

It is one thing (and still necessary) to insist on the determinate weight in society of those arrangements we call economic; it is another to believe that in doing so we have poked through the texture of signs and conventions to the bedrock of

---

[27] Elliott, *Althusser*, p. 132.

[28] Doreen Massey, "Review of *Social justice and the city*," *Environment & Planning A* 6 (1974) pp. 229–35. Massey's objection was hardly surprising, Harvey has since remarked, because his objective "was to ground urban studies in Marxian political economy and that meant reading *Capital* rather than *Reading Capital*": David Harvey, "Three myths in search of a reality in urban studies," *Environment & Planning D: Society and Space* 5 (1987) pp. 367–76; the quotation is from p. 369.

[29] Lefebvre, *Production*, p. 341.

matter and action upon it. Economic life – the "economy," the economic realm, sphere, level or what-have-you – is in itself a realm of representations. How else are we to characterize money, for instance, or the commodity form, or the wage contract?[30]

Harvey would, I think, insist rather more firmly than Clark on the *centrality* of money within capitalism's spaces of representation, conceiving of it as a crucial modality of power that not only "fuses the political and the economic into a genuine political economy" but also bridges "what are otherwise so frequently seen as separate worlds of political economy and cultural production." In Harvey's view, "money, time and space all exist as concrete abstractions framing daily life," and the task of his historico-geographical materialism is thus to tease out the interconnections through which those spaces of representation are configured – in effect, to "de-naturalize" them by showing how they are socially produced through an historico-geographical process. Indeed, I take that to be the central concern of *The condition of postmodernity*.[31]

Historicism and the history of space

Lefebvre wants to elucidate the specificity of the capitalist mode of production of space, to understand how the production of space came to be saturated with the tonalities of capitalism. He attempts to do so by sketching out, in different but overlapping texts, what he eventually called "the long history of space." Running through these sketches – obviously so in *La révolution urbaine*, more obliquely in *La production de l'espace* – is a definite historicism, and I have difficulty in seeing his work as quite the "shattering attack" on the "hegemonic historicism" of mainstream Marxism that Soja makes it out to be.[32] The dialectical progression that Lefebvre describes is, I think, imposed on his account by the peculiar purpose he wants it to fulfill. Any attempt to construct a history of the present – which is how I read Lefebvre's history of space – always risks imposing an ineluctable logic on its narrative as alternatives are closed off one by one in order to arrive at the designated terminus. I put so much weight on this precarious historiography because some critics have characterized Lefebvre's approach as a highly speculative one that is cut off from the substantive force and sheer complexity of human history.

[30] T. J. Clark, *The painting of modern life: Paris in the art of Manet and his followers* (Princeton: Princeton University Press, 1984) p. 6.

[31] Harvey, *Condition of postmodernity*, p. 102; idem, "Postmodern morality plays"; see also his "Money, time, space and the city" in *Consciousness*, pp. 1–35.

[32] Henri Lefebvre, *La révolution urbaine* (Paris: Gallimard, 1970); idem, *Production*; Edward Soja, *Postmodern geographies: The reassertion of space in critical social theory* (London: Verso, 1989) p. 50.

In his early writings, for example, Castells argued that the paths that Lefebvre sought to open up in *La révolution urbaine* were washed away "in the flood of a metaphilosophy of history"; when he later warmed to Lefebvre's work, after his own break with Althusser, he still admonished him for failing to provide practical "instruments of research, given the speculative character of his philosophical perspective."[33] In fact, Lefebvre was at pains to distance himself from interpretations of this sort. Far from being an exercise in unrestrained speculation – which would have been a purely utopian project – his intention was to disclose tendencies embedded in the history of the present whose potential realization was absent from our anticipations of the future. *La révolution urbaine* was not an urban history (or an historical geography) in any conventional sense, therefore, but neither was it a predictive mysticism: It was, rather, an exercise in what Lefebvre called "transduction" or, more simply, a "reflection on the possible." Accordingly, he warned his readers not to confuse the tendency with the established fact. The urban revolution of his title lay not in the distant past, as it did for scholars preoccupied with the origins of urbanism, but in the immanent future: It prefigured what Lefebvre saw as the *possible* and *positive* constitution of "urban society," which had to be thought of not as an accomplished reality but as "a horizon, a materializing virtuality."[34] The same attempt to bring a possible future into political focus animates *La production de l'espace*, in which Lefebvre proposed a new sequence of spatialities – absolute space, historical space, abstract space – that again had to be read as the announcement of a *project* rather than the recitation of a chronology. Lefebvre said he advanced this sequence as a strategic hypothesis, pregnant with political implications, which "straddles the breach between science and utopia, reality and ideality." I will have much more to say about this in due course, but in outline, abstract space – the space of contemporary capitalism – was supposed to carry within itself "the seeds of a new kind of space": a *differential* space in which differences would be respected rather than planed into the homogeneities typical of the space of the commodity. This intersects with Lyotard's project, of course, but in Lefebvre's case it is given an historicist inflection by locating the present conjuncture (it should be remembered that he was writing in the early 1970s) on the cusp between abstract space and differential space; Lefebvre placed its bifurcations and contradictions within a long historical process whose "origins lie very far away" and whose "goal" was "still far distant." This is not a determinism; it is not shaped by any linear logic; but it is nonetheless an historicism.[35]

---

[33] Castells, *Urban question*, p. 94; *idem, The city and the grassroots: A cross-cultural theory of urban social movements* (Berkeley: University of California Press, 1983) p. 300.

[34] Lefebvre, *Révolution*, pp. 9, 12, 27, 184.

[35] *Idem, Production*, pp. 48–50, 52, 60, 409.

Harvey has also called for "the urbanization of revolution" – which may seem equally speculative to his critics – but his concerns seem to me to be shaped by a much closer imbrication of the theoretical and the practical than appears in Lefebvre's writings. Although most of his work no longer involves the quarrying of archival materials – at least, not in the narrow sense that many historians would still understand their craft – Harvey nevertheless deals in abstractions that are much more "concrete" than Lefebvre's. By "concrete abstractions" Harvey means concepts that are available to us in everyday speech and drawn upon in the conduct of everyday life. Their identification – and the elucidation of their taken-for-granted "embeddedness" in social practices – is an essential moment in the process of theory construction that, according to Harvey, involves representing and integrating concrete abstractions within a coherent system of thought. To disclose those integrations, he continues, requires the construction of a parallel series of abstract concepts that are usually *not* part of our everyday understandings, concepts that are qualitatively different from concrete abstractions like money but having the power to reveal what is happening beneath the surface of the commonsense facticity of the world about us.[36] This depth model evidently breaks with the theoreticism of structural Marxism, but it is not (quite) what Lefebvre had in mind when he described his own project as a metaphilosophy. His intention was to open up philosophy to both the "real" and the "possible": to provide concepts at such a high level of generality that they could transcend the fragmentary sciences in order to grasp the mosaic – the absent, virtual, open totality – which he took to be immanent in their shards. His purpose was not so much to establish foundations, so he said, as to provide an "orientation" that would "open up paths" and disclose a "horizon" that was concealed by conventional categories of thought.[37] Harvey's project, by contrast, *is* intended to expose the foundations of capitalism as a mode of production: to identify its "laws of motion," its "underlying systema-ticities," its stubborn and persistent "logic." To be sure, there is nothing inevitable about the outcomes (though they are shaped and structured in various determinate ways) and Harvey offers fewer glimpses of a possible future than Lefebvre; by and large, his work fastens on an explanatory-diagnostic moment and rarely broaches the anticipatory-utopian moment that animates Lefebvre's project. But when Harvey does sketch a possible scenario, I think it significant that his vision of the future is the negative of Lefebvre's positive image of a prospective "urban society":

---

[36] David Harvey, "Introduction" to his *The urban experience* (Oxford: Blackwell Publishers, 1989) pp. 1–16; this is a new introduction to a selection of essays from his *Studies in the history and theory of capitalist urbanization*. See also David Harvey and Allen Scott, "The practice of human geography: Theory and empirical specificity in the transition from Fordism to flexible accumulation," in Bill McMillan (ed.), *Remodelling geography* (Oxford: Blackwell Publishers, 1989) pp. 217–29.

[37] Lefebvre, *Révolution*, pp. 89–90, 92; *idem*, *Production*, p. 368.

The urbanization of capital on a global scale charts a path toward a total but also violently unstable urbanization of civil society. The urbanization of consciousness intoxicates and befuddles us with fetishisms, rendering us powerless to understand let alone intervene coherently in that trajectory. The urbanization of capital and of consciousness threatens a transition to barbarism in the midst of a rhetoric of self-realization.[38]

I suggest that *The condition of postmodernity* attempts to open a series of windows on exactly that transition, which Harvey sees as immanent within the contemporary conjunction of flexible accumulation and postmodernism: a *noir* scenario that he conjures up most dramatically in his account of a dystopian Los Angeles in Ridley Scott's *Blade Runner*. This ghastly prospect is made visible by reading the signs of a new conjuncture that has been installed since Lefebvre offered his utopian reading of urban society.[39]

Humanism and the politics of space

Lefebvre's problematic of spatiality has sometimes been seen as the deployment of an unrestrained humanism. When Castells was writing under the sign of structural Marxism, he treated the thesis set out in *La révolution urbaine* as a "spontaneism": "It indicates that space, like the whole of society, is the ever-original work of that freedom of creation that is the attribute of Man, and the spontaneous expression of his desire."[40] But this is far too glib. Lefebvre's activism and his expansive sense of human agency undoubtedly explain part of Harvey's attraction to (and affection for) his work, but Lefebvre's interventions cannot be reduced to a liberal humanism.

There are two points I wish to make. First, Lefebvre's is a *critical* humanism that resides in the sphere of "everyday life." This term has a specific meaning and purpose within Lefebvre's corpus; it is a double-sided concept which serves to some degree as an elaboration of Marx's supposedly incomplete thematization of alienation. On the one side, "everyday life" parallels what the young Lukács had previously described as *Alltäglichkeit*: "the 'trivial life' of the human being, indistinguishable from the world of objects – the dreary, mechanical and repetitive unfolding of the everyday."[41] But

---

[38] Harvey, *Consciousness*, p. 276.

[39] Harvey, *Condition of postmodernity*, pp. 308–14. Harvey dates the "sea-change" to 1972–73.

[40] Manuel Castells, "From urban society to urban revolution," in his *Urban question*, pp. 86–95; the quotation is from p. 92. For a more affirmative reading of Lefebvre's humanism, together with a critical commentary on Castells, see Peter Saunders, "The urban as ideology," in his *Social theory and the urban question* (London: Hutchinson, 1986; 2nd ed.) pp. 152–82.

[41] Michel Trebitsch, "Preface" to Henri Lefebvre, *Critique of everyday life*. Vol. 1: *Introduction* (London: Verso, 1991) pp. ix–xxviii; the quotation is from p. xvii. This volume was originally published as *Critique de la vie quotidienne* (Paris: Grasset, 1947) and was republished a decade later with a new foreword (Paris: L'Arche Éditeur, 1958).

Lefebvre, who constructed his concept independently of Lukács, gave his version of it an historical and a geographical inflection. Translated into the problematic of the production of space, "everyday life" is the domain that is enframed, constrained, and colonized by the space of the commodity and the territory of the state; it is a product of modernity, and it is within its sphere, for example, that the routinized spatial practices that are so relentlessly diagrammed by time-geography take place. On the other side, however, "everyday life" also contains traces and memories of spatial practices that were untouched by modernity's estrangements. Lefebvre has a romantic, almost nostalgic vision of those traditional practices, seeing in them "an innocent life," now "impoverished and humiliated." But even in this vestigial form he thinks they hold out the promise of recovering a lost creativity and the prospect of achieving a common self-realization. Lefebvre glimpses the figure of what he calls "the *total man*" – what Marx called "man in the whole wealth of his being, man richly and deeply mentally alive" – as "but a figure on a distant horizon, beyond our present vision." Again, its realization requires both an historical and a geographical sensibility, and Lefebvre uses a passage in which Marc Bloch reads the agrarian history of France out of its present-day landscapes to provide a moving reflection on the disclosure and, indeed, the redemptive power of the everyday. For Bloch, the commonplace and the everyday were never beneath the attention of the serious historian; quite the reverse, and Lefebvre has the great *Annaliste* open his readers' eyes to the richness and significance of "humble, familiar, everyday objects":

Our search for the human takes us too far, too "deep," we seek it in the clouds or in mysteries, whereas it is waiting for us, besieging us on all sides.... "The familiar is not necessarily the known," said Hegel. Let us go farther and say that it is in the most familiar things that the unknown – not the mysterious – is at its richest, and that this rich content of life is still beyond our empty, darkling consciousness, inhabited as it is by imposters, and gorged with the forms of Pure Reason, with myths and their illusory poetry.[42]

Putting these two sides together, one might say that for Lefebvre everyday life "is both a parody of lost plenitude and the last remaining vestige of that plenitude," so that his *critique* of everyday life is "at once a rejection of the inauthentic and the alienated, and an unearthing of the human which still lies buried therein."[43]

Second, this "unearthing" turns Lefebvre's project into an archaeology of sorts: but it is certainly no spontaneism. It is true that from time to time Lefebvre celebrated the creative energies released during the "festivals"

---

[42] Lefebvre, *Critique*, pp. 66, 132, 210.
[43] Trebitsch, "Preface," p. xxiv.

of the Paris Commune, the Liberation, and May 68, all moments in which, so he said, the everyday was dislocated, social space was reclaimed, and the possibility of revolutionary self-realization seemed at hand. In 1967 he had in fact called for (and perhaps even anticipated) a permanent cultural revolution:

The Festival rediscovered and magnified by overcoming the conflict between everyday life and festivity and enabling these terms to harmonize in and through urban society ... [is] the final clause of the revolutionary plan.[44]

But in every case, including the events of May 68, those hopes were dashed. According to Lefebvre, such fleeting moments of opportunity were stilled by the production of ever more abstract spatialities that reached ever further into the intimacies of everyday life. Capitalism "has subordinated everything to its own operations by extending itself to space as a whole," he wrote in the wake of May 68, and in doing so had established itself at the level of – had in effect colonized – the everyday.[45] In the face of such a powerful and pervasive series of encroachments, he continued, spontaneism, festivals, peripheral mobilizations would never be enough to turn the tide. Instead, it was necessary to come to terms with the modalities through which space had been *centralized* – "fixed in a political centrality" – and at the same time "distributed into peripheries that are hierarchized in relation to the center." This had a particular salience in France, but the point was not peculiar to the Fifth Republic: Lefebvre claimed that it was through the processes of centralization, fragmentation and hierarchization that colonization was generalized. "Around the centers there are nothing but subjected, exploited and dependent spaces: neo-colonial spaces."[46] While Lefebvre could accept the importance of Foucault's ascending analytics of power, therefore, he also insisted that its preoccupation with the periphery resulted in a parochialism:

The issue of prisons, of psychiatric hospitals and antipsychiatry, of various converging repressions, has a considerable importance in the critique of power. And yet this tactic, which concentrates on the peripheries and only on the peripheries, simply ends up with a lot of pinprick operations which are separated from each other in time and space.[47]

---

[44] Henri Lefebvre, *Everyday life in the modern world* (New York: Harper and Row, 1971) pp. 32–33, 37, 206; this was first published as *La vie quotidienne dans le monde moderne* (Paris: Gallimard, 1968). It was, in part, a summary of the first two volumes of Lefebvre's critique of everyday life – *Critique*, and *Fondements d'une sociologie de la quotidienneté* (Paris: L'Arche Éditeur, 1961) – and an anticipation of the third volume: *De la modernité au modernisme* (*Pour une métaphilosophie du quotidien*) (Paris: L'Arche Éditeur, 1981).

[45] Henri Lefebvre, *La survie du capitalisme* (Paris: Éditions Anthropos, 1973), translated as *The survival of capitalism* (London: Allison and Busby, 1976) pp. 38, 58.

[46] *Ibid.*, pp. 84–85.

[47] *Ibid.*, p. 116. Although Lefebvre does not mention Foucault by name, he must surely have in mind Foucault's work with the *Groupe d'Information sur les Prisons* and the *Groupe Information Santé*.

Their collective force would always be weakened by their inability to pierce the center and centrality.

This comprehensively missed Foucault's point, which was that modernity was characterized by an anonymous and dispersed system of power that could not be brought within the "descending" analytics of the traditional juridico-political model, but Lefebvre's response was to propose an altogether different "archaeology" and "genealogy." His archaeology took the form of a spatial architectonics, which was to uncover the prehistory of social space by establishing the biomorphic, more or less primal relations between the human being and space. It would focus on the body, on its implication in and constitution of a "sensory-sensual space" that had never been erased by the production of subsequent spatialities.

> In space, what came earlier continues to underpin what follows. The preconditions of social space have their own particular way of enduring and remaining actual within that space.... The task of architectonics is to describe, analyze and explain this persistence, which is often invoked in the metaphorical shorthand of strata, periods, sedimentary layers and so on.[48]

This is a matter of conceptual excavation, of bringing into consciousness human capacities and creativities which form the matrix of an authentic everyday life, and it does not involve any appeal to the deep structures of structuralism. In Lefebvre's view, these "layers" are buried beneath a constellation of power-knowledge that has installed and sanctioned a distinctive regime of truth that he calls (ironically) "true space." The task of his genealogy is thus to provide a history of space that will show how this constellation of power-knowledge – this supposedly "true space" – is an artificial construction that privileges mental space, marginalizes social space and compromises lived experience. The hyperinflation of mental space installs as a cultural dominant what Lefebvre calls "the space of reductions, of force and repression, of manipulation and co-optation, the destroyer of nature and the body." This "true space" is the discursive space of philosophy and epistemology, and of the sciences of abstract space (including urban and regional planning and spatial science), in which dissident constructs that might be subversive of modernity are suppressed or swallowed up by the hegemonic representations of abstract space.[49] Lefebvre characterizes abstract space in different ways, often as a geometric-visual-phallic

---

Perhaps he was also mindful of the reissue (and recasting) of his *Folie et déraison: histoire de la folie à l'âge classique* (Paris: Gallimard, 1972) and *Naissance de la clinique: Une archéologie du regard médical* (Paris: Presses Universitaires de France, 1972). Although *Surveillir et punir* was not published until 1975, Foucault gave courses on the prison and the penal system at the Collège de France in 1971 and 1972.

[48] Lefebvre, *Production*, p. 228.

[49] *Ibid.*, pp. 354, 398.

space, but what is particularly important for my present purposes is that
it is supposed to be a space from which previous histories have been erased
and in which, in consequence, what Lefebvre calls "the time needed for
living" – a sense of historicity, of human finitude – "eludes the logic of
visualization and spatialization":

With the advent of modernity time has vanished from social space.... Our time,
this most essential part of lived experience, this greatest good of all goods, is no
longer visible to us, no longer intelligible.[50]

Against the mystifications and deceptions of this "true space," as a way
of recovering and reengaging with the authenticity and plenitude of everyday life – a language that would have been quite foreign to Foucault –
Lefebvre poses "the truth *of* space" (my emphasis). This will emerge, so
he claims, through the *articulation* of "mental space" and "social space." He
proposes a strategy that will restore a proper balance between the two by
recognizing what they have in common *but also respecting the differences between
them.* Lefebvre seeks to achieve this conceptual doubling with the figure of
centrality, through which both spaces pass in ceaseless animation. Here
Lefebvre trades on both the metaphorical and material implications of
"centrality" in order to rework and reinscribe the concept of totality within
his problematic of the production of space. In his early writings he had
argued that the self-realization that was immanent within the critique of
everyday life required a dialectical concept of totality: otherwise it would
indeed issue in a spontaneism, and the project would collapse into either
"radical contingency and pure relativism" or "a traditional theology."[51] In
La production de l'espace, however, he substituted the figure of centrality.

The notion of centrality replaces the notion of totality, repositioning it, relativizing
it, and rendering it dialectical. Any centrality, once established, is destined to suffer
dispersal, to dissolve or explode from the effects of saturation, attrition, outside
aggressions and so on. This means that the "real" can never become completely
fixed, that it is constantly in a state of mobilization. It also means that a general
figure (that of the center and of "decentering") is in play which leaves room for
both repetition and difference, for both time and juxtaposition.[52]

This is not quite the break with Hegel that it seems to be, for it is in this
figure that Hegel meets Marx and Nietzsche: "the revolutionary road of
the human and the heroic road of the superhuman meet at the crossroads
of space."[53]

---

[50] *Ibid.*, pp. 95–96.

[51] Lefebvre, *Critique*, p. 77.

[52] Lefebvre, *Production*, pp. 299–300, 399.

[53] *Ibid.*, p. 400.

I am not sure what Harvey would make of these claims. Certainly, his critique of Foucault is much the same as Lefebvre's. Although he is critical of the arguments of many humanistic geographers, he is also dubious about any radical decentering of the subject that would devalue everyday experience. He portrays himself as a "restless analyst" whose "rigorous theory building" requires a continuous dialogue with the everyday. Like Baudelaire's *flâneur*, he says, he wanders the city streets, watching and listening in order to remain open "to the unpredictable collision of experience and imagination that always lies at the root of new insight."[54] I doubt that this distances Harvey from Foucault as much as he would like to think, but it is certainly a far cry from the antihumanism of Althusser (who was in fact Foucault's instructor at the École Normale Superieure). In any event, Harvey is wary of both humanism and antihumanism: One of the most striking features of the debates over humanist Marxism and structural Marxism – those famous "arguments within English Marxism" – is that Harvey stayed out of them. He is also openly skeptical of the attempts to construct a post-Marxist theory of structuration in which these oppositions might be transcended, since he clearly believes that Marx's own writings remain the most important source for clarifying the relations between human agency and social structure: "People struggle to make history (and geography), not just as they please and not under conditions of their own choosing."[55]

For much the same set of reasons, Harvey also agrees with Lefebvre that Foucault fails to grasp the systematicity of capitalism, and he objects to the way in which Foucault's "micro-politics of power" leads him to the conclusion that its sites and practices are independent of and irreducible to "any systematic strategy of class domination." This theoretical error (as Harvey sees it) has practical consequences inasmuch as Foucault's political involvements were therefore "dedicated to the cultivation and enhancement of *localized* resistance to the institutions, techniques and discourses of organized repression" (my emphasis).

Foucault evidently believed that it was only through such a multifaceted and pluralistic attack upon localized practices of repression that any global challenge to capitalism might be mounted without replicating all the multiple repressions of capitalism in a new form.... Yet it leaves open, particularly so in the deliberate rejection of any holistic theory of capitalism, the question of the path whereby

---

[54] Harvey, *Consciousness*, pp. xiv–xv. Harvey goes on to recall Lefebvre's characterization of the city as "the place of the unexpected" and suggests that this is why they are both drawn back to it again and again.

[55] The original phrasing, of which this is a reformulation, appears in the opening paragraphs of Marx's *The Eighteenth Brumaire of Louis Bonaparte*. Harvey says he sees "absolutely no point" in discarding Marx's version for the "looser and vaguer language" of structuration theory: David Harvey, "Three myths in search of a reality in urban studies," *Environment and Planning D: Society and Space* 5 (1987) pp. 367–76; the quotation is from pp. 367–68.

such localized struggles might add up to a progressive, rather than regressive, attack upon the central forms of capitalist exploitation and repression.[56]

Harvey prefers a more inclusive ("global") theorization of capitalism as a mode of production, and most particularly of its mode of production of space, and when he convenes modernism and postmodernism as opposing tendencies within capitalism conceived as an historico-geographical totality, his strategy has something in common with Lefebvre's figure of centrality:

Within this matrix of internal relations, there is never one fixed configuration, but a swaying back and forth between centralization and decentralization, between authority and deconstruction, between hierarchy and anarchy, between permanence and flexibility.... The sharp categorical distinction between modernism and post-modernism disappears, to be replaced by an examination of the flux of internal relations within capitalism as a whole.[57]

It is not difficult to see the shadows of Hegel as well as Marx in all this, but what of Nietzsche? While it is true that Harvey rejects both Nietzsche's aestheticism and his nihilism, he nevertheless carries over some of Nietzsche's most extraordinary images of modernity – its "creative destruction and destructive creation," its spaces crashing in "a sea of forces, flowing and rushing together" – to suggest "the continuity of the condition of fragmentation, ephemerality, discontinuity and chaotic change" between modernism and postmodernism. But he is also very clear that "behind all the ferment" it is possible to discern "some simple generative principles that shape an immense diversity of outcomes": and here Marx is still his model.[58]

A history of space

Let me now try to make all this more concrete by juxtaposing Lefebvre's history of space with Harvey's historico-geographical materialism. Two overlapping narratives run through Lefebvre's history of space. The first is a more or less positive account that charts the horizon of "urban society" in which the project of self-realization is supposed to be accomplished. The second is a much more negative account that traces a "de-corporealization" of space in the West. I will describe these trajectories in turn, and in doing so Lefebvre will necessarily guide the discussion; but on the way I will also identify some of the parallels and disjunctures between his project and

---

[56] Harvey, *Condition of postmodernity*, pp. 45–46.
[57] *Ibid.*, pp. 339–42.
[58] *Ibid.*, pp. 15–17, 44, 274, 344.

Harvey's. My main focus will be on what I have been calling the history of the present: with identifying processes in the past which in some way bear on the constitution of the present. This implies nothing about "evolutionary" or "revolutionary" models of historical eventuation, still less a commitment to historicism on my part, but I do think it necessary to insist on the importance of a history of the production of space for any critical *politics* of space.

But what "history"? Or, rather, *whose* history? Both Lefebvre and Harvey reconstruct what is often described as "the rise of the West," a set-piece of historical hubris, and since the same focus will become intrusively apparent as my discussion proceeds, I should say at once that Lefebvre does not intend this as a celebration:

The West is ... responsible for what Hegel calls the power of the negative, for violence, terror and permanent aggression directed against life. It has generalized and globalized violence – and forged the global level itself through that violence. Space as locus of production, as itself product and production, is both the weapon and the sign of this struggle.[59]

And yet this reveals Lefebvre's (I think characteristic) ambivalence toward Hegel. On the one side, he suggests that the weakness of both Hegelianism and its critique is that neither "clearly perceived the violence at the core of the accumulation process" and, by extension, its central role "in the production of a politico-economic space" that was "the birthplace and cradle of the modern state." The state was thus not the site of Reason, as it was for Hegel; or, if it was, then it was a Reason corroded by violence, stained by "the very lifeblood of this space." On the other side, however, Lefebvre comes close to repeating Hegel's hypostatization of Europe as the subject of History. What systematic violence produced, Lefebvre declares, inside the process of capital accumulation, was "Western Europe – *the space of history*, of accumulation, of investment, and the basis of imperialism by means of which the economic sphere would eventually come into its own" (my emphasis).[60] This does not deny the intrinsic violence of imperialism, of course, but it locates the architects of "history" at a self-proclaimed center.

Lefebvre is on firmer ground in his discussion of gender and sexuality (certainly firmer than Harvey). He makes it very clear that, in his view, this violence is embodied in a masculine will-to-power. With the first stirrings of capitalism, he writes:

---

[59] Lefebvre, *Production*, p. 109.
[60] *Ibid.*, pp. 276–79. On Hegel and Europe as the subject of History, see Robert Young, *White mythologies: Writing history and the West* (London: Routledge, 1990) especially pp. 1–20.

The prestigious Phallus, symbol of power and fecundity, forces its way into view by becoming erect. In the space to come, where the eye would usurp so many privileges, it would fall to the Phallus to receive or produce them.... A space in which this eye laid hold of whatever served its purposes would also be a space of force, of violence, of power restrained by nothing but the limitations of its means. This was to be the space of the triune God, the space of kings, no longer the space of cryptic signs but rather the space of the written word and the rule of history. The space, too, of military violence – and hence a *masculine* space.[61]

The gendering of space is thus not confined to premodern societies. On the contrary, Lefebvre insists that it achieves a new intensity within the quintessentially *abstract* spaces of modernity in which "phallic brutality" precisely does *not* remain abstract because "it is the brutality of political power, of the means of constraint."[62] If this seems unduly metaphorical, too far removed from the oppressions and brutalities experienced by women in their daily lives, or insufficiently nuanced, prone to an essentialism, one should remember that Lefebvre was writing these lines long before the development of a feminist geography, and before feminist theory of any sort had a substantial presence within the academy. The same cannot be said of Harvey and *The condition of postmodernity*.

Urban society and the contradictions of urbanism

Lefebvre's most optimistic account of the passage to an "urban society" is set out in *La révolution urbaine*. My summary will center on this text, but I will need to make reference to some of his other writings which will, in places, add a darker coloration to the discussion. For the most part, I will refrain from critical comment since my purpose is not to provide a detailed historiography, but on occasion it will be necessary for me to qualify or amplify some of Lefebvre's claims. In some respects, his original sketch foreshadows the genealogy of the urban condition set out in *Social justice and the city*, but Harvey did not encounter Lefebvre's writings until he had completed the essays in his own volume and it was only when he came to compose his concluding reflections that he realized the continuities and differences between them.

In outline, Lefebvre's schema is straightforward enough. It describes a progressive distancing from "nature" through the production of a "second nature" which reaches its fullest expression in "the urban." In a later passage Lefebvre provides a particularly economical formulation of the process that I have summarized in figure 27.

---

[61] Lefebvre, *Production*, p. 262.
[62] *Ibid.*, p. 287.

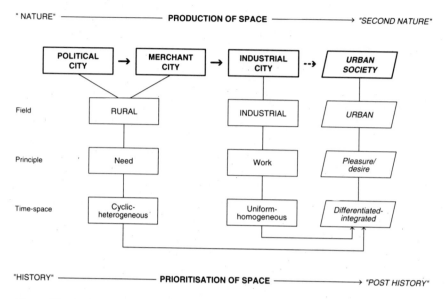

Figure 27 Lefebvre's urban revolution.

Nature, destroyed as such, has already had to be reconstructed at another level, the level of "second nature," i.e. the town and the urban. . . . The town, antinature or nonnature and yet second nature, heralds the future world, the world of the generalized urban. Nature, as the sum of particularities which are external to each other and dispersed in space, dies. It gives way to produced space, to the urban.[63]

The origins of urbanism are complex and various, but Lefebvre cuts into the dense fabric of argument with the formation of the *political city*. This is something of a portmanteau category, including cities as different as Babylon, Athens, and Rome, but it derives its cogency from the way in which the modalities of power of all these cities were inscribed in writing and enforced through politico-military means. The development of writing, no matter how circumscribed and fragile, permitted both the concentration and the extension of power over spans of time and space far greater than those available to purely oral cultures. "The city began as writing on the ground; writing prescribed and signified the city's power, its administrative capacity, its political and military sway; writing imposed the law of the city on village and countryside."[64]

---

[63] Lefebvre, *Survival*, p. 15. Although he has a number of criticisms of Lefebvre, Smith constructs a concept of socially produced space – of spatiality – in a similar way: see Neil Smith, *Uneven development: Nature, capital and the production of space* (Oxford: Blackwell Publishers, 1984; 2nd ed., 1990).

[64] Lefebvre, *Everyday life*, p. 154. This is a highly simplified sketch, of course, and the superimposition of the power of the city was never complete and often relied on more direct displays of physical force than the written word.

The city dominated subject populations that were scattered across multiple countrysides and extracted a surplus from their local ecologies. In places, Lefebvre attributes an essential unity to town and country. With the emergence of the classical city-state, for example, he proposes that

from this moment on, the vastness of pre-existing space appears to come under the thrall of a divine order. At the same time the town seems to gather in everything which surrounds it, including the natural and the divine, and the earth's evil and good forces. As image of the universe (*imago mundi*) urban space is reflected in the rural space that it possesses and indeed in a sense contains.[65]

This may approximate the ideal of the Greek *polis*, which Lefebvre describes as an absolute space "at once indistinguishably mental and social" in which is inscribed "the rational unity of Logos and Cosmos"; but such a model is less cohesive than it appears. In many other agrarian societies the town remained an outsider, whose shadows fell periodically across local communities without ever becoming a substantial, permanent presence in the day-to-day lives of their inhabitants. Lefebvre fractures the image of unity himself when he draws attention to its gendering; thus the free (male) citizen may have "envisioned the order of the world as spatially embodied and portrayed in *his* city" (my emphasis), but "women, servants, slaves, children – all had their own times, their own spaces."[66] Although Lefebvre is talking specifically about the cities of the Roman Empire, the same could be said of the Greek *polis*. In fact, any fuller account of the classical ideal of democracy and its embodiment in the physical space of the city, in the sites and institutions of the so-called "public sphere," would have to unravel its multiple imbrications in the geographies of patriarchy, slavery, and Athenian imperialism. But Rome does have a particular importance for Lefebvre's argument, because the production of its imperial space was achieved "not under the sign of the Logos," so he claims, "but under the sign of the Law." What he has in mind, I take it, is the constitution of a juridico-political space that not only guaranteed and memorialized imperial power – exemplified most vividly by the construction of ever more elaborate fora in the heart of Rome by successive emperors – but also and more importantly framed and codified the private ownership of land (the *villa* and the *latifundia*). This property principle was vital, Lefebvre argues, because "by dominating space – and this in the literal sense of subjecting it to its *dominion* – [it] put an end to the mere contemplation of nature, of the Cosmos or of the world, and pointed the way towards the mastery which transforms instead of simply interpreting."[67]

[65] Lefebvre, *Production*, p. 235.
[66] *Ibid.*, pp. 238, 240, 244.
[67] *Ibid.*, pp. 246, 252–53.

By these means, "during the supposed emptiness of the late imperial or early medieval period," a new space was established, which was necessary "for the subsequent development of a historical space, a space of accumulation."[68]

But the most far-reaching form of property was not bound to the soil or fixed in place. In most agrarian societies, Lefebvre argues, nontributary exchange and long-distance trade were present only in the interstices – in "heterotopias" – and were not part of the dominant spatial structure. The political city was thus threatened by the market, by merchants, and most of all by the insurgent power of the most mobile of all property forms: money. "Money and commodities, still *in statu nascendi*, were destined to bring with them not only a 'culture' but also a space." In Lefebvre's view, the triumph of the market was most decisive in the West. "The space that emerged in Western Europe in the twelfth century, gradually extending its sway over France, England, Holland and Italy, was the space of accumulation – its birthplace and cradle."[69]

The details are complicated but Lefebvre is only interested in the general outcome: "the medieval revolution [that] brought commerce inside the town and lodged it at the center of a transformed urban space." With the rise of the *merchant city* in fourteenth-century Europe, so he says, the town was no longer superimposed on the countryside. A new symbiosis was forged in which local ecologies were knit together in extra-local and supra-personal webs of exchange. These new spatial divisions of labor were articulated by the market economy, mediated by the institutions of the state and often codified in a new imaginary. The discovery of linear perspective imposed a systematic and harmonious order upon representations of town and country alike. This was a remarkable moment, Lefebvre believes, which offered a fleeting glimpse of transcendence of deep-seated contradictions between town and country:

---

[68] *Ibid.*, p. 253; see also *idem*, *La pensée marxiste et la ville* (Paris: Casterman, 1972) pp. 37–41, 53–57. The passages from antiquity to feudalism were also crossed by the immense, translocal span of the Catholic Church:

> One single institution spanned the whole transition from Antiquity to the Middle Ages in essential continuity: the Christian Church. It was, indeed, the main, frail aqueduct across which the cultural reservoirs of the Classical World now passed to the new universe of feudal Europe, where literacy had become clerical.... No other dynamic transition from one mode of production to another reveals the same splay in superstructural development: equally, none other contains a comparable spanning institution.

See Perry Anderson, *Passages from antiquity to feudalism* (London: New Left Books, 1974) pp. 131, 137. Lefebvre comments on the cultural significance of the Church in his *Production*, pp. 253–62, where (as I will show presently) it plays an important part in his account of the decorporealization of space; but the significance of the Church in medieval Europe was as much juridico-political as it was cultural or, indeed, disciplinary.

[69] Lefebvre, *Production*, pp. 263, 265.

At this unique moment, the town appeared to found a history having its own inherent meaning and goal – its own "finality," at once immanent and transcendent, at once earthly (in that the town fed its citizens) and celestial (in that the image of the City of God was supplied by Rome, city of cities). Together with its territory, the Renaissance town perceived itself as a harmonious whole, as an organic mediation between earth and heaven.[70]

This is, I think, unduly fanciful: or, at least, in many places the ideology of harmony and mediation was threadbare, left in tatters by the violence that raged within many Renaissance towns and the wars that broke out between them. In any event, Lefebvre argues that the rough equipoise between town and country tilted during the sixteenth century, and that the balance shifted decisively in favor of the towns. As "the time and space of commodities and merchants gained the ascendancy," he remarks, so time and space were themselves "urbanized."

Time – the time appropriate to the production of exchangeable goods, to their transport, delivery and sale, to payment and to the placing of capital – now served to measure space. But it was space which regulated time, because the movement of merchandise, of money and of nascent capital, presupposed places of production, boats and carts for transport, ports, storehouses, banks and money-brokers.[71]

The merchant city consolidated its powers and liberties, both at home and abroad, but in the struggle between feudalism and capitalism its victory was short-lived. Merchant capital yielded to industrial capital and in the process, Lefebvre argues, circuits of capital accumulation transcended the limits of the city; surplus value could be realized far beyond the locality and the region, and successive "spirals of spatial abstraction" established the nation-state and the world-economy as the main arenas of power and profit. Their contours were drawn in and from Western Europe, and at the center of that space of accumulation, thickly populated and "piled high with the rich spoils of years of rapine and pillage," "industry would pitch its tent."[72]

For much of the eighteenth century the locus of large-scale industrialization was ex-urban, but gradually its rhythms and its routines penetrated every sphere of civil society and reached into the heart of the city itself. In the course of the nineteenth century the *industrial city* came to be represented as a "second nature," partly through the incorporation of a domesticated and manicured nature within the city (parks and gardens) and partly through the heightened domination of nature within the process of industrial production (water-wheels and steam-engines). "Theoretically na-

---

[70] *Ibid.*, p. 271.

[71] *Ibid.*, pp. 277–78; see also Lefebvre, *Révolution*, pp. 17–22.

[72] Lefebvre, *Production*, pp. 269, 275–76.

ture was distanced," Lefebvre observes, "but the signs of nature and the natural multiplied, replacing and supplanting real 'nature.'" As the nineteenth century turned into the twentieth, the accelerating pace of industrialization was maintained by the propulsive tension between two dialectically related processes – which Lefebvre variously terms implosion and explosion; localization and globalization; and differentiation and integration – that carried urbanism beyond the industrial city into the "critical zone": the threshold of *urban society*.[73]

This skeleton account is unexceptional; its outlines trace those of more conventional urban histories, and its conceptual anatomy resembles other (more recent) bodies of social theory. Even Lefebvre's idealization of the unity of the classical city-state and the late medieval, early Renaissance city is not unusual: It repeats the humanist sympathies of Lewis Mumford's master-work, *The city in history*, which also "opens with a city that was, symbolically, a world" and "closes with a world that has become, in many practical aspects, a city."[74] What is distinctive about *La révolution urbaine*, I think, is its systematization. As figure 27 shows, in Lefebvre's original account this sequence of urban forms is described as a succession of "fields": the rural, the industrial, and the (generalized) urban. These fields are not defined in uniquely morphological terms but include experiences and perceptions, times and spaces, images and concepts, languages and rationalities, theories and practices. This is an extraordinary jumble, but each of the fields is dominated by a different principle and demarcated by a distinctive representation of time and space.[75]

All social practices supposedly articulate three basic principles – need, work, and pleasure – but these assume different forms in different societies and their importance varies systematically over time and space. Thus the rural is animated by what Lefebvre sees as the overwhelming imperative to satisfy "need" – however that may be socially defined – and to wrest the means of subsistence from a recalcitrant and volatile nature; there is no aspect of social life that is set apart from this elemental priority. As systems of industrial production come to dominate nature in new and intensified ways, however, so the industrial constitutes "work" as a separate sphere of social life with its own internal logic and places a premium on the pursuit of productivity (and profit) and hence on the enhanced appropriation of nature. The realization of the urban promises to transcend these basal concerns and to usher in what is sometimes called a "post-scarcity" society in which for the first time in history human beings will be free to realize themselves in what Marx once celebrated as "the whole wealth of

---

[73] Lefebvre, *Révolution*, pp. 23–34, 38–40.

[74] Lewis Mumford, *The city in history* (London: Penguin Books, 1987) p. 7; this was first published in 1961.

[75] Lefebvre, *Révolution*, p. 41.

[their] being." Lefebvre usually calls this principle *jouissance* and sometimes *désir*, but the English translations – "pleasure" or "desire" – are only pallid substitutes for the active, creative, and often sexual connotations of the French.[76]

Conceptions of time and space parallel these progressions. The rural is identified with the cyclical and the heterogeneous, whose rhythms and differences are embedded in the "natural economy" and transposed into ritual and myth, whereas the industrial is identified with the uniform and the homogeneous, whose regularities and disciplines are imposed by a series of instrumental rationalities which override the cadences of the seasonal calendar and the particularities of local ecology. The urban revolution then incorporates and transcends these representations to produce a time-space that is *simultaneously* differentiated and integrated: what Lefebvre calls "the differential."[77] In consequence, he wonders, "perhaps we are entering... into a period which is no longer that of history, where the homogeneous struggles against the heterogeneous"; a period, too, in which human beings are freed from the furrows of the previous fields – the determinisms and constraints of need and work – and are able to express the free impulses of *jouissance* and *désir*. If these oppositions are indeed being transcended within the contemporary then, in this sense, the realization of urban society promises to inaugurate what Lefebvre calls "a post-history."[78]

I do not propose to comment on the details of Lefebvre's scheme but two general riders are necessary. First, the sequence of urban forms is recognizably the same as that identified by Harvey in his early essays on urbanism, but the processes that shape them are conspicuously different. Lefebvre's sequence of urban forms does, I think, rest on a succession of modes of production (although he defines them in a somewhat looser sense than Marx), whereas Harvey's early discussion of urbanism depends on a series of modes of exchange: in particular, on a transition from reciprocity through redistribution to market exchange. It is only in his later writings that Harvey elects to build directly on Lefebvre's account. He distinguishes the mobilization of surpluses in the merchant city, the production of surpluses in the industrial city, and then – in a major elaboration of Lefebvre's thesis – the absorption of surpluses in the Keynesian city and the competition for surpluses in the post-Keynesian city.[79]

[76] *Ibid.*, pp. 46–47.

[77] *Ibid.*, pp. 47–57; cf. *idem, Le manifeste différentialiste* (Paris: Gallimard, 1971).

[78] *Ibid.*, p. 58; cf. *idem, La fin de l'histoire: épilégomènes* (Paris: Ed. Minuit, 1970). This is a very particular meaning of *post-histoire*, and (as Lefebvre himself makes clear) it is scarcely Hegelian in content: Lefebvre is acutely critical of the hypostatization of the modern state. For a discussion see André Vachet, "*De la fin de l'histoire à l'analyse différentielle: La révolution urbaine,*" *Dialogue* 11 (1972) pp. 400–419.

[79] Cf. Harvey, *Social justice*, pp. 195–284 and *idem, Urbanization*, pp. 191–226. I describe these last two categories as an elaboration of Lefebvre's thesis simply because they engage more directly than is usual in Harvey's work with the state, and that has come to be one of Lefebvre's major concerns.

Second, the trajectory of Lefebvre's project evidently requires him to move the city to the center of society until it *becomes* society, but this obscures many other important structural features of premodern social systems which cannot be reduced to the configuration of town-country relations. The scheme is, at bottom, an historicism. This is most obvious in the dialectical progression of the three fields and their conceptions of time and space – a sequence of thesis, antithesis, and synthesis which Hegel explicitly cautioned against – but it also becomes intrusive in the closing stages of the scenario. There Lefebvre represents industrialization and urbanization as a double movement in the course of which the second term in the couplet ineluctably supplants the first. His argument here is frankly impressionistic – as one might perhaps expect of an exercise in transduction, Lefebvre paints with bold brush-strokes and little attention to detail – but his claim (and in many respects his thesis) hinges on the emergence of two distinctive circuits of accumulation within capitalism. Marx was naturally preoccupied with the first of these, the circuit of industrial capital, since he was living through the convulsions of nineteenth-century industrial capitalism. To be sure, his Victorian contemporaries were fascinated by what Robert Vaughan called their "age of great cities," but the generalization of the urban (in Lefebvre's sense) was then barely on the horizon. As the configuration of capitalism changed, however, the outlines of other circuits became clearer. Lefebvre draws attention to the existence of a second circuit of "speculative capital," directed toward investment in the built environment, which he claims assumes a particular importance in the twentieth century. His argument is that during a crisis there is a pervasive tendency for capital to switch from the first circuit into the second. Extrapolating:

As the proportion of global surplus value formed and realized in industry diminishes, so the proportion formed and realized in speculation and in the construction of the built environment grows. The second circuit supplants the first. From being incidental, it becomes essential.[80]

To the extent that this occurs, then Lefebvre asserts that the principal contradiction of capitalism will no longer be located between town and country but within the urban itself.[81] It is this thesis, I suggest, that has provided the mainspring for much of Harvey's subsequent work.

Speculations and crises

At first, Harvey was openly skeptical about Lefebvre's thesis. Although urbanism and property speculation were intimately related to each other,

[80] Lefebvre, *Révolution*, p. 212.
[81] *Ibid.*, p. 225.

he could not accept that the circuit of speculative capital had replaced the circuit of industrial capital.[82] But that was in 1972–1973. The date is significant because the "sea-change" that Harvey subsequently detected within the regime of capital accumulation – the transition from Fordism to flexible accumulation – supposedly had its origins in the same period.[83] At the time, however, his misgivings were both theoretical and empirical, but he moved quickly to set them aside.

Theoretically, he objected, Lefebvre had provided no mechanism that would account for the switching of capital from one circuit to another. Harvey soon worked out a partial explanation centering on the ability of property speculators and developers to exact "class-monopoly rent." Ricardo and the classical economists had assumed that absolute rent could only be levied on an island where all resources were in use and where an absolute scarcity existed "in nature," so to speak, but Harvey argued that urbanization creates its own "islands" through the *social* production of space: partly through the imposition of a formal grid of property relations and zoning laws but partly – and more importantly – through "the informally structured absolute spaces of [property] submarkets." The crucial point for Harvey's analysis was the existence of interdependencies between these submarkets, because this immediately generates the possibility of powerful multiplier effects as rents derived from one submarket are realized in another. In Baltimore, for example, Harvey showed that capital was being steadily withdrawn from the inner-city housing market and switched into suburban real estate. The circuits between the two submarkets were put in place by finance capital and the switches between them were being pulled by a series of financial institutions. On this basis Harvey could now argue that the emergence of a "finance form of capitalism" – in response to "the inherent contradictions in the competitive industrial form" – "treats money as a 'thing-in-itself' and thereby constantly tends to undermine the production of value in pursuit of the form rather than the substance of wealth." Insofar as the operation of finance capital is cut loose from the production of value and takes flight on the wings of speculation and realization, then Harvey thought it was now possible to provide "a comprehensible inner logic for Lefebvre's 'ensemble of transformations' through which industrial society comes to be superceded by urban society."[84]

Empirically, however, Harvey had considered Lefebvre's thesis to be at best "premature." This was Lefebvre's view too, of course, and I have already emphasized that he envisioned urban society as a "possible object" rather than an accomplished reality. In some considerable degree his thesis

---

[82] Harvey, *Social justice*, p. 313.

[83] Harvey, *Condition of postmodernity*, p. 188.

[84] Harvey, *Urbanization*, pp. 81–82, 88. The original essay, "Class-monopoly rent, finance capital and the urban revolution," was published in 1974.

was a response to the events of 1968, but as the 1970s advanced so that horizon of possibility came much closer. Far from being premature, Lefebvre's thesis now seemed to be remarkably prescient. Harvey accepted that there were indeed "serious grounds for challenging the adequacy of the urban-rural dichotomy . . . as a primary form of contradiction within capitalism."

Focusing on this dichotomy may be useful in seeking to understand social formations that arise in the transition to capitalism. . . . But in a purely capitalist mode of production . . . this form of expression of the division of labor . . . disappears within a general concern for geographical specialization in the division of labor.[85]

The complex spatial divisions of labor that were emerging within advanced capitalism – the vast and intricate mosaic of uneven development – entailed, as a necessary condition of their existence, what Lefebvre referred to as the generalization of the urban and what Harvey now called "the urbanization of capital." Where he had once asked how the laws of motion of the capitalist mode of production were written into the urban process, Harvey now refined (if not quite reversed) the question: "how does capital become urbanized?" The reason for the change was simple. It had become clear to him that:

Capital flow presupposes tight temporal and spatial coordinations in the midst of increasing separation and fragmentation. It is impossible to imagine such a material process without the production of some kind of urbanization as a "rational landscape" within which the accumulation of capital can proceed.[86]

And Harvey was convinced that the contemporary landscape of accumulation, unstable, volatile and deeply fractured, was the preserve of a capricious finance capital whose flows subjected the space-economy to more or less continuous transformation.

Urbanism has . . . been transformed from an expression of the production needs of the industrialist to an expression of the controlled power of finance capital, backed by the power of the state, over the totality of the production process. . . . Concomitantly, the urban realm becomes the locus for the controlled reproduction of the social relations of capitalism.[87]

These ideas are enlarged in *The limits to capital* where Harvey develops a detailed discussion of the emergence of finance capital. He begins with

---

[85] *Ibid.*, pp. 14–15.
[86] *Ibid.*, pp. 185, 190.
[87] *Ibid.*, p. 88. That last sentence dovetails with one of Lefebvre's central concerns (although Harvey does not address it directly).

what he calls a "first-cut" theory of crisis, which depends upon Marx's dissection of the instability of capitalist production and in particular of its chronic tendency to produce both a surplus of capital (overaccumulation) and a surplus of labor power (a strategic moment in devaluation). Harvey then deepens this analysis by making a "second cut" to show how these surpluses can be absorbed through new forms of circulation and in particular through the construction of new financial and monetary arrangements which are themselves structurally implicated in financial and monetary crises. This enables him to expose the connections between long-term crises of capital accumulation and the increasing socialization of capital. State intervention is of considerable importance in securing these connections, as Lefebvre's later writings have also made plain, and this prompts Harvey to move beyond a consideration of the purely temporal dynamics of capital accumulation and circulation. Accordingly, he wires the territoriality of the state to a final, "third-cut" theory of crisis that examines the possibility of providing a "spatial fix" for the convulsions of the capitalist economy. This scenario requires the surpluses of capital and labor power to be inscribed in the built environment or absorbed through the formation of new geopolitical configurations. But Harvey concludes that these can only ever be temporary solutions. They buy time through the production of space, but they cannot resolve the ever-present tension between "fixity and motion" – between the construction and deconstruction of successive landscapes of accumulation – which is inherent in the circulation of capital and which has been so very considerably aggravated by the hegemony and hypermobility of finance capital.[88]

Harvey builds still further on these basal propositions in *The condition of postmodernity*, where finance capital holds the key to the instabilities of flexible accumulation: or, at any rate, where flexible accumulation is supposed to look "more to finance capital as its coordinating power than did Fordism."[89] But Harvey now locks these instabilities into the contemporary crisis of representation. He begins by considering changes in the mode of representation of value.

---

[88] Harvey, *Limits*, pp. 316–29, 398, 422–31. This is a highly general account of the emergence of finance capital that pays little attention to what Lefebvre saw as one of its most significant features: its *uneven* character. "The fusion of industrial and bank capital was accomplished differently in Japan, the United States and France": Henri Lefebvre, *L'irruption: de Nanterre au sommet* (Paris: Anthropos, 1968), translated as *The explosion: Marxism and the French Revolution* (New York: Monthly Review Press, 1969) p. 43. This is a commentary on the events of May 1968 written while Lefebvre was teaching at Nanterre. The terminus of Lefebvre's argument about finance capital is much the same as Harvey's. Ten years later he too suggested that "the contradiction between capital bound to a territory, controlled and directed by the state" and "nonterritorialized" capital was becoming steadily more obvious and likely to lead to open conflict: *idem, De l'état. Tome 4: Les contradictions de l'état moderne* (Paris: UGE, 1978) p. 232.

[89] Harvey, *Condition of postmodernity*, p. 164.

None of the shifts in the experience of time and space would make the sense or have the impact they do without a radical shift in the manner in which value gets represented as money.... Since 1973, money has been "dematerialized" in the sense that it no longer has a formal or tangible link to precious metals ... or ... to any other tangible commodity. Nor does it rely exclusively upon productive activity within a particular space. The world has come to rely, for the first time in its history, upon immaterial forms of money.... The effect is to render the spaces that underpin the determination of value as unstable as value itself.[90]

He then connects the "dematerialization" of money to a more general crisis of representation: more general because it takes place within a world where virtually every sphere of social life has been commodified and, as a direct result of the hegemony of finance capital, radically destabilized; a world, too, where the hypermobility of finance capital has set in motion waves of time-space compression of unprecedented intensity. It is this conjuncture – what Harvey calls this "highly problematic intersection of money, time and space" – that constitutes (in the most active of senses) the *condition* of postmodernity.

The breakdown of money as a secure means of representing value has itself created a crisis of representation in advanced capitalism. It has also been reinforced by, and added its very considerable weight to, the problems of time-space compression . ... The rapidity with which currency markets fluctuate across the world's spaces, the extraordinary power of money capital flow in what is now a global stock and financial market, and the volatility of what the purchasing power of money might represent, define, as it were, a high point of that highly problematic intersection of money, time, and space as interlocking elements of social power in the political economy of postmodernity.[91]

Lefebvre regards that intersection as highly problematic too, though he rarely bothers to distinguish "postmodernity" as such. Even in his latest account he insists that it is wrong to oppose modernity to postmodernity. The project of modernity has not stood still, of course, but Lefebvre is adamant that its historical progression cannot be comprehended through an oppositional grid of this sort.[92] Whatever this most recent phase of modernity is to be called, however, Lefebvre argues that it is the product of a series of transformations that have wrenched capitalism out of the frame that Marx had constructed in *Capital*. "The power of money, so forcefully denounced by Marx, has supposedly become stronger since his time," he remarks, and yet "the relation of money to political power is not clearly visible."[93] In order to reveal that connective imperative – to

---

[90] *Ibid.*, pp. 296–97.
[91] *Ibid.*, p. 298.
[92] Lefebvre, *De la modernité*, p. 52.
[93] Lefebvre, *Explosion*, p. 16.

prise apart Harvey's "interlocking elements of social power" – Lefebvre examines the crisis of representation and, like Harvey, concludes that *abstraction is a leitmotiv of capitalist modernity.* The victory of abstract signs is announced over the rubble of concrete referentials, he declares, and most insistently through the floating signs of finance capital: what he calls *les jeux d'écriture bancaires et monétaires.*[94]

The violence of abstraction and the decorporealization of space

In his later writings, and most particularly in *La production de l'espace* and a fragment from *De l'état*, Lefebvre fills in some of the details missing from his first outline and reworks some of its framing assumptions.[95] He pays particular attention to the production of a sequence of spaces that culminates in the hegemony of an abstract space (figure 28). Although this sequence is connected to a parallel succession of modes of production, Lefebvre insists that the one cannot be collapsed directly into the other.

Each mode of production has its space; but the characteristics of space do not amount to the general characteristics of the mode of production; medieval symbolism is defined neither by the rents paid by the peasants to the lords of the manor nor by the relations between town and country. The reduction of the aesthetic, of the social and the mental to the economic was a disastrous error.[96]

The schema is largely descriptive and Lefebvre says very little about the struggles that were involved in the transitions he identifies. It would not be difficult to show that these struggles had their symbolic as well as their material dimensions – many of them involved competing conceptions of space, demarcations of territory, and grids of property; they often mobilized spatial representations to advance their claims and counterclaims – and it should be possible to accord them a central place in a more analytical schema. But Lefebvre's primary purpose is still to provide a history of the present: All he seeks to do here is to sketch a history of space within the horizon of his own metaphilosophy. Seen like this, one of the most significant aspects of this new typology is its emphasis on what I want to call the *decorporealization* of space. This strikes me as one of the central achievements

---

[94] Lefebvre, *De la modernité*, p. 52.

[95] Lefebvre, *Production*; idem, *De l'état*, pp. 282–92.

[96] *Ibid.*, p. 292. Cf. Mario Rui Martins, "The theory of social space in the work of Henri Lefebvre," in Ray Forrest, Jeff Henderson, and Peter Williams (eds.), *Urban political economy and social theory: Critical essays in urban studies* (Aldershot: Gower, 1982) pp. 160–85, who regards this as an auto-critique: as an explicit recognition of "the major error of his previous approach" (p. 176). I am not so sure.

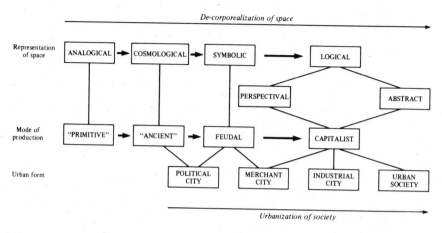

Figure 28 The production of abstract space.

of Lefebvre's problematic: to establish an essential connection between the history of the body and the history of space and, more particularly, to comprehend that shift "from the space of the body to the body-in-space" whose momentum carries over into "the spiriting-away or *scotomization* of the body."[97] Since Lefebvre made these suggestions, there have been several attempts to elucidate the history of the human body, but these have been independent of his work and have rarely attempted to connect bodies and spaces in any systematic fashion.[98] For this reason, but also because Lefebvre's account is often impressionistic and sometimes frankly enigmatic, I will flesh out some of his more important propositions. In doing so, however, I make no claims for completeness, and neither do I mean to imply that these remarks are unimpeachable.

As figure 28 shows, Lefebvre argues that the representation of space in "primitive" societies is dominated by *analogical space*. The physical form of the dwelling and the village itself typically represent and reproduce a divine body that is itself a projection, often in distorted or exaggerated form, of the human body. One of the most celebrated demonstrations is Bourdieu's analysis of the Kabyle house in Algeria. Its internal space is at once corporealized and gendered:

[97] Lefebvre, *Production*, pp. 196, 201.

[98] The exception, ironically, is Foucault, especially in his *Discipline and punish: The birth of the prison* (London: Allen Lane, 1975) and several of the essays included in his *Power/knowledge: Selected interviews and other writings, 1972–1977* (Brighton: Harvester, 1980). For a discussion, see Felix Driver, "Power, space and the body: A critical assessment of Foucault's *Discipline and punish*," *Environment and Planning D: Society and Space* 3 (1985) pp. 425–46. For a sample of other, more recent histories of the human body, see *Fragments for a history of the human body*, which appeared in three special issues of *Zone*, 3–5 (1989), edited by Michael Feher with Romona Neddaff and Nadia Tazi.

The low, dark part of the house is...opposed to the upper part as the female to the male. Not only does the division of labor between the sexes (based on the same principle of division as the organization of space) give the woman responsibility for most of the objects belonging to the dark part of the house, the carrying of water, wood and manure, for instance; but the opposition between the upper part and the lower part reproduces, within the internal space of the house, the opposition between the inside and the outside, between female space – the house and its garden, the place *par excellence* of *haram*, i.e. the sacred and forbidden – and male space. The lower part of the house is the place of the most intimate secret within the world of intimacy, that is, the place of all that pertains to sexuality and procreation.[99]

These distinctions are inscribed within the very materials of which the house is made; the master beam, extending the protection of the male part of the house to the female, is identified with the master of the house, and it is made to rest upon a forked trunk that is identified with his wife. These oppositions are then repeated and reversed beyond the house, for it turns out that the orientation of the house is a rotation of the orientation of external space: that "the inner space is but the inverted image or mirror reflection of male space." And, as Bourdieu subsequently makes clear, this is not a purely "logical" operation – or, rather, its logic is not abstract, the product of a series of exclusively mental operations, but resides in – is literally, physically embodied in – the human body. Terms like "displacement" and "rotation" must be given their practical senses as movements of the body in order to understand the integration of body space with cosmic space.[100]

Analogical space often inscribes ideologies and practices of gender, though there is considerable variation from one society to another and by no means all of them cast women in a subordinate position. Tuan universalizes these spatialities by claiming that all vocabularies of spatial organization have terms in common that "are ultimately derived from the structure and values of the human body." He also insists on their particular salience in "traditional" societies – like the Kabyle – in which "the earth is the human body writ large."[101] This speaks directly to Lefebvre's spatial architectonics, but what Tuan fails to make clear, unlike Bourdieu and Lefebvre, is that these conceptions of space are not "ordering frameworks"

---

[99] Pierre Bourdieu, "The Kabyle house, or the world reversed," in his *Algeria 1960* (Cambridge: Cambridge University Press, 1979), pp. 133–53; the quotation is from pp. 137–38. See also *idem*, *The logic of practice* (Cambridge, England: Polity Press, 1990) pp. 271–83.

[100] Bourdieu, "Kabyle house," pp. 139, 153; *idem*, *Outline of a theory of practice* (Cambridge: Cambridge University Press, 1977) pp. 90–91, 117.

[101] Yi-Fu Tuan, "Body, personal relations and spatial values," in his *Space and place: The perspective of experience* (London: Edward Arnold, 1977) pp. 34–50; the quotations are from pp. 37, 89. See also Daphne Spain, *Gendered spaces* (Chapel Hill: University of North Carolina Press, 1992) pp. 17–64.

but rather *ways of being-in-the-world.* This is considerably clarified in Mitchell's rereading of the Kabyle house. He argues that the kind of order that is disclosed there departs from modern Western conceptions of order in four particulars:

First, it is not concerned with order as a framework, whose lines would bring into existence a neutral space in terms of which things were to be organized. Second, such ordering does not work by determining a fixed boundary between an inner world and its outside. Third, it is not concerned with an order set up in terms of an isolated subject, who would confront the world as his or her object. Nor, finally, is it concerned with meaning as a problem for this individual subject of fixing the relations between the world and its plan of representation; or with truth as the certainty of such representation.[102]

This matters so much, to Mitchell and I think to Lefebvre too, because "what we [modern] inhabitants of the world-as-exhibition ordinarily take for granted as the elements of any order – framework, interior, subject, object and an unambiguous meaning or truth – remain problematized and at play in the ordering of the Kabyle house."[103]

This attentiveness to the world is extended and elaborated still further in the ancient world, where the built form of the political city inscribes a *cosmological space* whose elements and configurations are supposed to express the architecture of the cosmos. One of the richest elucidations of this close parallelism between the mathematically expressible regimes of the heavens (the movements of "heavenly bodies") and the biophysically determined rhythms of life on earth is Paul Wheatley's *The pivot of the four quarters.* Wheatley's primary purpose is to treat the ancient Chinese city as what he calls a "cosmo-magical symbol," but for my present purposes the term is an awkward one: While the cosmo-magical force of the ancient city is unquestionable, I will need to suggest that its *symbolic* character is considerably more ambiguous.

There is no difficulty in accepting Wheatley's initial premise: religions believing that human order was brought into being at the creation of the world tend to dramatize the cosmogony "by reproducing on earth a reduced version of the cosmos" and, in particular, by sacralizing space around a central point at which the creative force of the gods is focused. The basic elements of this "astrobiological" system (as Berthelot called it) are the geomantic identification of the sacred site; the delimitation of the sacred space, usually by the construction of a wall; the orientation of the processional axis along the celestial meridian; and the raising of an *omphallos* at the center, the point of transition at which divine power enters the world

---

[102] Timothy Mitchell, *Colonizing Egypt* (Cambridge: Cambridge University Press, 1988) pp. 50–51; Bourdieu, *Outline,* p. 116.
[103] *Ibid.,* p. 51.

and flows out through four cardinally oriented gates, and the *axis mundi* around which the kingdom revolves.[104] These practices establish and guarantee the reality of cosmological space or, as Eliade puts it, "reality" is here a function of the imitation of a celestial archetype. The delimitation of sacred space thus "provided the framework within which could be conducted the rituals necessary to ensure that intimate harmony between the macrocosmos and the microcosmos without which there could be no prosperity in the world of men."[105] It usually *was* a world of men, of course, and while Wheatley says nothing about the procreative significance of the *omphallos* raised at the sacred center other scholars are less cautious. Thus Kostof describes the stepped *ziggurat* in Mesopotamia "thrust[ing] upward like a solid prayer," "marrying in its shape the dark cave below and the dome of heaven above," and, indeed, as a bed "for the pure virgin whom the male god would have chosen for himself and whose union with him would bring about, for one more year, fertility and abundance in the land."[106] There were significant variations from place to place and Wheatley describes the ways in which these principles and parallels were refracted through different cultural lenses. But he concludes that the structural regularities that underwrote them were visible not so much in their architectural configurations or the particulars of their spatial forms but rather through "the manner in which the whole assemblage that we designate a city was believed to *function*."[107] This sense of functionality or, better, of practical implication is, so Lefebvre claims, a characteristic of absolute space more generally:

There is a sense in which the existence of absolute space is purely mental and hence "imaginary." In another sense, however, it also has a social existence and hence a specific and powerful "reality." The "mental" is "realized" in a chain of "social" activities, because in the temple, in the city, in monuments and palaces the imaginary is transformed into the real.[108]

Expressed like this, I think it is possible to grasp what Lefebvre intends analogical and cosmological space to have in common, and thus what

---

[104] Paul Wheatley, *The pivot of the four quarters: A preliminary inquiry into the origins and character of the ancient Chinese city* (Edinburgh: Edinburgh University Press, 1971) pp. 417–68; see also Robert Sack, *Conceptions of space in social thought: A geographic perspective* (London: Macmillan, 1980) pp. 144–64.

[105] Wheatley, *Pivot*, p. 418. For a detailed illustration of the ways in which social practices were embedded in and constituted through cosmological space, see James Duncan, *The city as text: The politics of landscape interpretation in the Kandyan kingdom* (Cambridge: Cambridge University Press, 1990) pp. 119–53.

[106] Spiro Kostof, *A history of architecture: Settings and rituals* (Oxford: Oxford University Press, 1985) pp. 57–58.

[107] Wheatley, *Pivot*, p. 451; my emphasis.

[108] Lefebvre, *Production*, p. 251.

differentiates them from the symbolic space that is supposed to succeed them: namely, iconicity. The term derives from Peirce, who proposed a distinction between the naturalness of iconic signs and the arbitrariness of symbols. It has to be said that the distinction is far from satisfactory or secure. Our recognition of the resemblance between a sign and its object is often based on cultural conventions, so that most semioticians would now accept that iconicity is at best a complex and heterogeneous property.[109] But I think Mitchell captures the essence of what Lefebvre has in mind in the following passage:

There is nothing symbolic in this world. Gall is not associated with wormwood because it symbolizes bitterness. It occurs itself as the trace of bitterness. The grain does not represent fertility, and therefore the woman. It is itself fertile, and duplicates in itself the swelling of a pregnant woman's belly. Neither the grain nor the woman is merely a sign signifying the other and neither, it follows, has the status of the original, the "real" referent or meaning of which the other would be merely the sign. These associations, in consequence, should not be explained in terms of any symbolic or cultural "code," the separate realm to which we imagine such signs to belong.[110]

One might say the same of the pivot of the four quarters: Its geometry was not abstract but existential and, to repeat, "reality" was a *function* of the imitation of a celestial archetype.

Lefebvre argues that it was not until the emergence of feudalism in Western Europe that a properly *symbolic space* was constituted, and even then its representations of the church in particular maintained close affinities with corporeal imagery through the crucified body of Christ.[111] Christianity set great store by its tombs, shrines, and relics, and Lefebvre refers to this ritualization and solemnization of death as the consecration of a "cryptic space" – a play on words working equally well in French and English – which was the subterranean locus of absolute space, of the "world," during the decline of the Roman Empire and far beyond. But in the twelfth century a spectacular inversion occurred, and the urban landscape turned this space on its head:

What do the great cathedrals say? They assert an inversion of space as compared with previous religious structures. They concentrate the diffuse meaning of space onto the medieval town. They "decrypt" in a vigorous (perhaps more than a rigorous) sense of the word: they are an emancipation from the crypt

---

[109] John Lyons, *Semantics* (Cambridge: Cambridge University Press, 1977) p. 102.

[110] Mitchell, *Colonizing Egypt*, p. 61. The echoes of Foucault are unmistakable: see Michel Foucault, *Les mots et les choses* (Paris: Gallimard, 1966) translated as *The order of things* (London: Tavistock, 1970) pp. 17–30.

[111] Lefebvre, *De l'état*, p. 284.

and from cryptic space. The new space did not merely "decipher" the old, for, in deciphering it, it surmounted it; by freeing itself it achieved illumination and elevation.[112]

"Illumination and elevation": the pinions of the Gothic cathedral. Within its precincts, whose towers soared into the vastness of space and whose aisles were pierced by shafts of light, medieval worshippers were presented with a symbol of the kingdom of God on earth. Its modulated proportions and sequences, sanctioned by Augustinian aesthetics, moving ever upward through a series of horizontal levels, were intended to lead the mind from the world of appearances to the contemplation of the divine order and its transcendent unity; similarly, light was conceived as the form that all things have in common, "the simple that imparts unity to all," and seen as an intimation of the perfection of the world to come.[113] As Lefebvre knows very well, however, this symbolic space was not merely a mental space but also a fully social space, suffused with the tonalities of earthly power. The Gothic idiom began in the royal abbey of St. Denis in the Ile de France, where Charlemagne had been crowned and where the Capetian kings were buried. Its first original constructions were all cathedrals in the royal domain – Chartres, Amiens, Reims, Bourges – and its early diffusion coincided with the extension of royal power. The filiations between Church and Crown in these buildings were associative rather than formal; Lefebvre declares that the façades of the cathedrals rose "in affirmation of prestige" and that their purpose "was to trumpet the associated authorities of Church, King and city to the crowds flocking towards the porch." The same lesson could be read from "the verticality and political arrogance of the towers," and I think more directly from the stained glass windows: At Chartres, for example, the royal household and the noble houses of the Ile de France were all memorialized.[114] In sum, then, Lefebvre believes that symbolic space was constituted through the advance of a "visual logic" (the term is Panofsky's) "in collusion on the one hand with abstraction, with geometry and logic, and on the other with authority."[115]

Lefebvre describes the advance of visualization in a language of menace. It is at once "an intense onslaught" and "a threatening gambit," whose

---

[112] Lefebvre, *Production*, pp. 256–57.

[113] Here I follow Otto von Simson, *The Gothic cathedral: Origins of Gothic architecture and the medieval concept of order* (Princeton: Princeton University Press, 1988). Lefebvre himself uses Erwin Panofksy, *Gothic architecture and scholasticism* (New York: New American Library, 1976; first published in 1951) but only as a foil to contrast his own views with what he takes to be Panofsky's "idealism and spiritualism."

[114] Kostof, *History of architecture*, pp. 329–30, 339–40; Lefebvre, *Production*, p. 261. Political involvements were confined neither to the Gothic nor to France, of course, and cathedrals and churches throughout medieval and postmedieval Europe were studded with signs of political power and patronage.

[115] Lefebvre, *Production*, p. 261.

victim is the human body. Lefebvre argues that this strategy of visualization achieved a decisive victory with the installation of a *logical space* that had its origins, so he believes, in the discovery of linear perspective during the European Renaissance. The production of *perspectival space* involved the construction of symmetrical visual pyramids, the apex of one the "vanishing point" in the painting, the apex of the other the (single) eye of the painter or viewer. The geometricization of knowledge that this implies was focal to Renaissance humanism as a whole, and in one sense, as Cosgrove has shown in a series of marvelously suggestive essays, "at the center of Renaissance space, the space produced by perspective, was the human individual." But this was no humanism that Lefebvre could endorse, for it was a thoroughly bourgeois humanism in which, as Cosgrove also shows, "visual space is rendered the *property* of the individual detached observer, from whose divine location it is a dependent, *appropriated* object."[116] I will return to those two emphases in a moment; what matters for now is that, in quite another sense, the human body is supremely irrelevant to the production of this perspectival space. For across its geometric screens the emotional investment of the viewer is nullified: "The moment [of] erotic projection in vision – what St Augustine had anxiously condemned as 'ocular desire' – was lost as the bodies of the painter and the viewer were forgotten in the name of an allegedly disincarnated, absolute eye."[117] The Janus face of perspectival space, simultaneously installing and diminishing the human being, is central to Lefebvre's argument and he invokes it in this passage:

Space remained symbolic of the body and of the universe, while at the same time becoming measured and visual. The transformation of space towards visualization and the visual is a phenomenon of the utmost importance.... [Perspectival space] recaptures nature by measuring it and subordinating it to the exigencies of society, under the domination of the eye and no longer of the body as a whole.[118]

---

[116] Denis Cosgrove, "Prospect, perspective and the evolution of the landscape idea," *Transactions of the Institute of British Geographers* 1 (1985) pp. 45–62; the quotations are from pp. 48, 51 (my emphasis). See also *idem*, "The geometry of landscape: Practical and speculative arts in sixteenth-century Venetian land territories," in Denis Cosgrove and Stephen Daniels (eds.), *The iconography of landscape* (Cambridge: Cambridge University Press, 1988) pp. 254–76; *idem*, "Historical considerations on humanism, historical materialism, and geography," in Audrey Kobayashi and Suzanne Mackenzie (eds.), *Remaking human geography* (Boston: Unwin Hyman, 1989) p. 189–226.

[117] Martin Jay, "Scopic regimes of modernity," in Scott Lash and Jonathan Friedmann (eds.), *Modernity and identity* (Oxford: Blackwell Publishers, 1992) pp. 178–95; the quotation is from p. 181.

[118] Lefebvre, *De l'état*, p. 287. "Nature" is here recaptured at several removes, through a process of abstraction, and not by physically sensuous activity. In much the same way, human figures scarcely appear in the ideal townscapes of the late fifteenth-century school of Pierro della Francesca; and according to Cosgrove they have no need to, "Prospect," pp. 49–50, precisely because "the measure of man" is written into "the measured architectural façades and proportioned spaces of the city, an intellectual measure rather than sensuous human life."

In deference to the classical, geometrically inspired conception of artistic space and the projective techniques used by Ptolemy to map the curved surface of the earth on a flat plane, Edgerton speaks of the Renaissance *re*-discovery of linear perspective, but other scholars are unequivocal: linear perspective was invented by Filippo Brunelleschi around 1413. A range of techniques from classical and medieval science no doubt provided a background, a working horizon, but it seems unlikely that any of them were the source of Brunelleschi's ideas.[119] He demonstrated the basic principles himself but, as Lefebvre notes, it was his friend Leon Battista Alberti who elaborated and systematized them in what proved to be an extremely authoritative treatise, *Della Pittura*, which was published in 1435. For most of the fifteenth century, in consequence, art and aesthetics were no longer preoccupied with the transcendence of experience *but with the representation of visual experience itself.* Every effort was made to cultivate what Kearney calls the illusion of reality: the impression that what is depicted on the canvas is taking place before the spectator's eyes. To Leonardo, the painting of the *Quattrocento* was "the art of offering the eye perfect simulacra of natural objects."[120] And, as Leonardo's own astonishing range of accomplishments makes clear, the production of this perspectival space was more than an aesthetic accomplishment: "At the same time as the mathematically inspired concept of perspectival space began to appear in art, the physical image of the world came to be conceived as ordered in accordance with geometric principles."[121]

The "worlding" of perspectival space was closely bound up with the growth of commercial and banking capital in the city-states of Renaissance Italy, and I have already implied that linear perspective turned the visual field into a commodity: that space was rendered as the "property" of the spectator who "appropriated" it. Indeed, Berger suggests that Alberti's metaphor of the canvas as a "transparent window" could be replaced by that of a safe set into the wall "in which the visible has been deposited."[122] But Lefebvre is reluctant to explain its installation by economic and social conditions alone. Visual techniques of estimation and measurement are clearly useful to a mercantile society, and Cosgrove in fact suggests that the principles that underlay the theory of perspective "were the everyday skills of the urban merchant."[123] But this is hardly a sufficient explanation

---

[119] Samuel Edgerton, *The Renaissance rediscovery of linear perspective* (New York: Harper & Row, 1976); Martin Kemp, *The science of art: Optical themes in western art from Brunelleschi to Seurat* (New Haven: Yale University Press, 1990).

[120] Richard Kearney, *The wake of imagination* (London: Hutchinson, 1988) p. 136.

[121] Joan Gadol, *Leon Battista Alberti* (Chicago: University of Chicago Press, 1969) pp. 21–22, 103–104.

[122] John Berger, *Ways of seeing* (London: BBC Publications, 1972) p. 109. For a brutal view of the relations between linear perspective and mercantile capitalism, see Leonard Goldstein, *The social and cultural roots of linear perspective* (Minneapolis: MEP Publications, 1988).

[123] Cosgrove, "Prospect," p. 51; Michael Baxandall, *Painting and experience in fifteenth-century Italy* (Oxford: Clarendon Press, 1974).

since, as Kemp points out, there were other societies in Europe, particularly in the Netherlands, in which mercantile skills were cultivated with equal necessity *but without the corresponding development of a "measured" art.*[124] In fact, there is a complex cultural geography of perspective that radiates into far wider constellations of power and knowledge: perspective was not only a "visual ideology," to use Cosgrove's term, but also what Foucault would have called a technology of power. The same geometrical techniques were used by surveyors to map private estates, by cartographers to map political boundaries, and by engineers to calculate the trajectories of cannon.[125] Lefebvre suggests that together they constituted a "code of space," a language common to both theory and practice, that was repeated across the whole field of representation and across the landscape of the city itself. "It is probably the only time in the history of space that there was a single code across all levels of the hierarchy – the room, the building, the group of buildings, the quarter, the town, and its insertion into the surrounding space."[126]

Within this urban landscape, perspective and façade were interlaced in an intricate ideology.

Perspective established the line of façades and organized the decorations, designs and mouldings that covered their surfaces. It also drew on the alignment of façades to create its horizons and vanishing-points.[127]

But façade implies a distinction between what is shown and what is not shown, and for this reason Lefebvre suggests that it carries the implication of asymmetry, even of duplicity. What he has in mind, I think, is not so much Goffman's distinction between front and back regions (though that is not without relevance) but a contradiction between the figure to which façade and perspective appealed and the figure that was quietly effaced by their installation. In sixteenth-century Rome, for example, the appeal was direct:

The basic configuration of space applied equally to the whole and to each detail. Symbolism infused meaning not into a single object but rather into an ensemble of objects presented as an organic whole.... The prestigious dome [of St Peter's] represents the head of the Church, while the colonnades are this giant body's arms, clasping the piazza and the assembled faithful to its breast.[128]

[124] Lefebvre, *De l'état*, pp. 286–87; Kemp, *Science of art*, p. 335; my emphasis; see also Svetlana Alpers, *The art of describing: Dutch art in the seventeenth century* (Chicago: University of Chicago Press, 1983).
[125] Cosgrove, "Prospect," pp. 46, 54–55.
[126] Lefebvre, *Production*, pp. 64–65; idem, *De l'état*, pp. 288–89.
[127] Lefebvre, *Production*, p. 273.
[128] *Ibid.*, p. 274.

This organic, corporeal imagery naturalizes a particular constellation of power-knowledge through its appeal to the unity of the body. *But it also marks the erasure of the living body itself: this is a space dominated by the eye and the gaze.*

The final victory of decorporealization was the installation of the *abstract space* of twentieth-century capitalism.

By the time this process is complete, space has no social existence independently of an intense, aggressive and repressive visualization. It is thus – not symbolically but in fact – a purely visual space. The rise of the visual realm entails a series of substitutions and displacements by means of which it overwhelms the whole body and usurps its role.[129]

Lefebvre pays particular attention to that convulsive period when modernity was revealed in its purest, most sovereign form. The opening years of the twentieth century witnessed what he repeatedly describes as *la chute des référentiels*:

With the loss of absolute time and space – the space of Euclid and Newton – perceptible reality lost its stable referentials – a fact that was promptly translated into the sphere of aesthetics; perspective changed, the vanishing point, a token of geometric space, vanished.[130]

All of these foundational assumptions, which had been grounded in what had come to seem a permanent rationality, collapsed in Europe between 1905 and 1910. In the physical sciences, this crisis of meaning was articulated and aggravated by the development of quantum theory and relativity theory, which made it apparent, says Lefebvre, "that 'our space' was just one among many possible spaces." Intimations of this sort left a lasting impression on his own thought. Like many of his contemporaries, he had followed Bergson – not without misgivings – in thinking of time and space as quite separate:

With my surrealist friends, I walked around Paris for whole nights ... and we said, time and space are so separate in cities. The decisive element for me was a paper given by Einstein at the *Société française de philosophie*, where he confronted Bergson and where, in my opinion, he crushed him.... [He] introduced the close relationship between space and time.... From that moment my attention as a Marxist was drawn towards these phenomena and to the idea that there was a gap to be filled within Marxist thought.[131]

---

[129] *Ibid.*, p. 286. Lefebvre also identifies two other formants of abstract space: the geometric and the phallic.

[130] Lefebvre, *Everyday life*, p. 113; see also *idem, Survival*, p. 21.

[131] See "An interview with Henri Lefebvre," *Environment and Planning D: Society & Space* 5 (1987) pp. 27–38; the quotation is from p. 33.

But all this lay in the future. Lefebvre's own argument hinges on the crisis of representation in the arts, and particularly on the development of Cubism, which according to one commentator was the movement in modern painting that "most consciously destroyed the representational system that had prevailed since the Renaissance."[132] There were resonances between the sciences and the arts, but they were probably not direct.

Although Planck published his Quantum Theory in 1901, its implications were not understood until the 1920s at the earliest, and by that time all the Cubist innovations had been made. Nor is it likely that the Cubists read Einstein in 1905.[133]

The connections and homologies were thus intuitive:

The simultaneous overturning of fundamental notions in painting and in physics gave rise to comparisons that were rather vague but sometimes *intuitively* right.... [C]ubist painting arrived at a continuum of space-time unknown until then in painting.[134]

What interests Lefebvre is less the techniques than the *politics* of modernism, which he treats as a politics of space by virtue of its involvement in the production of a spatial imaginary whose assumptions and entailments extended far beyond the canvas. The opening decades of the twentieth century were racked by successive, overlapping crises: the crisis of liberalism, of imperialism, of the subject, of meaning itself. Modernism confronted and contributed to all of them, "in face of the shattered figures of a world in pieces, in face of a disjointed space, and in face of a pitiless 'reality' that [could not] be distinguished from its own abstraction, from its own analysis, because it [was] already an abstraction, already in effect an analytics."[135]

Modernism was thus at once complicit in and critical of the triumph of this ("contradictory") abstract space. The significance of Cubism lay not so much in the response of contemporary critics and publics to its showings, but in what its showings *showed*. Quite simply, Lefebvre believes that "Picasso's space *heralded* the space of modernity":

What we find in Picasso is an unreservedly visualized space, a dictatorship of the eye – and of the phallus.... Picasso's cruelty towards the body, particularly the

---

[132] Peter Bürger, *Theory of the avant-garde* (Minneapolis: University of Minnesota Press, 1984) p. 73. Although Lefebvre identifies a logic of visualization that spans perspectival space and abstract space – which constitutes a "logical space" – he also describes modernism as "the space of catastrophe" for linear perspective: *De l'état*, p. 289.

[133] John Berger, *The success and failure of Picasso* (New York: Pantheon, 1980) p. 69.

[134] Wolfgang Paalen, *Form and sense* (New York: Wittenborn, 1945) p. 27; my emphasis. See also Mark Roskill, *The interpretation of Cubism* (London and Toronto: Associated University Presses, 1985) p. 30.

[135] Lefebvre, *Production*, p. 302.

female body, which he tortures in a thousand ways and caricatures without mercy, is dictated by the dominant form of space, by the eye and by the phallus – in short, by violence. Yet this space cannot refer to itself – cannot acknowledge or admit its own character – without falling into self-denunciation. And Picasso, because he is a great and genuine artist, . . . inevitably glimpsed the coming dialectical transformation of space and prepared the ground for it.[136]

To Lefebvre, Picasso is representative of the privilege that modernism accorded to *opticality* as the field and ground for aesthetic practice – "it is only things seen that are knowledge for Picasso," Gertrude Stein declared; and for much the same reason Paul Éluard described his work as a project "to liberate the vision"[137] – but Lefebvre also believes that Picasso hypostatized a logic of visualization that caged and crushed the human body.[138]

I hope these remarks will not be misunderstood. Lefebvre is hardly unmasking Picasso as the angel of history, ushering abstract space to its geometric place. Instead, he is using the avant-garde as a way of figuring (and prefiguring) the emergent topologies of abstract space. He believes that Picasso's work, together with that of Klee and Kandinsky, registered the collapse of the referentials and revealed the space that was appearing in the ruins. More than this, they also prepared the ground for the Bauhaus, which by the early 1920s had in fact hired both Klee and Kandinsky and was already closely associated with the *Neue Sachlichkeit*, the "New Objectivity." Bauhaus projects in architecture and design were synchronized with the new industrial economy and meshed with modern techniques of mass production and standardization. I take this to be the (materialist) crux of Lefebvre's argument. Whereas avant-garde painters had *shown* that "all aspects of an object could be considered simultaneously," the Bauhaus was concerned with the global *production* of space in the same way. It was "no

---

[136] *Ibid.*, p. 302. Krauss presses the argument still further. To her, Picasso's use of collage created "a metalanguage of the visual" which could "talk about space without deploying it" and which thus presaged the development of a *post* modernist art. See Rosalind Krauss, "In the name of Picasso," *October* 16 (1981) pp. 5–22. The relationship between the visual and the carnal is also explored in *idem*, "The blink of an eye," in David Carroll (ed.), *The states of "theory": History, art and critical discourse* (New York: Columbia University Press, 1990) pp. 175–99.

[137] Gertrude Stein, *Picasso* (Boston: Beacon Press, 1959) pp. 22, 35; the memoir was originally published in Paris in 1938; Paul Éluard, *Pablo Picasso* (New York: Philosophical Library, 1947) p. 38.

[138] In a series of essays that address these themes Krauss argues that while modernism privileged "opticality" as the only way to ground or legitimate aesthetic practice, there was also what she calls an "antioptical or antivisual option that existed within the artistic avant-garde in the first half of the twentieth century." Her exemplar is Duchamp – who sought to "corporealize the visual (against the disembodied opticality of modernist painting)" – but this "underground" also included Georges Bataille, Salvador Dali, Max Ernst and others. See *idem*, "Im/pulse to see," in Hal Foster (ed.), *Vision and visuality* (Seattle: Bay Press, 1988) pp. 51–75; *idem*, "The story of the eye," *New Literary History* 21 (1990) pp. 283–97. Lefebvre would not be altogether out of place in this list.

longer a question of introducing forms, functions or structures in isolation," therefore, "but rather one of mastering global space by bringing forms, functions and structures together in accordance with a unitary conception."[139]

This was true of much of the work produced at the Bauhaus under the direction of Walter Gropius, but it became even more obvious after his replacement by Hannes Meyer in 1928. Significantly, Meyer's appointment delighted Otto Neurath, the doyen of the Vienna Circle, the cradle of logical positivism, and one historian describes the Bauhaus project as an attempt "to use scientific principles to combine primitive color relations and basic geometrical forms to eliminate the decorative and create a new antiaesthetic aesthetic that would prize functionality."[140] The close connections that this implies between the artists and architects of the Bauhaus and the philosophers and scientists of the Vienna Circle – political, intellectual, personal – are fully germane to Lefebvre's argument, because he thinks it hard to distinguish abstract space "from the space postulated by the philosophers" in their "fusion of knowledge with power." There is, of course, an immense irony in all this, for the members of the Bauhaus believed they were establishing a revolution (an early manifesto anticipated the building of "the cathedral of socialism"); their enemies in the Weimar Republic and the Third Reich saw them as Bolsheviks, and in fact closed the school in 1933; and yet, Lefebvre concludes, all the time they were introducing, theoretically and practically, the quintessentially abstract space of advanced capitalism.[141]

The eye of power

I now want to bring Lefebvre's two narratives together and juxtapose them with Harvey's account. It has to be remembered that in many respects Lefebvre's project pivots around the events of May 68 and that, by 1989, Harvey is responding to an altogether different conjuncture. Even so, the history of space he offers in *The condition of postmodernity* covers much of the same ground as Lefebvre.

As one might expect, Harvey focuses on the production of successive capitalist spatialities. During the European Renaissance, information yielded by the voyages of discovery was brought within a vision of the globe as

[139] Lefebvre, *Production*, pp. 124–25, 304–308.

[140] Peter Galison, "Aufbau/Bauhaus: Logical positivism and architectural modernism," *Critical Inquiry* 16 (1990) pp. 709–52; the quotation is from p. 711. See also Kenneth Frampton, *Modern architecture: A critical history* (London: Thames and Hudson, 1985) pp. 123–29, 130–41 and Andreas Faludi, "Planning according to the 'scientific conception of the world': The work of Otto Neurath," *Environment and Planning D: Society and Space* 7 (1989) pp. 397–418.

[141] Lefebvre, *De l'état*, p. 290; idem, *Production de l'espace*, p. vii; idem, *Production*, p. 308.

a knowable totality. Logics of visualization and geometricization, which were articulated by the various discourses of perspectivism, made possible new representations of space not only in architecture, cartography, and art but also in drama and poetry. Harvey argues that many of the emphases of this "Renaissance revolution in concepts of space and time" were carried forward, in heightened form, by the Enlightenment project of the eighteenth century. The production of space – its conquest and ordering – was now conducted under the sign of a supposedly *universal* rationality. "Universal" is a highly charged word, of course, and Harvey knows very well that these universals were ethnocentric; but he pays more attention to the means by which the rationalities that had been inscribed within the Renaissance imaginary, "islands of social practice within a sea of social activities" in which quite other, often mystical conceptions of time and space remained undisturbed, were *generalized*. In effect, Harvey seems to suggest that this was the counterpoint to Enlightenment ideals of emancipation, since it was private property in land and the buying and selling of space as a commodity – the advance of capitalism – that consolidated space as "universal, homogeneous, objective and abstract in social practice."[142]

Harvey argues that these absolutes were swept away during the crisis of European capitalism in the late 1840s. He makes much of the conjunction between this "first unambiguous crisis of capitalist overaccumulation" and the famous "springtime of peoples," the revolutions that spread like a bushfire across the map of Europe, and in many respects his account can be read as an elaboration of Lukács's spirited attempt to read the cultural forms of the European bourgeoisie from the recompositions of capital that were set in motion by the revolutions of 1848. But it is an elaboration not a repetition, because to Harvey one of the most extraordinary characteristics of this conjuncture was its speed and simultaneity, which had far-reaching consequences.

Events proved that Europe had achieved a level of spatial integration in its economic and financial life that was to make the whole continent vulnerable to simultaneous crisis formation. The political revolutions that erupted at once across the continent emphasized the synchronic as well as the diachronic dimensions to capitalist development. The certainty of absolute space gave way to the insecurities of a shifting relative space, in which events in one place could have immediate and ramifying effects in several other places.[143]

---

[142] Harvey, *Condition of postmodernity*, pp. 242–59. It should be emphasized that these processes were by no means confined to Europe and that they were indelibly woven into the violent threads of European colonialism and imperialism.

[143] *Ibid.*, p. 261. This is not to say that the events of 1848 were undifferentiated. They had their own geography, and any rigorous analysis would have to make a series of careful discriminations. But these differences do not militate against a parallel grid of structural interconnections, which is precisely Harvey's point. In the case of England and France, for example, Traugott has emphasized that it

How could this be? Harvey's answer is uncompromising, and consistent with his theorization of money, time, and space as the central elements in modern constellations of power.

European space was becoming more and more unified precisely because of the internationalism of money power. 1847–8 was a financial and monetary crisis which seriously challenged received ideas as to the meaning and role of money in social life.[144]

These developments ushered in a profound crisis of representation:

It is no accident that the first great modernist cultural thrust occurred in Paris after 1848. The brushstrokes of Manet that began to decompose the traditional space of painting and to alter its frame, to explore the fragmentations of light and color; the poems and reflections of Baudelaire that sought to transcend ephemerality and the narrow politics of place in the search for eternal meanings; and the novels of Flaubert with their peculiar narrative structures in space and time coupled with a language of icy aloofness; all of these were signals of a radical break of cultural sentiment that reflected a profound questioning of the meaning of space and place, of present, past and future, in a world of insecurity and rapidly expanding spatial horizons.[145]

Harvey argues that the response to these political and economic crises – the creation of new forms of temporal and spatial displacement through the production of space – only intensified the sense of insecurity. The crisis of 1873–1896, to which Harvey makes no direct reference, compounded those uncertainties. The internationalization of the grain market precipitated a slide into agricultural depression in Europe, where the restructuring of the industrial economy set off an accelerating, almost dizzying series of technical innovations. He argues that these changes had a major impact on the conventions of realism – in Zola's *La Terre* and Norris's *The octopus*, for example, farmers grapple with the simultaneity of the modern agricultural economy, but this is rendered in strikingly different ways in the two texts – and that "a fascination with technique, with speed and motion," with what Benjamin called "the work of art in an age of mechanical reproduction," was a leitmotiv of modernism. At the same time, Harvey continues, European colonialism, imperialism, and World War I – "the Cubist War" – fragmented the spatial imaginary still further.

is "only when the responses of these national economies are considered as aspects of an interactive system (rather than as independent or merely analogous reactions to a common stimulus) [that] the 'global' character of their simultaneous crises emerge[s]." See Mark Traugott, "The mid-nineteenth-century crisis in England and France," *Theory and Society* 12 (1983) pp. 455–68; the quotation is from Traugott's introduction to this special issue of the journal, p. 449.

[144] Harvey, *Condition of postmodernity*, p. 262.

[145] *Ibid.*, p. 263.

And yet he believes that it is nonetheless possible to distinguish two currents within the "confusions and oppositions" of contemporary social thought. On the one side was a revolutionary (and eventually "heroic") modernism, which explored and celebrated the new rationalities of relative space and which was committed to universalism and internationalism, a sensibility that "sought to show how the accelerations, fragmentations and imploding centralization (particularly in urban life) could be represented and thereby contained within a singular image." On the other side was a reactive (often a reactionary) modernism that celebrated the particularities of place and nation, and often enshrined them in an aggressive geopolitics and in the aestheticization of politics: "The identity of place was reaffirmed in the midst of the growing abstractions of space." It is thus within the dialectic of space and place that Harvey situates early twentieth-century modernism.[146]

There are some substantive differences between these histories of space – of spatialities – and their valuations of modernism, but it is the contrast in conceptual structure that I want to accentuate here. Although Harvey's history of space is (of necessity) radically simplified, he is determined to provide a causal account of the transitions involved. The subtitle of *The condition of postmodernity* is unambiguous: an enquiry into the *origins* of cultural change. Lefebvre, in contrast, is so conscious of the dangers of collapsing representations of space directly into modes of production that, apart from the simple sketches in *La révolution urbaine*, he more or less eschews causal analysis altogether. In *La production de l'espace* he identifies a terribly important thematic – the decorporealization of space – but he is unable to account for the succession of different spatialities within which it is inscribed, except by appealing to a transcendent logic of visualization. As I propose to show, however, the price of Harvey's achievement is a high one.

I make so much of differences in conceptual structure because Harvey's analysis of the condition of postmodernity rests, in large measure, upon his decoding of the spatialities of early twentieth-century Europe – that is to say, it hinges on early modernism rather than the "high modernism" from which postmodernism is supposed to depart – and the reasons for this are, I think, essentially theoretical. If "the changing experience of space and time" was somehow causally implicated in the emergence of modernism, Harvey argues, then it is reasonable to treat postmodernism as a response "to a new set of experiences of space and time."[147] Let me sketch the system of concepts on which this parallellism depends. Although Harvey does not say so, the oppositions between place and space that he identifies within early modernism mirror (and are presumably intended to mirror) what he has identified elsewhere as a series of contradictions within the landscape of capitalism (figure 29). At the level of the economy, he

---

[146] *Ibid.*, pp. 10–38, 272–80.
[147] *Ibid.*, p. 283.

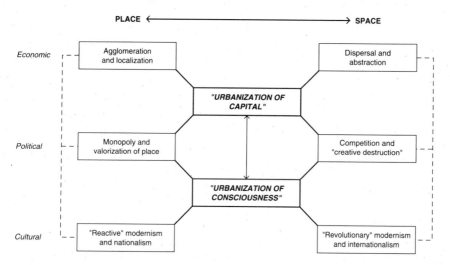

Figure 29   *Spatiality, capitalism, and modernity in early twentieth-century Europe.*

suggests, there is "a tension within the geography of accumulation between fixity and motion, between the rising power to overcome space and the immobile structures required for such a purpose": in other words, between agglomeration in place and dispersal over space. In *The limits to capital* Harvey used this couplet to frame Marx's fundamental distinction between concrete labor (which is inscribed in the particularities of place) and abstract labor (which depends upon the abstractions of space). Here, surely, is Lefebvre's "space of the commodity" writ large, but Harvey invokes Lefebvre more explicitly to mobilize the same opposition at a political level. There is, he says, a pervasive tension between intraregional coalitions whose interests usually coincide with the preservation or enhancement of the valorization of place, and interregional relations of competition and domination, which spasmodically threaten to restructure the space of capital accumulation through de-valorization and re-valorization – a process of "creative destruction." In *The condition of postmodernity*, Harvey transposes these oppositions into a third, cultural register in order to explore the tensions both within modernism and between modernism and postmodernism. He is plainly more sympathetic to the modernist project than Lefebvre, and he believes that it carried within it spaces of representation that were at once international in scope and yet subversive of the abstract spaces of global capitalism. The tragedy of modernism, for Harvey, is that in the end "the rule that form follows profit as well as function dominated everywhere."[148]

---

[148] This paragraph draws on David Harvey, "The geopolitics of capitalism," in Derek Gregory and John Urry (eds.), *Social relations and spatial structures* (London: Macmillan, 1985) pp. 128–63; Harvey, *Limits*, p. 375; idem, *Urbanization*, pp. 32–61, 125–64; idem, *Condition of postmodernity*, pp. 226, 281–83.

Set out like this, it is perhaps easy to see why Meaghan Morris is dismayed at "the neat little symmetries of [Harvey's] oppositional grids." I have described the oppositions between space and place that reappear in the economic, political, and cultural registers as "mirrors," and so they are. The topography of this figure simply does not allow for the complex web of overdeterminations that structures Althusser's social formation: on the contrary, Morris argues that *The condition of postmodernity* relies on "a reflection model of culture." In fact, she suggests that it installs two mirrors: one, the master-mirror, which displays "the truth of coherent reflection" (historical materialism) and the other, set up in opposition to it, displays the illusion of fragmentation (postmodernism) that is "created in the negative image of its own categories."

One mirror commands the universal, the other is mired in the local; one "grasps" (or "penetrates"), the other "masks" (or "veils"); one wants to know reality, the other worships the fetish; one quests for truth, the other lives in illusion.[149]

It will be apparent from this that Harvey is considerably less cautious than Lefebvre about making connections between modes of production and the production of space. He makes much more of the logic of capital than Lefebvre, though rather less of class than many of his critics seem to think. He says surprisingly little about the social space of modernism, and apart from Klimt, Schiele, and Loos "clinging together in the midst of a crisis of bourgeois culture, caught in its own rigidities but faced with whirlwind shifts in the experience of space and time" – a wonderfully evocative image – there are few passages in which Harvey so unambiguously maps the class location of the protagonists.[150] It is also true that he makes much less of the intersections of power, gender, and sexuality than Lefebvre. This is a major lacuna, and not one that can be filled by extrapolation from the labor theory of value or through marginal concessions to feminism. The gendering of modernism and its representation of space is as important as the class positions that it inscribes, and here Lefebvre's logic of visualization is likely to be of great importance (though it will need to be developed in close association with more recent feminist critiques of the gaze).

I want to press this possibility still further, and suggest that the price of Harvey's causal analysis is a high one in a quite literal sense: as I will try to show, his project privileges precisely that visual logic from which Lefebvre so forcefully dissents. My focus will be on the conceptual contrast between *time-space colonization* (Lefebvre) and *time-space compression* (Harvey).

---

[149] Meaghan Morris, "The man in the mirror: David Harvey's 'Condition' of postmodernity," *Theory, Culture and Society* 9 (1992) pp. 253–79; the quotations are from pp. 268–70.

[150] *Ibid.*, p. 275. The class coding of the early twentieth-century crisis of representation is made particularly clear in Donald Lowe, *History of bourgeois perception* (Chicago: University of Chicago Press, 1982) p. 110.

Modernity and time-space colonization

Lefebvre argues that by the middle of the twentieth century abstract space had been imposed on the concrete space of everyday life. Figure 30 summarizes his argument in its most schematic and general form as what I want to call (playfully but nonetheless seriously) "the eye of power."[151]

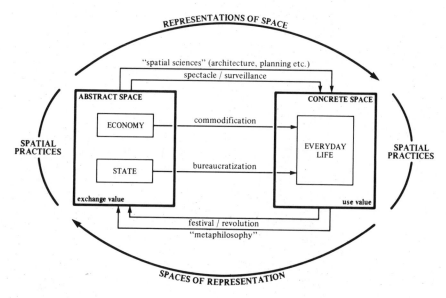

Figure 30 The eye of power.

Abstract space is produced through two major processes, each of them "doubled." First, modernity is shaped by an intensified *commodification of space*, which imposes a geometric grid of property relations and property markets on the earth, and an intensified *commodification through space*, which involves the installation of economic grids of capital circulation by means of which abstract space inscribes abstract labor and the commodity form. Second, modernity is shaped by a heightened *bureaucratization of space*, whereby each administrative system "maps out its own territory, stakes it out and signposts it," and a heightened *bureaucratization through space*, which involves the installation of juridico-political grids by means of which social life is subject to systematic surveillance and regulation by the state.[152] These

---

[151] I have borrowed the phrase from Michel Foucault, "The eye of power," in his *Power/knowledge*, pp. 146–65.
[152] Lefebvre, *Production*, pp. 341, 387; Lefebvre, *Everyday life*, p. 160.

processes reinforce each other to constitute abstract space as preeminently
the space of *exchange value*:

Capitalist and neocapitalist space is a space of quantification and growing homogeneity, a merchandized space where all the elements are exchangeable and thus
interchangeable; a police space in which the state tolerates no resistance and no
obstacles. Economic space and political space thus converge towards the elimination
of all differences.[153]

Concrete space is the space of everyday life, a space that is enframed,
constrained, and colonized by the economy and by the state; but, as I said
earlier, it is also the trace and memory of other spatialities and other ways
of being-in-the-world. The dominant spatial imaginary of early twentieth-
century Europe may have been dislocated by modernism and modern
physics, but according to Lefebvre,

everyday life consolidated itself, as the site of preservation of the old reality and
the old representations, deprived of external bearings but perpetuating themselves
in practice. "One" continued to live in Euclidean and Newtonian space even when
theoretical knowledge moved into the space of relativity.... Everyday life is
certainly not immutable and even modernity will modify it; but it affirms itself
as the place of continuity, escaping the strange cultural revolution.... From that
date onwards, thought and everyday life, and so theory and practice, separated, took
different and diverging paths.[154]

Lefebvre's purpose is to bring those two paths together again: to reclaim
and redeem everyday life. His intention is thus

to conceive everyday life in such a way as to retrieve it from its modern state of
colonization by the commodity form and other modes of reification. A critique
of the everyday can be generated only by a kind of alienation effect, insofar as it
is put into contact with its own radical *other*, such as an eradicated past ... or an
imagined future.[155]

What Lefebvre's history of space seeks to show, therefore, as a history of
the present that opens a window onto a possible future, is that in spite
of the violence of abstraction, concrete space is still "a space of 'subjects'
rather than of calculations": the space of *use value.*[156]

[153] Henri Lefebvre, "Space: social product and use value," in J. W. Freiburg (ed.), *Critical sociology:
European perspectives* (New York: Irvington, 1979) pp. 285–95; the quotation is from p. 293. Following
directly from these remarks, Lefebvre offers a characterization which speaks directly to one of
Harvey's most prescient insights: "The space of the commodity may be defined as a homogeneity
made up of specificities": *Production*, p. 387.

[154] Lefebvre, *De la modernité*, p. 49.

[155] Edward Ball, "The great sideshow of the Situationist International," *Yale French Studies* 73
(1987) pp. 21–37; the quotation is from p. 30; see above, pp. 362–63.

[156] Lefebvre, *Production*, pp. 356, 362, 381–82.

I am aware that figure 30 mimics the claims developed by Habermas in his theory of communicative action: most obviously, the colonization of the lifeworld by the system.[157] Indeed, at one point Lefebvre suggests that

everyday life [has] *replaced* the colonies. Unable to maintain the old imperialism, looking for new instruments of domination and having decided to bid on the internal market, capitalist leaders treated the everyday as they had treated colonial territories.[158]

But the parallel is far from exact. Habermas does ground his account in a history of the present, but it is a reconstructive history – and one that privileges the (Western) present as the high point, though not the end point, of an evolutionary succession – which is conspicuously silent about the production of space. For Lefebvre, "colonization" is more than a figure of speech, and he is deeply conscious of the implications of occupation, dispossession, and reterritorialization with which it is freighted. The imagery of colonization was often invoked by the situationists, with whom he was closely associated for a short space of time, and they were also sensitive to the physicality and violence carried within it: If everyday life was "a foreign country where everyone actually lives," this must have had a particular resonance in France in the years surrounding the Algerian struggle for independence.[159]

To Lefebvre, the colonization of everyday life means the superimposition and hyperextension of abstract space, and in figure 30 I have roughed in some of the modalities which bear most directly on his account. He frames them with three concepts:

1 *Spatial practices*, which refer to the time-space routines and the spatial structures – the sites and circuits – through which social life is produced and reproduced;
2 *Representations of space*, which refer to conceptions of space – or, more accurately perhaps, to constellations of power, knowledge, and spatiality – in which the dominant social order is materially inscribed (and, by implication, legitimized);
3 *Spaces of representation*, which refer to counterspaces, spatial representations that "arise from the clandestine or underground side of social life" and

---

[157] Jürgen Habermas, *The theory of communicative action*. Vol. 2: *The critique of functionalist reason* (Cambridge, England: Polity Press, 1987); the original was first published in Germany in 1981.

[158] In 1968 Lefebvre argued that "organized capitalism" had its colonies in the metropolis and was carrying "the colonial experience into the midst of erstwhile colonizing peoples": *Explosion*, p. 93. By the 1980s the message was even grimmer. "It could happen that one day...an army of bureaucrats...will treat everyday life not...as a semi-colony [but] quite simply as a conquered land": Lefebvre, *De la modernité*, p. 287.

[159] Greil Marcus, *Lipstick traces: A secret history of the twentieth century* (Cambridge: Harvard University Press, 1989) p. 145.

from the critical arts to imaginatively challenge the dominant spatial practices and spatialities.[160]

Among the modern representations of space to which Lefebvre attaches particular importance are the discourses of the spatial sciences (including mainstream architecture and urban planning) and the spectacularization of urban space – what Sorkin has marvelously called "variations on a theme park" – which have together done so much to sustain *l'illusion urbanistique*.[161] Its dual is what Karnouh has characterized, apparently independently of Lefebvre, as *l'illusion rustique*:

Everything urban people have lost: identity, direct relation to others, community of knowledge, and that consumerism cannot fulfill, lead[s] them to a search for the impossible completeness of a world exploded into artificial needs. The imaginary countryside brackets daily alienation and creates the momentary illusion of a rediscovered unity . . . [which is] part of a program preestablished by merchants of leisure within the decor of urban scenography.[162]

Within such a schema, Lefebvre argues, the eye has an extraordinary power:

[Thought] soars up into the abstract space of the visible, the geometric. The architect who designs, the planner who draws up master-plans, see their "objects," buildings and neighborhoods, from on high and from afar. . . . They pass from the "lived" to the abstract in order to project that abstraction onto the level of the lived.[163]

This way of seeing is by no means confined to the optics of expert systems; the gaze has, in a sense, been generalized, and Bell identifies an anxiety of particular relevance to the present discussion when he suggests that contemporary culture has become freighted with the characteristically visual obsessions of an urban society:

---

[160] Lefebvre, *Production*, pp. 33, 38–46. The English translation refers to spaces of representation as "representational spaces" but I have preferred the longer version, which is also Harvey's usage.

[161] Michael Sorkin (ed.), *Variations on a theme park: The new American city and the end of public space* (New York: Noonday Press, 1992).

[162] Claude Karnoouh, "The lost paradise of regionalism: The crisis of post-modernity in France," *Telos* 67 (1986) pp. 11–26; the quotation is from p. 20. Karnoouh explicitly anchors his critique around the events of 1968. It should also be noted that Lefebvre places the origins of the festival in the rural field: Lefebvre, *Everyday life*, p. 36.

[163] Lefebvre, *Révolution*, p. 241; cf. *idem*, *Production*, pp. 361–62. This echoes Merleau-Ponty's critique of what he called *pensée au survol*, "the high-altitude thinking which maintained the Cartesian split between a distant, spectatorial subject and the object of his sight." As Jay shows, this is part of a much wider critique of the gaze – or, less sharply, the interrogation of sight – in French social thought: Martin Jay, "In the empire of the gaze: Foucault and the denigration of vision in twentieth-century French thought," in David Couzens Hoy (ed.), *Foucault: A critical reader* (Oxford: Blackwell Publishers, 1986) pp. 175–204; the quotation is from p. 178.

To "know" a city, one must walk its streets; but to "see" a city, one must stand outside in order to perceive it whole. From a distance, the skyline "stands for" the city. Its massed density is the shock of cognition, its silhouette the enduring mark of recognition. This visual element is its symbolic representation.[164]

By these means, through the generalization and naturalization of these modalities, representations of space function as technologies of power, as disciplinary technologies that produce dispositions of useful and docile bodies. The phrasing is Foucault's, but the sentiment is also Lefebvre's: "Living bodies . . . are caught up not only in the toils of parcellized space, but also in the web of images, signs and symbols. These bodies are transported out of themselves, transferred and emptied out, as it were, via the eyes."[165]

But concrete space is also the site of resistance and active struggle, the origin of spaces of representation that provide counterdiscourses and create alternative spatial imaginaries. This is one of the responsibilities of Lefebvre's "metaphilosophy" which, in contrast to the conventional spatial sciences, is rooted in concrete space and seeks to reassert difference over homogeneity, unity over fragmentation, and equality over hierarchy. In doing so, it pays particular attention to the body, at once the victim of the history of the present and the locus of a new "history" of the future: "The body, at the very heart of space and of the discourse of power, is irreducible and subversive. It is the body which is the point of return."[166] For this reason, Lefebvre makes much of the festival, in opposition to the spectacle, as the site of participation and of the possibility of the poesis of creating new situations from desire [*jouissance*]. The revolutionary project thus requires and demands "the reappropriation of the body in association with the reappropriation of space."[167]

If the proximity of the eye of power to the events of May 68 is incontrovertible, it is by no means estranged from the present. I think, for example, of those critics who trace postmodernism back to the culture and counterculture of the sixties; who, like Huyssen, see the revolt against abstract expressionism in art, the exploration of new forms of immediacy, spontaneity, and participation in performance, and the exuberant rejection of "the congealed canon" in literature as moments in an avant-garde project to reintegrate art and everyday life: a project that reached back to early modernism but which in the late 1960s also anticipated the more liberating aims of some versions of postmodernism.[168] I think, too, of the intensely

[164] Daniel Bell, *The cultural contradictions of capitalism* (New York: Basic Books, 1976) pp. 104–108.

[165] Lefebvre, *Production*, p. 98; see also p. 308.

[166] Lefebvre, *Survival*, p. 89.

[167] Poster, *Existential marxism*, p. 256; Lefebvre, *Production*, pp. 166–67.

[168] Andreas Huyssen, *After the Great Divide: Modernism, mass culture and postmodernism* (London: Macmillan, 1988) pp. 163–68 and *passim*.

creative contemporary engagements between art, cultural studies, and social theory. A pertinent example is Martha Rosler's project, "*If you lived here...*" a series of events that took place in New York's Soho district between 1987 and 1989 and which combined theories about the social production of art with critical discourses about the formation of the city. Rosler was animated by basic questions – "How can one represent a city's 'buried' life, the lives in fact of most city residents? How can one show the conditions of tenants' struggles, homelessness, alternatives to city planning as currently practiced?" – which not only spoke directly to Lefebvre's abiding concerns: the recovery of history and social memory, the redemption of everyday life, the reclamation of concrete space, *le droit à la ville*. Lefebvre's ideas about the city as a contested terrain and about the production of a genuinely democratic, public space were also woven into the development of Rosler's project.[169]

I am not sure what Harvey would make of these parallels and extensions, except to say that (in relation to figure 30) he is also critical of spatial science and the spectacle and that he too calls for a revolution that will, through the union of theory and practice in popular struggle, strive for "a genuinely humanizing experience."[170] And yet he does not directly address the time-space colonization of everyday life, and the opposition he identifies between space and place cannot be mapped directly onto Lefebvre's "abstract space" and "concrete space." In fact, although Harvey sketches out and fills in Lefebvre's grid of spatial practices, representations of space and spaces of representation, he makes very little use of it.[171] The core concept in his discussion of modernity and its dominant spatial imaginary is time-space compression.

Modernity and time-space compression

Time-space compression is a hybrid concept, perhaps appropriately: It owes something to concepts of time-space convergence and time-space distanciation, but it is intended to have a more *experiential* dimension than either of them. Yet its theoretical armature is provided by Bourdieu rather than Lefebvre. Although Bourdieu's views on historical materialism are ambiguous – and a good deal more ambiguous than Harvey seems to realize – the attraction of his ideas is, I think, twofold.

[169] Here I draw on Martha Rosler, "Fragments of a metropolitan viewpoint" and Rosalyn Deutsche, "Alternative space," both in Brian Wallis (ed.), *If you lived here: The city in art, theory and social activism (A project by Martha Rosler)* (Seattle: Bay Press, 1991) pp. 15–43 and pp. 45–66; see also Rosalyn Deutsche, "Uneven development: Public art in New York City," *October* 47 (1988) pp. 3–52, which also draws on Lefebvre.
[170] Harvey, *Consciousness*, p. 276.
[171] Harvey, *Condition of postmodernity*, pp. 218–22.

In the first place, Harvey considers Lefebvre much too vague about the relations between spatial practices, representations of space and spaces of representation, and he believes that Bourdieu's concept of the *habitus* is capable of clarifying them in a more precise way.[172] I find it helpful – as a first approximation – to think of the span of the habitus in both a vertical and a horizontal dimension. "Vertically," Bourdieu offers the concept as a means of overcoming the dichotomy between what he once described as a subject-less structuralism and the philosophy of the subject. The habitus is a system of internalized *dispositions* that mediate between social structures and practical activity. In French, *disposition* connotes both "the result of an organizing action" and "a predisposition," and the recursive movement between result and predisposition within the flow of practical activity has something in common with the model of structuration developed within Anglophone social theory. In much the same way Bourdieu insists that the habitus is reducible neither to the imperative of structures nor to the intentionality of agents, and offers his theory of practice as a way of transcending that classical opposition. But it is not difficult to see why some of his critics diasgree: Honneth claims that the habitus must depend on a reductionist model, because it is in some substantial degree the lateral extension of social structures that shapes and underwrites the horizontal span of the habitus as "the locus of practical realization of the 'articulation' of fields." Bourdieu argues that the coherence of different social practices is the result of "the coherence which the generative principles constituting that habitus owe to the social structures of which they are a product and which they tend to reproduce."[173] By extension, therefore, and "horizontally," Bourdieu intends the habitus to harmonize and homologize social practices from one sphere of social life to another. Hence:

One of the fundamental effects of the orchestration of habitus is the production of a commonsense world endowed with the objectivity secured by consensus on

---

[172] *Ibid.*, p. 219. For much the same maneuver, see Rob Shields, *Places on the margin: Alternative geographies of modernity* (London: Routledge, 1990).

[173] My understanding of Bourdieu's work has been greatly helped by Rogers Brubaker, "Re-thinking classical theory: The sociological vision of Pierre Bourdieu," *Theory and Society* 14 (1985) pp. 745–75, and Axel Honneth, "The fragmented world of symbolic forms: Reflections on Pierre Bourdieu's sociology of culture," *Theory, Culture and Society* 3 (1986) pp. 55–56. See also Bourdieu, *Outline*, pp. 83, 97. My comments here refer to a "structural reductionism" – to the conditions and constraints determined by social structures of all kinds – and it should be clearly understood that reductionism in this general sense does not automatically collapse into *economic* reductionism in a more specific sense. But here too Bourdieu is ambiguous. Although he distinguishes between "economic capital" and "symbolic capital," on occasion insisting that they obey different logics, he also argues that "all practices, including those purporting to be disinterested or gratuitous, and hence non-economic, [can be treated] as economic practices directed towards the maximizing of material or symbolic profit": Bourdieu, *Outline*, p. 183. This is precisely Honneth's objection, and, I suspect, another reason for Harvey's attraction to his work.

the meaning ... of practices and the world. ... The homogeneity of habitus is what – within the limits of the group of agents possessing the schemes (of production and interpretation) implied in their production – cause practices and works to be immediately intelligible and foreseeable and hence taken-for-granted.[174]

The concept of the habitus is thus a way of elucidating the coherence of social life; its systematicity is always partial and precarious – always an achievement, something to be negotiated through social practice rather than imposed through a trans-situational logic – but it is nonetheless real.

Harvey uses the concept of the habitus to capture the coherence of social life in a particular frame: the experience of time and space. He claims to derive this directly from Bourdieu who, on his reading, suggests that "spatial and temporal experiences are primary vehicles for the coding and reproduction of social relations." Although Bourdieu is acutely aware of the significance of symbolic space, he thinks it more important in oral cultures than in literate cultures, and so Harvey has to generalize what one might call the spatial sedimentation of the habitus. In doing so, he relocates it within the framework of historico-geographical materialism:

The objectivity of time and space is given ... by the material practices of social reproduction, and to the degree that these latter vary geographically and historically, so we find that social time and social space are differentially constructed. Each distinctive mode of production ... [will] embody a distinctive bundle of time and space practices and concepts.[175]

And yet: Bourdieu's accent on the repetitive and taken-for-granted character of social life, like several other models of social reproduction, threatens to institute a closure that is extraordinarily difficult to prise open. Bourdieu certainly does not mean to imply that social life is an endless stasis; one of the primary concerns of his cultural sociology is to elucidate the essential *creativity* that is supposed to inhere in all practical activity. Still, if the "objective conditions" set limits to that creativity, as Bourdieu insists they do, if "every established order tends to produce 'the naturalization of its own arbitrariness,' expressed in the 'sense of limits' and the 'sense of reality' which in turn form the basis for an 'ineradicable adherence to the established order,' " as Harvey agrees they do, then how does one break out of the circle? If something "goes without saying," then how is critique even possible: How is "the undiscussed [brought] into discussion"?[176] Bourdieu answers these questions by inverting them.

---

[174] *Ibid.*, p. 80; see also *idem, Logic*, pp. 58–59. This means that it should also be possible to install a series of connections between the "horizontal" dimension of the *habitus* and the concept of time-space distanciation that is focal to Giddens's structuration theory.

[175] Harvey, *Condition of postmodernity*, p. 204; Bourdieu, *Outline*, p. 89.

[176] Bourdieu, *Outline*, pp. 167–68; Harvey, *Condition of postmodernity*, p. 345.

It is when the social world *loses* its character as a natural phenomenon that the question of the natural or conventional character ... of social facts can be raised.

This in turn requires the coherence of the habitus to be *dislocated* in some way, "breaking the immediate fit between the subjective structures and the objective structures" and thereby "destroy[ing] self-evidence practically."[177]

Bourdieu suggests two ways in which this can happen. The first possibility is through the juxtaposition of different cultures in the same space. "If the emergence of a field of discussion" – and of critique more generally – "is historically linked to the development of cities," he argues, this is because the heterogeneity of urban life "favors the confrontation of different cultural traditions, which tends to expose their arbitrariness *practically*, through first-hand experience, in the very heart of the routine of the everyday order, of the possibility of doing the same things differently."[178]

This is precisely what postmodernism and postcolonialism are about, of course, and in their different ways they can remind us of our *own* "otherness." But Harvey does not pursue this possibility; on the contrary, he warns against being "seduced by otherness" and hints at ways in which the construction and imbrication of microcultural geographies in many late twentieth-century cities "draw[s] a veil over real geography."[179] He is more interested in the second possibility: the "practical questioning" which is brought about by "political and economic crises correlative with class division."[180]

The expression of ... contradictions in the form of objective and materialized crises plays a key role in breaking the powerful link "between the subjective structures and the objective structures" and thereby lay[ing] the groundwork for a critique that "brings the undiscussed into discussion and the unformulated into formulation."[181]

Although Harvey looks to political and economic crises to provide the causal momentum for his history of space, however, class division dissolves and disappears into the shadows of his account. In fact, he treats the habitus as a way of conceptualizing *individual* spaces and times in social life, whereas Bourdieu emphasizes that "the singular habitus" has to be placed within a wider social field in which *class-positions* structure (though

---

[177] Bourdieu, *Outline*, pp. 168–69.

[178] *Ibid.*, p. 233, n. 16.

[179] Harvey, *Condition of postmodernity*, pp. 47, 87. Here I presume Harvey has in mind the commercial dramatization and "spectacularization" of cultural difference rather than the display and celebration of those differences as such.

[180] Bourdieu, *Outline*, p. 168.

[181] Harvey, *Condition of postmodernity*, p. 345.

they do not uniquely determine) the system of dispositions. Hence "each individual system of dispositions is a structural variant of the others, expressing the singularity of its position within the class and its trajectory."[182] Indeed, Bourdieu's entire analysis of late twentieth-century French society is based on this claim:

The habitus is not only a structuring structure, which organizes practices and the perception of practices, but also *a structured structure*: the principle of division into logical classes which organizes the perception of the social world is itself the product of internalization of the division into social classes.[183]

What is at issue here is not the acceptability (or otherwise) of Bourdieu's formulation. It is, rather, Harvey's selective appropriation of a concept in such a way that discriminations that it carries within it are erased.

In the second place, Harvey uses Bourdieu to connect the dislocation of the habitus to the disorientation that he claims is induced by *time-space compression*. He intends this term to signal "processes that so revolutionize the objective qualities of space and time that we are forced to alter, sometimes in quite radical ways, how we represent the world to ourselves." Lefebvre describes the logic of visualization in the language of menace, and Harvey does the same for time-space compression: alarming, disturbing, even threatening (rarely exhilarating), a "maelstrom" and a "tiger," it induces "foreboding," "shock," a "sense of collapse" and ultimately "terror."[184] These reactions speak in the same voice and derive, in some measure, from Bell's critique of the cultural contradictions of capitalism. Politically, the two are far apart and Harvey strenuously repudiates Bell's neoconservatism. Analytically, however, he concedes that Bell's account was probably more accurate than many radical attempts to understand what was happening.[185]

---

[182] *Ibid.*, pp. 214–16; Bourdieu, *Logic*, p. 60. Elsewhere Harvey makes it plain that "spatial practices derive their efficacy in social life only through the structure of social relations within which they come into play." In a capitalist society, therefore, spatial practices "become imbued with class meanings" (p. 223). So they do; but Harvey's discussion of the habitus is silent about those meanings and the emphasis is thrown exactly where I have marked it: on *individual* spaces and times.

[183] Pierre Bourdieu, *Distinction: A social critique of the judgement of taste* (Cambridge: Harvard University Press, 1984) p. 170; my emphasis.

[184] Harvey, *Condition of postmodernity*, p. 240; *idem*, "Between space and time: Reflections on the geographical imagination," *Annals of the Association of American Geographers* 80 (1990) pp. 418–34; the quotations are from pp. 426, 433.

[185] *Idem, Condition of postmodernity*, p. 352. This is not an exceptional view. Habermas – who, like Harvey, objects to Bell's neoconservatism – nonetheless accepts that Bell "is a good social theorist" and argues that "in his analysis of the causes of the cultural crisis he does not proceed in a neoconservative manner at all": Jürgen Habermas, "Neoconservative cultural criticism in the United States and Germany," in his *The new conservatism: Cultural criticism and the historians' debate* (Cambridge: MIT Press, 1989) pp. 22–47; the quotation is from p. 28.

Bell was greatly exercised by "the disjunctions of cultural discourse" and "the diremption of culture" which he thought had been brought about by the "eclipse of distance." Like Harvey, he argues that post-Renaissance Western aesthetics took as its central task (and, I think, trust) the establishment of "certain formal principles of art around the rational organization of space and time." It was these principles that gave social life its foundation, its structure, and its coherence. During the last fin de siècle, however, this quintessentially rational order was disrupted in two registers: Not only was "physical distance compressed" (the phrase is Bell's) by new systems of transportation and communication, but aesthetic distance was effaced by the modernist sensibility and in particular by what he takes to be its insurgent stress on "immediacy, impact, sensation and simultaneity."[186] Of course, Bell's critique is directed against modernism, Harvey's against postmodernism: but in a purely formal sense their concerns compound one another. Both authors warn against the deformation of the project of modernity; both of them conjoin this to the production of space. When Bell asserts that "the organization of space has become the primary aesthetic problem of mid–twentieth-century culture," therefore, Harvey is not far behind.[187] Viewed from such a perspective, one might say that the contemporary crisis of representation is implicated in "a characteristically 'de-centered' habitus."[188]

What Harvey wants to do, in contradistinction to Bell, is to wire this crisis of representation to a crisis of capital accumulation. As Habermas has accentuated, the concept of "crisis" is doubly coded. In classical aesthetics, for example, crisis signifies "the turning-point in a fateful process that, despite all objectivity, does not simply oppose itself from the outside and does not remain external to the identity of the persons caught up in it." The same is still true in medicine. And Habermas argues that the same ought to be true in social theory, where any concept of crisis, if it is to be of any use, "must grasp the connections between system integration and social integration."[189]

---

[186] Bell, *Cultural contradictions*, pp. 91, 106, 108 and *passim*.

[187] *Ibid.*, p. 107; Harvey cites this remark in *Condition of postmodernity*, p. 201.

[188] Scott Lash, *Sociology of postmodernism* (London: Routledge, 1990) p. 253.

[189] Jürgen Habermas, *Legitimation crisis* (London: Heinemann, 1976) p. 2. The problem is by no means straightforward since, as Habermas acknowledges, "social integration" and "system integration" are concepts that derive from radically different theoretical traditions:

> We speak of social integration in relation to systems of institutions in which speaking and acting subjects are socially related. Social systems are seen here as *life-worlds* that are symbolically structured. We speak of system integration with a view to the steering performances of a self-regulated *system*. Social systems are considered here from the point of view of their capacity to maintain their boundaries and their continued existence by mastering the complexity of an inconstant environment. Both paradigms, life-world and system, are important. The problem is to demonstrate their interconnection.

Habermas's most developed solution to the problem will be found in his theory of communicative action.

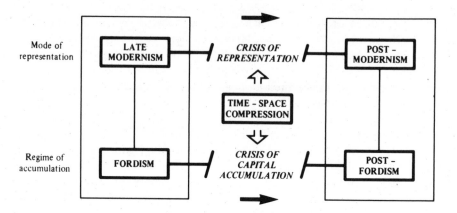

Figure 31   Transitions in late twentieth-century capitalism.

Harvey's critique of postmodernism incorporates both meanings: the contemporary crisis of modernity is at once "objective" (a crisis of capital accumulation) and "subjective" (a crisis of representation). In one sense this is a far less sophisticated scheme than Habermas's and the connections that Harvey makes between these twin crises are more direct than those to be found within Habermas's account. But in another sense Harvey's is the more radical formulation, because it relocates the classical problematic of crisis within the horizon of a geographical imagination. Like Habermas (in his early writings at any rate), Harvey also identifies the contemporary epicenter of politico-economic shock waves as a deep crisis within the regime of capital accumulation, but he draws on the work of the Regulation School to theorize this in considerably more detail and conceives of it as "in large part a crisis of temporal and spatial form."[190] Its resolution is thus marked not only by the transformation from Fordism to post-Fordism ("flexible accumulation") but also by the installation of new and radically unstable landscapes of accumulation. This "spatial fix" sets off waves of time-space compression that are directly and structurally implicated in the contemporary crisis of representation (figure 31).

Harvey insists that it is the *intensity* of the experience of time-space compression that distinguishes the condition of postmodernity. With the

[190] Harvey, *Condition of postmodernity*, p. 196. I have deliberately refrained from commenting on Harvey's construction of the political economy of late twentieth-century capitalism in any detail, but I must make two summary qualifications. First, I have theoretical reservations about the work of the Regulation School and in particular about the way in which it conceives of the relations between a regime of accumulation and a mode of regulation. Second, I think the empirical distinction between Fordism and Post-Fordism is oversimplified, and that a more nuanced historical geography of industrialization is required in order to capture the complexities of combined and uneven development.

transition to flexible accumulation, processes of production intensify and processes of transfer, exchange and consumption accelerate: the accent is on volatility, instantaneity, ephemerality, disposability. Harvey pays most attention to their "more general, society-wide" implications and to the ways in which these "postmodern ways of thinking, feeling and doing" bear down on the individual. He glides from society to the individual through a series of downward moves. "Money and commodities are themselves the primary bearers of cultural codes," he argues, and since these are "entirely bound up with the circulation of capital" it follows that "cultural forms are firmly rooted in the daily circulation process of capital" and that it is "with the daily experience of money and the commodity that we should begin." Through these reductions Harvey puts a critical gloss on what are, for most people (in the West), mundane experiences like having breakfast or watching television. In both these cases, he points out, as well as countless others, time-space compression has radically changed the mix of commodities and images that enter into the production and reproduction of everyday life. Local food systems have been incorporated into global commodity exchange, and local systems of meaning have been subsumed within global information and communication networks.

The general implication is that through the experience of everything from food, to culinary habits, music, television, entertainment and cinema, it is now possible to experience the world's geography vicariously, as a simulacrum. The interweaving of simulacra in daily life brings together different worlds (of commodities) in the same space and time. But it does so in such a way as to conceal almost perfectly any trace of origin, of the labor processes that produced them, or of the social relations implicated in their production.[191]

By putting these experiences in their historico-geographical context, Harvey argues, by treating the contemporary dislocation of the habitus

as part of a history of successive waves of time-space compression generated out of the pressures of capital accumulation with its perpetual search to annihilate space through time and reduce turnover time, we can at least pull the condition of postmodernity into the range of a condition accessible to historical materialist analysis and interpretation.[192]

In other words, it becomes possible to discern the social relations implicated in the production of this (postmodern) space. In doing so, however, the multiple and compound *geographies* of those processes seem to disappear: time-space compression, as Harvey describes it, concentrates

---

[191] *Ibid.*, pp. 299–300. Harvey derives the concept of simulacrum from Baudrillard, but he gives it a different inflection.

[192] *Ibid.*, pp. 306–307.

its furious energies on individuals and points. And yet the dislocation of the habitus varies through society – through compound topographies of class as well as gender, sexuality, and ethnicity, which together constitute a web of overlapping subject-positions – and *over space*. Within the nightmare world of postmodern abstraction, Harvey insists that places are revalorized: that capital produces, packages, and profits from differentiation. And yet the "sensitivity of capital to difference" finds little place in his own theorizations of cultural change.[193] In some ways, I think, time-space compression is the dual of time-space colonization. As I read it, the metaphor of time-space colonization has inscribed within it an *outward* movement: It connotes a congeries of processes that spread, invade, occupy. By implication, therefore, it imposes a responsibility to chart the shifting constellations within these changing local-global geographies, and it is not surprising that Lefebvre should be so painfully aware – even when he writes about France alone – of the violence of abstraction wreaked in the name of the West and throughout the world. The metaphor of time-space compression, by contrast, seems to carry within it an *inward* movement: the implication of a world collapsing in on itself. It is then no more surprising that Harvey should describe the process in terms of an undifferentiated, scrambled stream of commodities and images cascading toward the West: that "one of the prime conditions of postmodernity is that no one can or should discuss it as an historico-geographical condition."[194]

If, as Harvey clearly believes, a critical human geography is needed to challenge such a world-view, then it will have to recognize that different people in different places are implicated in time-space colonization and compression in different ways. For some, these processes undoubtedly present new opportunities and demand larger responsibilities, reveal wider horizons and enhance geographical imaginations; but for others they impose additional burdens and raise higher barriers, create further distinctions and diminish individual capacities. This means that a critical human geography must not only chart the differential locations and the time-space manifolds that are created through these processes – a project for which some of the concepts of spatial science might still be reclaimed – but also draw out the multiple, compound, and contradictory subject-positions that they make available. The production of space is not an incidental by-product of social life but a moment intrinsic to its conduct and constitution, and for geography to *make* a difference – politically and intellectually – it must be attentive *to* difference.

---

[193] *Ibid.*, pp. 293–96.

[194] *Ibid.*, p. 337. Harvey means this as a critique of postmodernism, and many of his critics have dismissed it as such: on the contrary, and against both houses, I think it an accurate summary of the way in which he himself constructs and works with the concept of time-space compression.

Dreams of liberty and wings of desire

Many of Harvey's critics treat the suppression of difference, the failure to make distinctions, that occurs in *The condition of postmodernity* as symptomatic of the "fantasy, denial and desire" that comes with totalizing claims to knowledge and inheres in what Meaghan Morris calls the "mythic space of meta-theory."[195] Let me explain what I think they mean by recalling Bell's image of the modernist reading of the city as a distant silhouette. Unlike Lefebvre, Harvey also accords a particular privilege to the distant view.

The relation between such a "god-like" vision of the city and the turbulence of street life is interesting to contemplate. Both perspectives, though different, are real enough. Nor are they independent. . . . The eye is never neutral and many a battle is fought over the "proper" way to see. Yet, no matter what the associations and aspirations, a special satisfaction attaches to contemplating the view from on high, for we have seen the city as a whole, taken it into our minds as a totality. Afterwards, the experience of street life cannot help acquiring new meaning.[196]

This goes to the very heart of Harvey's modernist sensibility. It is one that sees the prospect of "revolution" from the high peaks but must put in place a series of pitches and traverses down to the depths of everyday life. Whatever his critics might think, Harvey is not unaware of the precariousness of the project. In his commentary on Wim Wenders's *Wings of desire* he remarks that "only the angels have an overall view and they, when they perch on high, hear only a babble of intersecting voices and whispers and see nothing but a monochromatic world."[197] Casarino's reading of the same film sharpens the dilemma. Wenders himself claims that the angels "were only an artifice to enable him to represent and narrate the everyday" and to this extent, Casarino argues, they are the trace-signs of a modernism that has long since been superseded. The aerial perspective is thus not so much celebratory as elegiac. *Wings of desire* "represents the impossibility of representing a history of contemporary metropolitan experience." It is, in the most profound of senses, an aporetic vision:

Wings of desire raises the spectator to the transcendental and omniscient vision of the angels. In this ascent the film represents the deep-seated longing for a gnostic position from which to articulate the fragments of the everyday in a way which would reveal them . . . as being structured by and within the same socio-historical situation. The angels' wanderings and investigations textualize the otherwise radically invisible structure of a history in which all the fragments share. Between a

---

[195] Deutsche, "Boys town," pp. 9–11; Morris, "Man in the mirror," p. 275.
[196] David Harvey, *The urban experience* (Oxford: Blackwell Publishers, 1989) p. 1.
[197] Harvey, *Condition*, p. 317.

longing for representation and the representation of this longing *Wings of desire* unfolds as a utopian gesture.[198]

The task of a critical human geography – of a geographical imagination – is, I suggest, to unfold that utopian gesture and replace it with another: one that recognizes the *corporeality of vision* and reaches out, *from one body to another*, not in a mood of arrogance, aggression, and conquest but in a spirit of humility, understanding, and care. This is not an individualism; neither is it a corporatism. If it dispenses with the privileges traditionally accorded to "History," it nonetheless requires a scrupulous attention to the junctures and fissures between many different histories: a multileveled dialogue between past and present conducted as a history (or an historical geography) of the present.[199] In seeking to connect the history of the body with the history of space, it seems to me that Lefebvre was striving for the production of a genuinely human geography in exactly this sense.

---

[198] Cesare Casarino, "Fragments on *Wings of desire* (or, fragmentary representation as historical necessity)," *Social Text* 24 (1990) pp. 167–81; the quotations are from p. 173, 177, 178, 179.

[199] In fact, Wenders shares this concern with the Office of Metropolitan Architecture and, by implication and inclusion, Vriesendorp's *Dream of Liberty*. The founding project of OMA juxtaposed a divided Berlin with a delirious New York: see Fritz Neumayer, "OMA's Berlin: The polemic island in the city," *assemblage* 11 (1990) pp. 37–53.

Select Bibliography

This bibliography is confined to the major texts that I discuss, together with some of the more important ancillary materials.

Agnew, John and Duncan, James (eds.) 1989: *The power of place: Bringing together geographical and sociological imaginations* (Boston: Unwin Hyman).

Anderson, Perry 1976: *Considerations on Western Marxism* (London: New Left Books).

—— 1980: *Arguments within English Marxism* (London: Verso).

—— 1983: *In the tracks of historical materialism* (London: Verso).

—— 1984: "Modernity and revolution." *New Left Review*, 144, pp. 96–113.

Barnes, Trevor and Duncan, James (eds.) 1992: *Writing worlds: Discourse, text and metaphor in the representation of landscape* (London: Routledge).

Barrett, Michèle 1991: *The politics of truth: From Marx to Foucault* (Cambridge, England: Polity Press).

Benjamin, Walter 1968: *Illuminations: Essays and reflections* (New York: Schocken Books).

—— 1973: *Charles Baudelaire: A lyric poet in the era of high capitalism* (London: Verso).

—— 1986: *Reflections: Essays, aphorisms, autobiographical writings* (New York: Schocken Books).

Berman, Marshall 1982: *All that is solid melts into air: The experience of modernity* (London: Verso).

Bernstein, Richard 1983: *Beyond objectivism and relativism: Science, hermeneutics and praxis* (Oxford: Blackwell Publishers).

—— 1992: *The new constellation: The ethical-political horizons of modernity/postmodernity* (Cambridge: MIT Press).

Best, Steven and Kellner, Douglas 1991: *Postmodern theory: Critical interrogations* (New York: Guilford Press).

Bondi, Liz 1990: "Feminism, postmodernism and geography: Space for women?" *Antipode*, 22, pp. 156–67.

Bondi, Liz and Domosh, Mona 1992: "Other figures in other places: On feminism, postmodernism and geography." *Environment and Planning D: Society and Space*, 10, pp. 199–214.

Buck-Morss, Susan 1977: *The origin of negative dialectics: Theodor Adorno, Walter Benjamin and the Frankfurt Institute* (New York: Free Press).

—— 1981: "Walter Benjamin: Revolutionary writer." *New Left Review*, 128, pp. 50–75; 129, pp. 77–95.

Buck-Morss, Susan 1989: *The dialectics of seeing: Walter Benjamin and the Arcades Project* (Cambridge: MIT Press).

Callinicos, Alex 1989: *Against postmodernism: A Marxist critique* (Cambridge, England: Polity Press).

Carter, Paul 1987: *The road to Botany Bay: An essay in spatial history* (London: Faber).

Castells, Manuel 1975: *The urban question: A Marxist approach* (London: Edward Arnold).

—— 1983: *The city and the grass-roots: A cross-cultural theory of urban social movements* (Berkeley: University of California Press).

de Certeau, Michel 1984: *The practice of everyday life* (Berkeley: University of California Press).

Chakrabarty, Dipesh 1991: "Subaltern studies and the critique of history." *Arena*, 96, pp. 105–120.

—— 1992: "The death of history? Historical consciousness and the culture of late capitalism." *Public Culture*, 4, pp. 47–65.

—— 1992: "Postcoloniality and the artifice of history: Who speaks for 'Indian' pasts?" *Representations*, 37, pp. 1–26.

Clark, T. J. 1984: *The painting of modern life: Paris in the art of Manet and his followers* (Princeton: Princeton University Press).

Clifford, James 1986: "Introduction: Partial truths." In Clifford and Marcus, pp. 1–26.

—— 1988: *The predicament of culture: Twentieth-century ethnography, literature and art* (Cambridge: Harvard University Press).

—— 1992: "Travelling cultures." In Grossberg, Nelson and Treichler, pp. 96–112.

Clifford, James and Marcus, George (eds.) 1986: *Writing culture: The poetics and politics of ethnography* (Berkeley: University of California Press).

Comaroff, Jean and Comaroff, John 1991: *Of revelation and revolution: Christianity, colonialism and consciousness in South Africa* (Chicago: University of Chicago Press).

Cosgrove, Denis 1984: *Social formation and symbolic landscape* (London: Croom Helm).

—— 1985: "Prospect, perspective and the evolution of the landscape idea." *Transactions of the Institute of British Geographers*, 10, pp. 45–62.

—— 1988: "The geometry of landscape: Practical and speculative arts in sixteenth-century Venetian land territories." In Cosgrove and Daniels, pp. 254–76.

—— 1989: "Historical considerations on humanism, historical materialism and geography." In Kobayashi and Mackenzie, pp. 189–226.

Cosgrove, Denis and Daniels, Stephen (eds.) 1988: *The iconography of landscape* (Cambridge: Cambridge University Press).

Crary, Jonathan 1990: *Techniques of the observer: On vision and modernity in the nineteenth century* (Cambridge: MIT Press).

Davis, Mike 1990: *City of Quartz: Excavating the future in Los Angeles* (London: Verso).

Dear, Michael 1986: "Postmodernism and planning." *Environment and Planning D: Society and Space*, 4, pp. 367–84.

—— 1991: "The premature demise of postmodern urbanism." *Cultural Anthropology*, 6, pp. 535–48.

Deutsche, Rosalyn 1988: "Uneven development: Public Art in New York City." *October*, 47, pp. 3–52.

—— 1991: "Boys town." *Environment and Planning D: Society and Space*, 9, pp. 5–30.

—— 1991: "Alternative space." In Brian Wallis (ed.), *If you lived here: The city in art, theory and social activism (A project by Martha Rosler)* (Seattle: Bay Press) pp. 45–66.

Dews, Peter 1987: *Logics of disintegration: Post-structuralist thought and the claims of critical theory* (London: Verso).

Dhareshwar, Vivek 1990: "The predicament of theory." In Martin Kreiswirth and Mark Cheetham (eds.), *Theory between the disciplines: Authority/vision/politics* (Ann Arbor: University of Michigan Press) pp. 231–50.

Driver, Felix 1985: "Power, space and the body: A critical assessment of Foucault's *Discipline and Punish.*" *Environment and Planning D: Society and Space*, 3, pp. 425–46.

—— 1988: "The historicity of human geography." *Progress in human geography*, 12, pp. 497–506.

—— 1992: "Geography's empire: Histories of geographical knowledge." *Environment and Planning D: Society and Space*, 10, pp. 23–40.

Duncan, James 1989: "The power of place in Kandy, Sri Lanka, 1780–1980." In Agnew and Duncan, pp. 185–201.

—— 1990: *The city as text: The politics of landscape interpretation in the Kandyan kingdom* (Cambridge: Cambridge University Press).

Duncan, James and Duncan, Nancy 1988: "(Re)reading the landscape." *Environment and Planning D: Society and Space*, 6, pp. 117–26.

—— 1992: "Ideology and bliss: Roland Barthes and the secret histories of landscape." In Barnes and Duncan, pp. 18–37.

Duncan, James and Ley, David 1982: "Structural Marxism and human geography: A critical assessment." *Annals of the Association of American Geographers*, 72, pp. 30–59.

Eagleton, Terry 1978: *Criticism and ideology: A study in Marxist literary theory* (London: Verso).

—— 1981: *Walter Benjamin, or towards a revolutionary criticism* (London: Verso).

—— 1983: *Literary theory: An introduction* (Oxford: Blackwell Publishers).

—— 1986: *Against the grain: Essays 1975–1985* (London: Verso).

—— 1990: *The significance of theory* (Oxford: Blackwell Publishers).

—— 1990: *The ideology of the aesthetic* (Oxford: Blackwell Publishers).

Foster, Hal (ed.) 1988: *Vision and visuality* (Seattle: Bay Press).

Foucault, Michel 1970: *The order of things: An archaeology of the human sciences* (London: Tavistock).

—— 1977: *Discipline and punish: The birth of the prison* (London: Penguin Books).

—— 1978: *The history of sexuality: An introduction* (London: Penguin Books).

—— 1980: *Power/knowledge: Selected interviews and other writings, 1972–1977* (Brighton, England: Harvester Press).

Frisby, David 1981: *Sociological impressionism: A reassessment of Georg Simmel's social theory* (London: Heinemann).

—— 1984: *Georg Simmel* (London: Tavistock).

—— 1985: *Fragments of modernity* (Cambridge, England: Polity Press).

—— 1992: *Simmel and since: Essays on Georg Simmel's social theory* (London: Routledge).

Geertz, Clifford 1973: *The interpretation of cultures* (New York: Basic Books).

—— 1983: *Local knowledge: Further essays in interpretive anthropology* (New York: Basic Books).

Giddens, Anthony 1976: *New rules of sociological method: A positive critique of interpretative sociologies* (London: Hutchinson).

—— 1977: *Studies in social and political theory* (London: Hutchinson).

—— 1979: *Central problems in social theory: Action, structure and contradiction in social analysis* (London: Macmillan).

Giddens, Anthony 1981: *A contemporary critique of historical materialism.* Vol. 1: *Power, property and the state* (London: Macmillan).

—— 1984: *The constitution of society: Outline of the theory of structuration* (Cambridge, England: Polity Press).

—— 1985: *A contemporary critique of historical materialism.* Vol. 2: *The nation-state and violence* (Cambridge, England: Polity Press).

—— 1987: *Social theory and modern sociology* (Cambridge, England: Polity Press).

—— 1990: *The consequences of modernity* (Stanford: Stanford University Press).

—— 1991: *Modernity and self-identity: Self and society in the late modern age* (Cambridge, England: Polity Press).

Grossberg, Lawrence; Nelson, Cary and Treichler, Paula (eds.) 1992: *Cultural studies* (New York: Routledge).

Guha, Ranajit and Spivak, Gayatri Chakravorty (eds.) 1988: *Selected subaltern studies* (New York: Oxford University Press).

Habermas, Jürgen 1971: *Toward a rational society* (London: Heinemann).

—— 1972: *Knowledge and human interests* (London: Heinemann).

—— 1976: *Legitimation crisis* (London: Heinemann).

—— 1987: *The theory of communicative action.* Vol. II: *The critique of functionalist reason* (Cambridge, England: Polity Press).

—— 1987: *The philosophical discourse of modernity: Twelve lectures* (Cambridge, England: Polity Press).

—— 1989: *The new conservatism: Cultural criticism and the historians' debate* (Cambridge: MIT Press).

Haraway, Donna 1989: *Primate visions: Gender, race and nature in the world of modern science* (New York: Routledge).

—— 1991: *Simians, cyborgs and women: The reinvention of nature* (London: Routledge).

—— 1992: "The promises of monsters: A regenerative politics for inappropriate/d others." In Grossberg, Nelson, and Treichler, pp. 295–337.

Harley, J. B. 1988: "Maps, knowledge and power." In Cosgrove and Daniels, pp. 277–312.

—— 1992: "Deconstructing the map." In Barnes and Duncan, pp. 231–47.

—— 1992: "Rereading the maps of the Columbian encounter." *Annals of the Association of American Geographers*, 82, pp. 522–42.

Harvey, David 1969: *Explanation in geography* (London: Edward Arnold).

—— 1973: *Social justice and the city* (London: Edward Arnold).

—— 1982: *The limits to capital* (Oxford: Blackwell Publishers).

—— 1985: *Consciousness and the urban experience* (Oxford: Blackwell Publishers).

—— 1985: *The urbanization of capital* (Oxford: Blackwell Publishers).

—— 1985: "The geopolitics of capitalism." In Derek Gregory and John Urry (eds.) *Social relations and spatial structures* (London: Macmillan) pp. 128–63.

—— 1987: "Three myths in search of a reality in urban studies." *Environment and Planning D: Society and Space*, 5, pp. 367–76.

—— 1989: *The condition of postmodernity: An enquiry into the origins of cultural change* (Oxford: Blackwell Publishers).

—— 1989: "From models to Marx: Notes on the project to 'remodel' geography." In Bill Macmillan (ed.) *Remodelling geography* (Oxford: Blackwell Publishers) pp. 211–16.

—— 1989: *The urban experience* (Oxford: Blackwell Publishers).

—— 1990: "Between space and time: Reflections on the geographical imagination." *Annals of the Association of American Geographers*, 80, pp. 418–34.

—— 1992: "Postmodern morality plays." *Antipode*, 24, pp. 300–326.

Harvey, David and Scott, Allen 1989: "The practice of human geography: Theory and empirical specificity in the transition from Fordism to flexible accumulation." In Bill Macmillan (ed.) *Remodelling geography* (Oxford: Blackwell Publishers) pp. 217–29.

Huyssen, Andreas 1986: *After the Great Divide: Modernism, mass culture and postmodernism* (London: Macmillan).

Jameson, Fredric 1972: *The prison-house of language: A critical account of structuralism and Russian formalism* (Princeton: Princeton University Press).

—— 1983: "Postmodernism and consumer society." In Hal Foster (ed.) *The anti-aesthetic: Essays on postmodern culture* (Seattle: Bay Press) pp. 111–25.

—— 1988: *The ideologies of theory: Essays 1971–1986*. Vol. 2: *Syntax of history* (Minneapolis: University of Minnesota Press).

—— 1990: *Late Marxism: Adorno or the persistence of the dialectic* (London: Verso).

—— 1991: *Postmodernism, or the cultural logic of late capitalism* (Durham: Duke University Press).

—— 1991: "Conversations on the New World Order." In Robin Blackburn (ed.) *After the fall: The failure of communism and the future of socialism* (London: Verso) pp. 255–68.

—— 1991: "Thoughts on the late war." *Social text*, 28, pp. 142–46.

Jay, Martin 1984: *Marxism and totality: Adventures of a concept from Lukács to Habermas* (Cambridge, England: Polity Press).

—— 1984: *Adorno* (Cambridge: Harvard University Press).

—— 1986: "In the empire of the gaze: Foucault and the denigration of vision in twentieth-century French thought." In David Couzens Hoy (ed.), *Foucault: A critical reader* (Oxford: Blackwell Publishers) pp. 175–204.

—— 1988: *Fin-de-siècle socialism and other essays* (London: Routledge).

—— 1991: "The textual approach to intellectual history." *Strategies*, 4/5, pp. 7–18.

—— 1992: "Scopic regimes of modernity." In Scott Lash and Jonathan Friedman (eds.), *Modernity and identity* (Oxford: Blackwell Publishers) pp. 178–95.

Jennings, Michael 1987: *Dialectical images: Walter Benjamin's theory of literary criticism* (Ithaca: Cornell University Press).

Kern, Stephen 1983: *The culture of time and space, 1880–1918* (Cambridge: Harvard University Press).

Kobayashi, Audrey and Mackenzie, Suzanne (eds.) 1989: *Remaking human geography* (Boston: Unwin Hyman).

Laclau, Ernesto and Mouffe, Chantal 1985: *Hegemony and socialist strategy: Towards a radical democratic politics* (London: Verso).

Lefebvre, Henri 1947: *Critique de la vie quotidienne* (Paris: Grasset).

—— 1961: *Fondements d'une sociologie de la quotidienneté* (Paris: L'Arche Éditeur).

—— 1969: *The explosion: Marxism and the French revolution* (New York: Monthly Review Press).

—— 1970: *La révolution urbaine* (Paris: Gallimard).

—— 1970: *La fin de l'histoire: Épilégomènes* (Paris: Minuit).

—— 1971: *Everyday life in the modern world* (New York: Harper & Row).

—— 1971: *Le manifeste différentialiste* (Paris: Gallimard).

—— 1972: *La pensée marxiste et la ville* (Paris: Casterman).

—— 1975: *Le temps des méprises* (Paris: Stock).

—— 1976: *The survival of capitalism: Reproduction of the relations of production* (London: Allison and Busby).

Lefebvre, Henri 1978: *De l'état.* Vol. 4: *Les contradictions de l'état moderne* (Paris: UGE).

—— 1979: "Space: Social product and use value." In J. W. Freiburg (ed.), *Critical sociology: European perspectives* (New York: Irvington) pp. 295.

—— 1981: *De la modernité au modernisme (Pour une métaphilosophie du quotidien)* (Paris: L'Arche Éditeur).

—— 1991: *The production of space* (Oxford: Blackwell Publishers).

—— 1991: *Critique of everyday life.* Vol. 1: *Introduction* (London: Verso).

Ley, David 1989: "Fragmentation, coherence and the limits to theory in human geography." In Kobayashi and Mackenzie, pp. 227–44.

—— 1989: "Modernism, post-modernism and the struggle for place." In Agnew and Duncan, pp. 44–65.

Ley, David and Mills, Caroline 1992: "Can there be a postmodernism of resistance in the urban landscape?" In Paul Knox (ed.) *The restless urban landscape* (Englewood Cliffs, N.J.: Prentice Hall) pp. 255–78.

Lunn, Eugene 1985: *Marxism and modernism: An historical study of Lukács, Brecht, Benjamin and Adorno* (London: Verso).

Lyotard, Jean-François 1971: *Discours, figure* (Paris: Éditions Klincksieck).

—— 1973: *Dérive à partir de Marx et de Freud* (Paris: Union Générale d'Éditions).

—— 1974: *Économie libidinale* (Paris: Minuit).

—— 1983: *Le différend* (Paris: Minuit).

—— 1984: *The postmodern condition: A report on knowledge* (Minneapolis: University of Minnesota Press).

—— 1988: *Peregrinations: Law, form, event* (New York: Columbia University Press).

—— 1991: *The inhuman: Reflections on time* (Stanford: Stanford University Press).

Marcus, George and Fischer, Michael 1986: *Anthropology as cultural critique: An experimental moment in the human sciences* (Chicago: University of Chicago Press).

Massey, Doreen 1985: *Spatial divisions of labor: Social structures and the geography of production* (London: Macmillan).

—— 1991: "Flexible sexism." *Environment and Planning D: Society and Space,* 9, pp. 31–57.

Meinig, Donald 1983: "Geography as an art." *Transactions of the Institute of British Geographers,* 8, pp. 314–28.

Mitchell, Timothy 1988: *Colonizing Egypt* (Cambridge: Cambridge University Press).

—— 1989: "The world-as-exhibition." *Comparative Studies in Society and History,* 31, pp. 217–36.

—— 1990: "Everyday metaphors of power." *Theory and society,* 19, pp. 545–77.

Morris, Meaghan 1992: "The man in the mirror: David Harvey's 'Condition' of postmodernity." *Theory, culture and society,* 9, pp. 253–79.

Nicholson, Linda (ed.) 1990: *Feminism/Postmodernism* (London: Routledge).

O'Hanlon, Rosalind 1988: "Recovering the subject: Subaltern Studies and histories of resistance in colonial South Asia." *Modern Asian Studies,* 2, pp. 189–224.

Olsson, Gunnar 1980: *Birds in egg/Eggs in bird* (London: Pion).

—— 1991: *Lines of power/limits of language* (Minneapolis: University of Minnesota Press).

Pollock, Griselda 1988: *Vision and difference: Feminity, feminism and histories of art* (London: Routledge).

Poster, Mark 1975: *Existential marxism in post-war France: From Sartre to Althusser* (Princeton: Princeton University Press).

—— 1990: *The mode of information: Poststructuralism and social context* (Chicago: University of Chicago Press).

Pratt, Mary Louise 1985: "Scratches on the face of the country; or, What Mr Barrow saw in the land of the Bushmen." *Critical inquiry,* 12, pp. 119–43.

—— 1992: *Imperial eyes: Travel writing and transculturation* (London: Routledge).

Pred, Allan 1981: "Social reproduction and the time-geography of everyday life." *Geografiska Annaler,* 63B, pp. 5–22.

—— 1984: "Place as historically contingent process: Structuration theory and the time-geography of becoming places." *Annals of the Association of American Geographers,* 74, pp. 279–97.

—— 1986: *Place, practice and structure: Social and spatial transformation in southern Sweden 1750–1850* (Cambridge, England: Polity Press).

—— 1990: *Making histories and constructing human geographies* (Boulder, Colo.: Westview Press).

—— 1990: *Lost words and lost worlds: Modernity and the language of everyday life in late nineteenth-century Stockholm* (Cambridge: Cambridge University Press).

Pred, Allan and Watts, Michael 1992: *Reworking modernity: Capitalisms and symbolic discontent* (New Brunswick, N.J.: Rutgers University Press).

Rabinow, Paul 1986: "Representations are social facts: Modernity and post-modernity in anthropology." In Clifford and Marcus (eds.), pp. 234–61.

—— 1989: *French modern: Norms and forms of the social environment* (Cambridge: MIT Press).

—— 1992: "Artificiality and enlightenment: From sociobiology to biosociality." In Jonathan Crary and Sanford Kwinter (eds.), *Incorporations* (New York: Zone) pp. 234–52.

Rajchman, John 1985: *Michel Foucault: The freedom of philosophy* (New York: Columbia University Press).

—— 1991: *Philosophical events: Essays of the 80s* (New York: Columbia University Press).

Rose, Gillian 1990: "The struggle for political democracy: Emancipation, gender and geography." *Environment and Planning D: Society and Space,* 8, pp. 395–408.

—— 1991: "On being ambivalent: Women and feminisms in geography." In Chris Philo (ed.), *New words, new worlds: Reconceptualising social and cultural geography* (Aberystwyth, Wales: IBG Social and Cultural Geography Study Group) pp. 156–63.

Ross, Andrew (ed.) 1988: *Universal abandon? The politics of postmodernism* (Minneapolis: University of Minnesota Press).

Ross, Kristin 1988: *The emergence of social space: Rimbaud and the Paris Commune* (Minneapolis: University of Minnesota Press).

Rouse, Joseph 1987: *Knowledge and power: Toward a political philosophy of science* (Ithaca: Cornell University Press).

Said, Edward 1984: *The world, the text and the critic* (London: Faber and Faber).

—— 1985: *Orientalism* (London: Penguin Books).

—— 1985: "Orientalism reconsidered." *Cultural critique* 1, pp. 89–107.

—— 1989: "Representing the colonized: Anthropology's interlocutors." *Critical Inquiry,* 15, pp. 205–25.

Slater, David 1992: "On the borders of social theory: Learning from other regions." *Environment and Planning D: Society and Space,* 10, pp. 307–28.

Smith, Neil 1990: *Uneven development: Nature, capital and the reproduction of space* (Oxford: Blackwell Publishers).

Smith, Paul 1988: *Discerning the subject* (Minneapolis: University of Minnesota Press).

Soja, Edward 1989: *Postmodern geographies: The reassertion of space in critical social theory* (London: Verso).

Soja, Edward 1990: "Heterotopologies: A remembrance of other spaces in the Citadel-LA." *Strategies*, 3, pp. 6–39.

—— 1992: "Inside exopolis: Scenes from Orange County." In Michael Sorkin (ed.) *Variations on a theme park: The new American city and the end of public space* (New York: Noonday Press) pp. 94–122.

Spivak, Gayatri Chakravorty 1988: "Can the subaltern speak?" In Cary Nelson and Lawrence Grossberg (eds.) *Marxism and the interpretation of culture* (Urbana: University of Illinois Press) pp. 271–313.

—— 1988: *In other worlds: Essays in cultural politics* (London: Routledge).

—— 1991: "Identity and alterity: An interview." *Arena*, 97, pp. 65–76.

—— 1991: "Feminism in decolonization." *Differences: A Journal of Feminist Cultural Studies*, 3, pp. 139–70.

—— 1991: "Women in difference: Mahasweta Devi's 'Delouti the Bountiful.'" In Andrew Parker, Mary Russo, Doris Sommer, and Patricia Yeager (eds.), *Nationalisms and sexualities* (New York: Routledge) pp. 96–117.

Stoddart, David R. 1986: *On geography and its history* (Oxford: Blackwell Publishers).

Thompson, E. P. 1968: *The making of the English working class* (London: Penguin Books).

—— 1975: *Whigs and Hunters: The origin of the Black Act* (London: Allen Lane).

—— 1978: *The poverty of theory and other essays* (London: Merlin).

Tuan, Yi-Fu 1971: "Geography, phenomenology and the study of human nature." *Canadian Geographer*, 15, pp. 181–92.

—— 1976: "Humanistic geography." *Annals of the Association of American Geographers*, 66, pp. 266–76.

—— 1977: *Space and place: The perspective of experience* (London: Edward Arnold).

—— 1989: "Surface phenomena and aesthetic experience." *Annals of the Association of American Geographers*, 79, pp. 233–41.

—— 1989: *Morality and imagination: Paradoxes of progress* (Madison: University of Wisconsin Press).

Wolin, Richard 1982: *Walter Benjamin: An aesthetic of redemption* (New York: Columbia University Press).

Young, Robert 1990: *White mythologies: Writing History and the West* (London: Routledge).

Name Index

Subject Index